Intestinal Microbiota in Health and Disease: Modern Concepts

Preface

Mucosal surfaces are home to a complex microbiota, which rapidly colonizes the tissues at birth and bestows important nutritional, metabolic and protective functions which benefit the host. The numerous and diverse microbial communities residing at the vast intestinal mucosal surface have a particularly important role to play in human health, any *dysbiosis* of this microbiota being an underlying factor in a variety of clinical diseases. As such, the intestinal microbiota is the subject of intense and rapidly advancing research. Recent progress in molecular microbiology, bioinformatics and ecology in particular, are providing new insight into the role of the intestinal microbiota in health and revealing hitherto unexplored means to prevent or treat certain diseases.

Through this book, we sought to provide the reader with an appraisal of the most recent advances in the field. To this end, we have invited recognized experts from the academic and medical communities and from the pharmaceutical and food industry, to provide a treatise on a range of topics that address host-microbe interactions in the gut and the consequences of these interactions for the host.

The opening chapters of the book discuss how microbes interact with host cells in specific intestinal niches, become established members of the intestinal microbiota and impact the development and function of a healthy immune system. The next chapters describe how pathogens overcome host defences and exploit the underlying mechanisms to infect the host. Chapters in the second half of the book then address the characteristics and the mechanisms leading to *dysbiosis* of the microbiota and how this culminates in the development of inflammatory bowel disease, obesity, infection or stress. In this context, authors also discuss how manipulation of gut ecology and the use of probiotic microorganisms are credible therapeutic strategies to treat diseases and conditions affecting both the inside and the outside the gastrointestinal tract.

We are confident that the authors have provided a balanced review of the field and that the contents of this book will offer new information to students and experienced researchers alike.

We sincerely thank each author for the outstanding quality of their respective contributions and we trust that you, like us, will find the contents of this book informative, fascinating and a pleasure to read.

Eduardo J. Schiffrin, Medical Director
Philippe Marteau, Professor
Dominique Brassart

Contents

1

Commensal Intestinal Microbiota and Mucosal Immune System Development and Function

Katarina Radulovic[1] and *Jan Hendrik Niess*[1,2,*]

Introduction

Instead of living as one individual organism, different species coexist in complex ecological niches constantly influencing each other. Humans are no exception from the symbiotic way of living since every healthy human individual coexists with an enormous number of microorganisms. The mutually dependent "life together" of two or more species is called symbiosis (Black 1996). Symbiotic relations of humans and microbial species always have positive outcomes for at least one member and this includes the relationships of mutualism and commensalism. Mutualism is a "win-win" situation in which both members benefit from the relationship (Black 1996). In commensalism the situation is "win-zero" since one member of the relationship benefits without helping or harming the other one (Black 1996). When considering the interactions between microbial communities,

[1]Department of Internal Medicine I, University Hospital Ulm, Ulm, Germany.
E-mail: katarina.radulovic@uni-ulm.de
[2]Department of Visceral Surgery and Medicine, Inselspital, Bern, Switzerland.
*Corresponding author: niess108@yahoo.com

some authors also include the relations with the negative outcome to one member into the symbiosis (Faust and Raes 2012). Such relations include parasitism ("win-lose" situation) amensalism ("zero-loose" situation) and competition ("loose-loose" situation) between the microbes (Faust et al. 2012; Faust and Raes 2012).

Humans start their development in a sterile intrauterine environment, but from the very moment of birth all the epithelial surfaces in direct contact with the environment (skin, respiratory, gastrointestinal and urogenital tract) are colonized by microorganisms (Levy 2007). We are living with these microorganisms in mutualistic or commensal relationships. We refer to the collection of all of these microorganisms as the human microbiome, microflora or commensal microbiota. Rising interest in investigations of how microflora influences the human health, as well as great improvement in microbiological methods have given valuable data about the composition of our microbiological partners and their interactions with the human body. Now we know that healthy individuals remarkably differ in the composition of the microbiota (The human microbiome project consortium 2012b). We also know that in the highly coevolved and bidirectional relationship with the human host, microflora have profound effects on the human health (The human microbiome project consortium 2012a).

This chapter will introduce the influence of microflora on immune system development and function in the gut, summarizing the results obtained on rodents and humans. We will describe the composition of intestinal microflora and the basics of intestinal immune system functioning. We will also discuss how the delicate balance between microflora and underlying immune system cells is maintained and how disturbances in microflora composition influence the function of intestinal immune cell subtypes.

The Composition of Intestinal Microbiota

As the largest body surface directly exposed to the influences of environment, the human intestine is populated with enormous number of microorganisms. More than 400 m^2 of the intestinal epithelial surface is colonized by approximately 100 trillion of microbial cells which is tenfold of the total number of cells in the human body (Backhed et al. 2005; Qin et al. 2010; Stephani et al. 2011). Bacteria are the main type of microbes present in the human gut, but other types like protozoa, fungi and viruses can be found, as well (Macpherson and Harris 2004; O'Hara and Shanahan 2006). The concentration of the bacterial cells gradually increases along the intestinal tract ranging from 10^3 cells/gram (g) of luminal content in the stomach and duodenum, to 10^4 to 10^7 cells/g in the small intestine and up to 10^{11} to 10^{14} cells/g in the colon (Sekirov et al. 2010; Stephani et al. 2011). Most of the intestinal bacteria are non-cultivable (Kuwahara et al. 2011) which created difficulties for the scientists who investigated them. The recent advance in culture-independent techniques, such as 16S ribosomal RNA sequencing, has made it possible to identify the majority of the bacteria living in the human intestinal tract and also to compare the microflora composition of different individuals and species. Up to 40.000 bacterial species are identified in the human intestine (Frank and Pace 2008) that are distributed in over 50 phyla from which 2 are dominant: Firmicutes and Bacteroidetes (Sekirov et al. 2010; The human microbiome project consortium 2012b). These two phyla represent the core of human intestinal microbiota including more than 80% of all the intestinal bacterial species, but greater inter-individual differences exist on the species level (Tap et al. 2009; The human microbiome project consortium 2012b). Some of the other phyla also represented in the human intestinal microbiome are Proteobacteria, Actinobacteria, Verrucomicrobiota, Fusobacteria and Cyanobacteria (Sekirov et al. 2010; Tap et al. 2009; The human microbiome project consortium 2012b).

Early Life Events Shape the Intestinal Microflora

As already discussed, intrauterine development of humans and all the other mammals, including rodents, is sterile due to efficient maternal immune mechanisms that keep the uterine cavity sterile (Levy 2007). Microbial colonization of the intestine starts at birth and bacteria can be found in faeces within the first few hours of life (Kaplan et al. 2011; Ouwehand et al. 2002). The intestinal flora of newborns is relatively unstable and beside the genotype of the host, efficient colonization depends on many extrinsic factors: the composition of maternal microflora, the mode of delivery, environmental hygiene conditions, and diet and medication (Fanaro et al. 2003; Fouhy et al. 2012). The pattern of initial colonization is very important since it shapes the gut microbiota composition in adulthood. Normally, the first intestinal habitants are the facultative anaerobic bacteria including *Escherichia coli* and the Firmicutes species form the genera *Streptococcus*, *Staphylococcus* and *Enterococcus* (Kaplan et al. 2011; Park et al. 2005). These bacteria take advantage of the oxygen abundance, but at the 1–2 week of age the oxidation-reduction potential decreases and obligate anaerobic bacteria of the genera *Bacteroides*, *Clostridium* and *Bifidobacterium* become dominant (Fanaro et al. 2003; Kaplan et al. 2011; Park et al. 2005; Scholtens et al. 2012). Weaning from the mother's milk to solid food further favours obligate anaerobes, leading to the final stabilisation of microflora composition that at the age of 1 year resembles the microflora of adults (Scholtens et al. 2012; Sekirov et al. 2010). In the adult gut, dominant bacterial genera are *Bacteroides*, *Clostridium*, *Fusobacterium*, *Eubacterium*, *Ruminococcus*,

Peptococcus, *Peptostreptococcus* and *Bifidobacterium*, while *Escherichia*, *Klebsiella*, *Lactobacillus* and *Enterobacter* are less represented (Fanaro et al. 2003; Vedantam and Hecht 2003).

Newborns delivered by caesarean section do not have the contact with maternal microflora during the process of birth and their intestinal microbiota composition highly depends on the environmental bacterial load. In these infants, the establishment of stabile microflora is delayed and instead of *Bifidobacterium*, the frequently isolated bacterial genera are *Klebsiella*, *Clostridium* and *Enterobacter* (Adlerberth et al. 2007; Fanaro et al. 2003; Kaplan et al. 2011). Also the intestine of formula-fed babies is not rich in *Bifidobacteria*, but rather in other bacterial genera such as *Enterobacter*, *Escherichia*, *Clostridium*, *Klebsiella* and *Bacteroides* (Fanaro et al. 2003; Kaplan et al. 2011). Preterm born infants are often hospitalized in the neonatal intensive care units and subjected to medication for a certain time period and this treatment crucially influences the composition of intestinal microbiota. In these infants, bacterial diversity is decreased and *Bifidobacteria* are rare while *Klebsiella*, *Enterobacter* and *Staphylococcus* are predominant (Arboleya et al. 2012; Fanaro et al. 2003; Kaplan et al. 2011).

These early life events and conditions have profound effects on the intestinal microflora composition and therefore on the health status of the individual during the whole life. Namely, an increased prevalence of atopic and allergic diseases in industrialized countries has been attributed to the increased hygiene and vaccination in the neonatal period and during childhood (Macpherson and Harris 2004; Ouwehand et al. 2002; Strachan 1989). Studies showed that children born and raised in farming conditions, rich in microbial antigen load, are protected from the development of asthma (Douwes et al. 2008; Ege et al. 2011) while avoidance of antigens during prenatal and neonatal period can increase the risk of allergy (Falth-Magnusson and Kjellman 1992; Woodcock et al. 2004). The two hypotheses trying to explain these observations, the "hygiene hypothesis" and the "fetal programming hypothesis", both point towards the importance of microbial-derived factors (Kaplan et al. 2011). It is considered that increased hygienic conditions ("hygiene hypothesis") reduce the exposure to microbial antigens, especially in the critical period during the intrauterine and neonatal stage ("fetal programming hypothesis") leading to the aberrant development of the immune system (Kaplan et al. 2011; Macpherson and Harris 2004). In general, the increased risk of atopic diseases is linked with the presence of *Enterobacteria*, *Clostridium*, *Bacteroides* and *Staphylococcus* and reduction of *Bifidobacteria* (Penders et al. 2007). These findings strongly indicate that the balanced composition of intestinal microflora is crucial for the host health.

Mutualism in the Intestine

Life in symbiosis with intestinal microorganisms brings mutual benefits for both partners. The gastrointestinal tract is rich in molecules that can serve as nutrients for the microorganisms and therefore represent a secure habitat in which microbes can survive and multipy (Costello et al. 2012; Sekirov et al. 2010). In return, the genome of gut microbiome, which contains ≥ 100 times the number of genes in the human genome, endows us with the functional features that we have not had to evolve ourselves (Backhed et al. 2005). Germ free (GF) animals provide an excellent model system for studying the contribution of microflora to various body functions. These animals are born and bred in special isolators that are ventilated with sterile filtered air and by using sterile bedding, food and water for their breeding in order to make them completely devoid of microorganisms (Macpherson and Harris 2004). Studies on GF animals showed that intestinal microbiota have nutritive, metabolic, protective and immunogenic functions (Fanaro et al. 2003; Rossi et al. 2011).

The intestinal flora assists in the digestion of otherwise indigestible energy substrates (e.g., plant-derived pectin, cellulose). In the line with this, it is shown that GF animals need about 30% more calories than conventional (CV) animals to keep a stable body weight (Wostmann et al. 1983). Through the production of short-chain fatty acids (SCFA), microflora stimulates the proliferation and differentiation of the intestinal epithelial cells (IECs) assuring the efficient nutrient absorption (O'Hara and Shanahan 2006). A prominent characteristic of GF animals is the pronounced enlargement of the caecum which can be explained by the accumulation of the luminal content, in particular mucus and urea that are normally degraded by microflora (Gustafsson et al. 1970; Juhr and Ladeburg 1986). The metabolic role of microbiota is also seen in the production of 1–20% of the total plasma level of lysine, the indispensable amino acid (Metges 2000) and regulation of the iron and copoper metabolism (Reddy et al. 1965a; Reddy et al. 1965b). Furthermore, intestinal microflora represents an important source of the vitamins that cannot be produced by the host body (Rossi et al. 2011). Vitamin K and the vitamins of B group are mainly produced by the bacterial genera *Bacteroides* and *Eubacterium* (Hill 1997; Rossi et al. 2011). Through the competitive occupation of the attachment sites and the consumption of nutrients, microflora protect the host from the pathogen invasion (Sekirov et al. 2010). The protection role of microflora is also achieved by the active interactions with the host immune system. The fact that lethal oral dose of the *Listeria monocytogenes* in GF mice is one million fold less than in CV animals (Zachar and Savage 1979) shows the importance of the microflora in host protection.

The Intestinal Immune System Coexist with the Enteric Microflora

Despite the symbiotic nature of the host-microbiota relationship and numerous beneficial effects of the intestinal microflora, presence of such a high non-self antigen burden in the intestinal lumen represents an immense challenge to the underlying mucosal immune system. Approximately 80% of all body's immune cells are located in the intestine (Smith and Garrett 2011). These cells are organized in the complex network of the intestinal immune system which faces a specific problem: it has to preserve the beneficial enteric microbiota without starting an immune response against them, but at the same time it has to efficiently eliminate pathogenic agents that can invade the body through the intestinal route. Closely associated together, microbiota and intestinal immune system interact in a bidirectional manner to ensure the maintenance of homeostasis. On one hand, the immune system evolved various adaptations to control the composition and distribution of the microflora without threatening the existence of these beneficial microorganisms. On the other hand, microflora has profound influence on the development and functioning of the immune system. Therefore, microflora and the intestinal immune system should be viewed and studied as a single entity.

The past few decades have seen some major advances in understanding the complex host-microbiota relationships due to the development of new model systems and methods in both microbiology and immunology. As already mentioned, the development of new culture-independent microbiological methods that rely on PCR, sequencing and DNA microarray techniques was a break-through in understanding the microflora composition and features. Still, further advance of the culture methods, especially of the anaerobic bacterial species, would be helpful for the development of the tools for manipulation of the intestinal microbiota. Much has been learned about host-microbial interactions from the experiments with GF animals that are shown to be a very suitable model for detailed investigation of how microflora interact with the host cells and regulate the functions of various host organ systems. Furthermore, GF animals served as the basis for development of gnotobiotic animals that become the important tool for investigation of host-microbiota interactions. Gnotobiotic animals are former GF animals colonized with the single microbioal specie or a defined mixture of species (Gordon and Pesti 1971). Nowadays, almost all the animal strains can be bred in GF and gnotobiotic conditions allowing the studies of how different genetic alterations in the host genome influence the host-microbiota symbiosis. Experiments of monoassociation of GF mice with the single bacterial species can explain the role of these specific strains in the immune system development and functioning. However, in normal conditions the complex interactions between different bacterial species in the intestine may be crucial for host-microbiota relationship. This is why some authors consider that the animals with reduced microflora due to the antibiotic treatment are better model systems than GF mice for studying the interactions between microbiota and host (Dimmitt et al. 2010). On the other hand, antibiotic treatment is changing the metabolism of intestinal macrobiota altering on that way the host-microbiota interactions (Beaugerie and Petit 2004).

The data obtained by using all these experimental systems and techniques showed the important role of microbiota in shaping every aspect of the intestinal immune system. It contains some different immune cell subtypes along with the cells with a unique and specific properties and functions. The next section will describe the development, organization and function of this complex immune system with regard to the role of microbiota in these processes.

Organization of the Intestinal Immune System

The mucosal immune system of the intestine consists of innate and adaptive immune components. Innate immunity is first activated by microbes and it serves to rapidly prevent, control or eliminate the infection (Abbas et al. 2007). On the other hand, cells of the adaptive immune component, lymphocytes, are activated after the innate immunity and they are able to start the efficient antigen-specific immune responses and create the immune memory. Adaptive immune responses are governed by the innate immunity since the lymphocyte require signals from the antigen presenting cells (APC) of the innate immunity for their full activation (Abbas et al. 2007).

Furthermore, the intestinal immune system is divided into inductive and effector sites. The organized lymphoid tissues of mesenteric lymph nodes (MLN), Peyer's patches (PP), isolated lymphoid follicles (ILF) in the small intestine and colonic patches in the large intestine are responsible for the induction phase of immune responses (Garrett et al. 2010; Makita et al. 2007; Nagler-Anderson 2001). These tissues consist of activated follicles of B lymphocytes surrounded by T cells and dendritic cells (DC). On inductive sites DC present the antigens to the naïve lymphocytes and start the adaptive immune responses (Makita et al. 2007; Mowat 2003). Intestinal epithelium and underlying lamina propria (LP) represent the effector sites of the intestinal immune system. Here the effector cells of both innate and adaptive immune component actively fight the invading pathogen (Garrett et al. 2010; Koboziev et al. 2010).

Studies on GF animals showed that absence of microflora induce the general underdevelopment of the intestinal immune system. In GF animals, fully formed ILF and colonic patches cannot be found while the number, size and cellularity of PP and MLN are reduced (Chinen and Rudensky 2012; O'Hara and Shanahan 2006; Ouwehand et al. 2002). The effector sites are also hypocellular in the absence of microflora since the numbers of innate and adaptive immune cells in both intraepithelial and LP effector compartments are significantly reduced (Dimmitt et al. 2010; O'Hara and Shanahan 2006; Williams et al. 2006). Colonization of GF mice with the mixture of bacteria efficiently reversed these effects (Cebra et al. 1998).

Similar to GF mice, human and murine infants are born with an immature intestinal immune system (Levy 2007; Ouwehand et al. 2002; Sekirov et al. 2010; Steege et al. 1997). Normally, in early neonatal life, lymphoid tissues complete the formation and both innate and adaptive immune cells proliferate and acquire the mature phenotype (Steege et al. 1997; Williams et al. 2006). However, in GF neonatal mice, intestinal immune systems do not develop much further with age (Williams et al. 2006). These data suggest that interactions with intestinal microflora are absolutely necessary for normal development and functioning of the intestinal immune system.

The Recognition of Microbiota by the Intestinal Immune System

The intestinal immune system communicates with the microflora by recognizing the molecules constituting the microbial cells or the products of microbial metabolism. It is considered that only the cells of the innate immunity can directly recognize the microbiota by recognition of the structures common to the group of the related microbes, the so called pathogen associated molecular patterns (PAMP). PAMPs include different microbial constitutes such as proteins, lipids, sugars and nucleic acids that are recognized by pattern recognition receptors (PRRs) (Abbas et al. 2007; Kinnebrew and Pamer 2012). PRRs are expressed on many different cell types and that ensures efficient and regulated microbial recognition and response to them. All the cells in the intestinal epithelium express PRRs and are the first ones that come in contact with the intestinal microbiota present in the lumen. Phagocytes in the intestine also express PRRs and certain types of them (CX_3CR1^+ cells) can directly sample the luminal bacteria using the transepithelial cellular processes (Niess et al. 2005). Therefore, luminal antigens can be presented to the immune cells by different types of intestinal epithelial cells, such as normal and villous M-cells, enterocytes and Goblet cells, but also by CX_3CR1 mononuclear phagocytes (Fig. 1). All of these cell types integrate the microbial signals through PRRs and coordinate the subsequent immune responses to them. Therefore PRRs represent the crucial component of the immune system in the host-microbiata interactions.

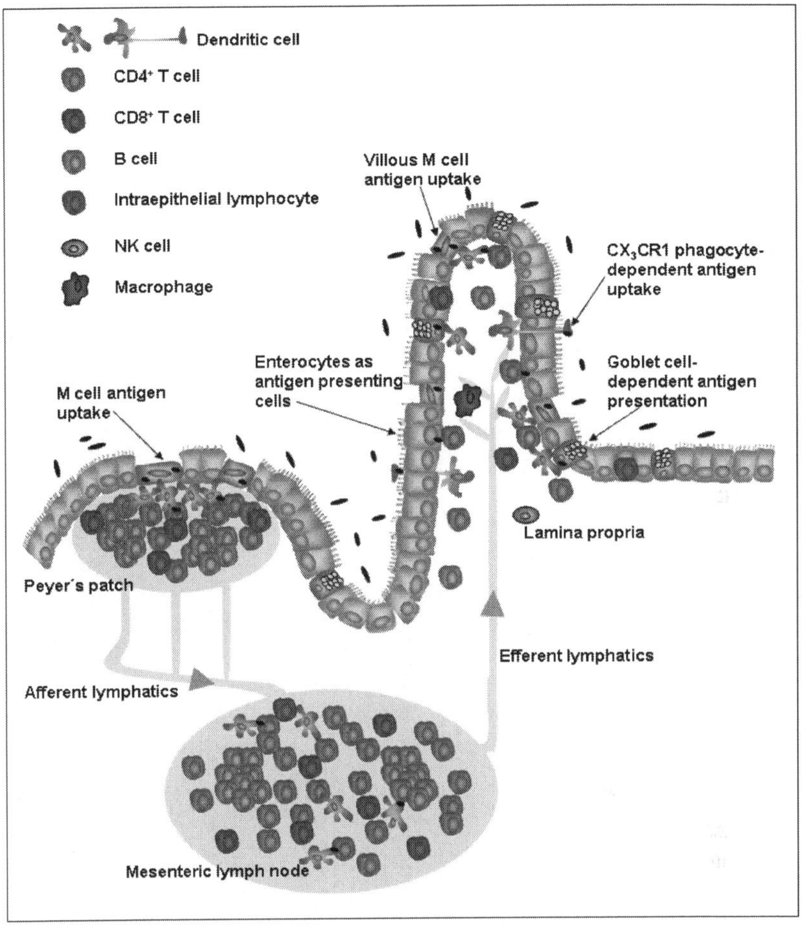

Figure 1. Antigen Recognition in the Intestine. Various cell types in the intestinal epithelium are able to uptake the luminal antigens and present them to the underlying immune system cells. Specialized antigen uptaking cells in the intestine are normal and villous micro-fold cells (M-cells) that transfer the antigen to the dendritic cells of the intestine. Enterocytes can function as the antigen presenting cells as they are able to process the luminal antigens and present them directly to the lymphocytes at the basal side. Goblet cells transfer the luminal antigens to the dendritic cells, while CX^3CR1^+ phagocyte can directly sample the luminal content using the transepithelial cellular processes.

Color image of this figure appears in the color plate section at the end of the book.

However, PRRs cannot distinguish between the pathogenic and beneficial microbes, since all of them express the same PAMPs. It is considered that the expression pattern of PRRs by intestinal epithelial cells can control their engagement by different types of microbes and also induction of different intracellular signalling pathways.

Pattern recognition receptors (PRRs)

PRRs include Toll-like receptors (TLRs), nucleotide oligomerization domain (NOD)-like receptors (NLRs) and C-type lectin receptors (CLRs) (Kinnebrew and Pamer 2012). Each of these receptors is specialized for the recognition of a specific microbial molecule or more of them, but the activation of all of them generally leads to the antimicrobial response that ensure the efficient and rapid microbial elimination. However, the products of normal intestinal microbita also engage PRRs, so beside the defence from the pathogens, these receptors have an important role in intestinal homeostasis.

Toll-like receptors (TLRs)

There are 10 functional TLRs in mice and 12 in humans (Kinnebrew and Pamer 2012). They are located either on the cell surface or on the endosomal cell compartment and they recognize microbial lipids, proteins and nucleic acids (Takeuchi and Akira 2010). Activation of TLRs associate them with adaptor proteins, such as myeloid differentiation factor 88 (MyD88) or Tool/interleukin (IL)-1 receptor domain-containing adapter inducing interferon (IFN)-β (TRIF), that can initiate the signalling cascade (Akira and Takeda 2004; Yamamoto et al. 2002). TLRs located on the plasma membrane induce pro-inflammatory cytokine, chemokine and antimicrobial peptides (AMPs) production (Akira and Takeda 2004) while the ones within the endosome activate IFN-β transcription (Schmitz et al. 2007). Studies showed that intestinal microbita induces epithelial cell proliferation in a TLRs/Myd88-dependent manner (Rakoff-Nahoum and Medzhitov 2007) and maturation of ILF by TLRs-, MyD88- and TRIF-depended pathway (Chinen and Rudensky 2012).

NOD-like receptors (NLRs)

Located in the cytosol, NLRs recognize the microbial products that have entered the cell by direct infection with intracellular pathogens, cytoplasmatic injection by extracellular bacteria or those escaping from the endosom after active phagocytosis (Kinnebrew and Pamer 2012). Both the NOD1 receptor expressed on epithelial cells and the NOD2 receptor predominately expressed by immune cells, recognize intracellular bacterial peptidoglycan and through the direct signalling cascade activate nuclear factor kappa B (NFkB) (Kinnebrew and Pamer 2012). They are very important for intestinal homeostasis since mutations in NOD2 gene are associated with inflammatory bowel disease (IBD) in humans (Abraham and Cho 2009), while recent study showed that NOD1 signals are crucial for microbiota-induced ILF development (Bouskra et al. 2008). Some other NLRs like NLRP1, NLRP3, NLRC4 and NLRP6 are associated with inflammasomes, signalling multiprotein complexes that serve to activate caspase-1 enzyme (Davis et al. 2011; Franchi et al. 2012). Once activated, caspase-1 proteolytically cleaves pro-IL-1β and pro-IL-18 releasing their active forms and enabling their secretion but also induce the specific type of cell death called pyroptosis by poorly understood mechanisms (Elinav et al. 2011; Franchi et al. 2012; Strowig et al. 2012). IL-1β and Il-18 are potent pro-inflammatory cytokines and therefore the activity of inflammasomes needs to be tightly regulated to avoid excess inflammatory response. The role of inflammasome-associated NLRs in the control of microbiota composition is also suggested since mice lacking NLRP3 or NLRP6 showed the intestinal enteropathy associated with overgrowth of Bacteroides (Henao-Mejia et al. 2012).

C-type lectin receptors (CLRs)

This receptor family contains a highly conserved C-type lectin like domain (CTLD) and includes both soluble and transmembrane proteins (Osorio and Reis e Sousa 2011). These proteins have various immunogenic and non-immunogenic functions, but the ones that have innate immune receptor function predominately recognize carbohydrate motifs of bacterial and fungal cell wall (Kinnebrew and Pamer 2012). The CLRs signalling cascade usually lead to NFkB activation and robust inflammatory response (Kinnebrew and Pamer 2012). Until now, there are no data on the role of CLRs in host-microbiota interactions.

The Innate Immune System in the Intestine

Innate immunity is a highly evolutionary conserved system that mediates the initial protection against infection. Innate immune cells use the common antigen-unspecific effector mechanisms against the related microbes (Abbas et al. 2007). Innate cells in the intestine have different origin, so they can be hematopoietic immune and somatic non-immune cells of the intestinal mucosa. Non-immune cells of the innate immune system in the intestine are the epithelial cells that form the mucosal barrier with the role to block the penetration of luminal microbes into the sterile inner body tissues. Different types of effector innate lymphoid cells (ILCs) as well as myeloid cells (macrophages and DC) belong to the immune cells of the intestinal innate immunity.

Mucosal Barrier

The mucosal barrier represents a dynamic system composed of mucus and a single-cell layer of intestinal epithelial cells (Fig. 2). The epithelium must absorb the dietary nutrients but also prevent the access of potentially harmful agents. The absorption is mainly done by enterocytes, while the other cell types present in the intestinal epithelium preferably, contribute to the immune barrier function. Being in the closest contact with the luminal microflora, intestinal epithelial cells mediate majority of the interactions between microflora and the host immune system. Beside the stem cells that ensure the regular renewal of the epithelium by the ability to differentiate into all the other cell types, there are five cell types in the intestinal

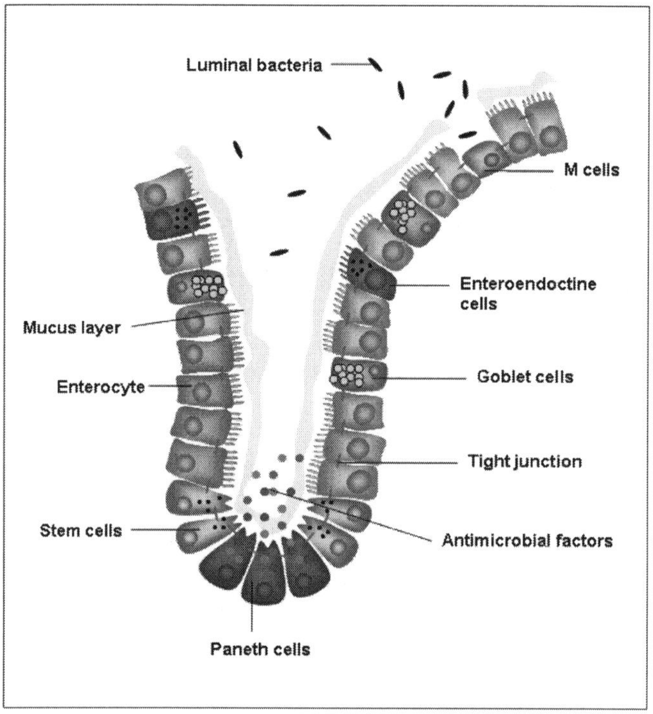

Figure 2. Mucosal Barrier in the Intestine. The intestinal epithelium consist of a single cell layer made of five cell types. All these cell types originate from stem cells located in the basis of intestinal crypts. The most numerous cells in the intestinal epithelium are enterocytes that are connected with the tight junctions. Intestinal epithelial cells are covered at the luminal side with mucus produced by Goblet cells. Mucus contains the antimicrobial factors secreted by Paneth cells. The mucus layer above micro-fold cells (M-cells) is disrupted allowing controlled luminal antigen uptake by these cells. Enteroendocrine cellssecrete the hormons that regulate the function of the intestinal epithelium.

Color image of this figure appears in the color plate section at the end of the book.

epithelium: Goblet cells, Paneth cells, enterocytes, enteroendocrine cells and microfold cells (M-cells).

Goblet cells

These cells produce and secrete large, heavily glycosylated mucin proteins that are assembled into a thick, complex bilayer network of mucus that covers the luminal side of epithelium (McGuckin et al. 2011) (Fig. 2). The mucus layer is the thickest in the stomach and colon (about 700 μm) while it is the thinnest in the small intestine (150–300 μm) (Atuma et al. 2001). This is probably connected to the primary absorption role of the small intestine. The mucus in the colon, but not the small intestine, consists of two layers: a thinner but more firm inner layer which is sterile and a thicker outer layer which is usually occupied by the commensal microorganisms (Hooper et al. 2012; McGuckin et al. 2011). Mucus is continuously degraded by the commensals and replaced by the Goblet cells (McGuckin et al. 2011). Furthermore, Gobleth cells also secrete other proteins that contribute the protective role of mucus layer, such as AMPs that ensure the sterility of the inner mucus layer and trefoil factors whose role is not fully understood but it appears that they can alter the biophysical properties of mucus (Gill et al. 2011; McGuckin et al. 2011; Thim et al. 2002). Mucus has to be porous to allow the passage of absorbed and secreted macromolecules but still it provides a very effective biophysical barrier to the microbes. The importance of mucus for the intestinal homeostasis is best observed in mice deficient in MUC2 mucin which lack the Goblet cells and the mice develop spontaneous colitis (McGuckin et al. 2011; Xavier and Podolsky 2007). In the cecum of GF, mice the number of Goblet cells is reduced (Table 1). There are evidences that both intestinal microflora and the immune system regulate the production of mucus. As already discussed, GF mice have reduced mucus degradation and accumulate it in the caecum (Gustafsson et al. 1970). Microbial metabolite butyrate contributes to the strengthening of the whole mucosal barrier and among other effects it also induces the mucus and AMPs production (Vanhoutvin et al. 2009). On the other side, cytokines of both innate and adaptive immune responses regulate the production and composition of the mucus (McGuckin et al. 2011). Recent data indicate that beside the role in innate immunity, Goblet cells directly regulate the adaptive immune responses by presenting the luminal antigens, preferably to the tolerogenic DC in intestinal LP (McDole et al. 2012).

Table 1. Intestinal Microflora Regulates the Number of Intestinal Immune Cells subtypes.

Cell type	**Cell nr. in intestine of GF compared to CV mice**	**Reference**
Mucosal barrier cells		
Goblet cells	↓in cecal mucosa	(Kandori et al. 1996)
Paneth cells	↓antimicrobial peptide production	(Cash et al. 2006; Hooper et al. 2003)
Entereocytes	Not determined	
Enteroendocrine cells	Not determined	
M-cells	Not determined	
Innate immune cells		
NK cells	No difference in spleen Not determined in intestine	(Sawa et al. 2011)
NKT cells	↑	(Olszak et al. 2012)
LTi cells	No difference	(Sawa et al. 2011)
ILC22	↓IL-22 expression	(Sanos et al. 2009; Satoh-Takayama et al. 2008; Spits and Di Santo 2011)
ILC17	Not determined	
Macrophages ($F4/80^+$ $CD11b^+$)	No difference	(Strachan and Sanders 1989)
Total DC ($CD11c^+$)	No difference	(Niess and Adler 2010; Strachan and Sanders 1989)
$CD11c^+$ $CD103^+$ DC	No difference	(Niess and Adler 2010)
CX_3CR1^+ phagocytes	↓	(Niess and Adler 2010)
Adaptive immune cells		
Total CD4 T lymphocytes	↓	(Mazmanian et al. 2005; Sawa et al. 2011)
Th1 cells	↓	(Mazmanian et al. 2005; Niess et al. 2008)
Th2 cells	↑	(Mazmanian et al. 2005; Niess et al. 2008)
Th17 cells	↓	(Niess et al. 2008; Round et al. 2011)
Treg cells	No difference	(Ishikawa et al. 2008)
LP CD8 T lymphocytes	↓	(Strachan 1989)
CD8 IEL	↓	(Strachan 1989)
B lymphocytes	↓IgA level	(Hapfelmeier et al. 2010; Macpherson and Harris 2004)

↓—reduced; ↑—increased

Paneth cells

The major AMP producing cells in intestine are Paneth cells (Fig. 2). These specialized secretory cells are located close to the stem cells deep in the intestinal crypts and are characterized by the intracellular granules containing a wide range of AMP molecules (Kinnebrew and Pamer 2012; Stappenbeck 2009). These AMPs are evolutionary ancient innate immune effectors that interact directly with the microbial membrane and disrupt it, but they cannot distinguish between commensal and pathogen microbes (Gill et al. 2011; McGuckin et al. 2011). Secretion of AMPs prevents the pathogenic organisms entering the underlying LP, limits the translocation of intestinal microflora and regulates the bacterial growth in lumen (Kinnebrew and Pamer 2012). Intestinal microbita regulate the function of Paneth cells as the expression of AMPs is reduced in GF animals and can be triggered by microbita colonization (Table 1). For full production of AMPs, complex microflora is needed but some single species can also induce the production of certain AMPs, e.g., *Bacteroides thetaiotomicron* induce the secretion of angiogenin-4 (Hooper et al. 2003; Sekirov et al. 2010). Signalling through PRRs was shown to be necessary for the induction of AMPs. TLRs- and NOD2-derived signals induce the secretion of α-defensins (Foureau et al. 2010; Kobayashi et al. 2005), small bactericidal peptides that reach the lumen and shape the composition of intestinal microflora (Hooper et al. 2012). MyD88-dependent signals are necessary for the induction of RegIIIγ expression (Vaishnava et al. 2011; Vaishnava et al. 2008) that is reactive against epithelium-associated Gram-positive bacteria (Hooper et al. 2012; Kinnebrew and Pamer 2012). Of course, the endogenous immune signal also controls the expression of AMPs, with IL-22 produced by both innate and adaptive immune cells being the most potent inducer of their expression (Spits and Di Santo 2011).

Enterocytes

The most numerous cells in the intestinal epithelium are enterocytes (Fig. 2). Their main role is nutrient absorption, so the apical surface of these columnar cells is covered with dense microvillus structures to increase the absorption surface (McGuckin et al. 2011). Enterocytes are closely connected together by the intercellular tight junctions (Fig. 2) that prevent the paracellular passage of even very small molecules and thus contribute to the mucosal barrier (Madara 1998). Contributing to the mucosal barrier function, intestinal microflora can strengthen the tight junctions between the enterocyes by butyrate secretion that directly induce the expression of tight junction proteins (Vanhoutvin et al. 2009). Furthermore, enterocytes produce mucin proteins and some AMPs (Miron and Cristea 2012).

Enterocytes also express PRRs and therefore they can sense the luminal intestinal microbiota (Miron and Cristea 2012). In response to microbial molecules, enterocyte can secrete various cytokines and chemokines that direct the innate and adaptive immune responses (Miron and Cristea 2012; O'Hara and Shanahan 2006). It is also shown that enterocytes express major histocompatibility complex (MHC) molecules and they can serve as APC therefore directly regulating the lymphocyte responses (Miron and Cristea 2012). Direct recognition of microbes by enterocytes is important for the prevention of microbial dissemination as it helps the activation of intestinal lymphocytes. This means that microbiota shape the epithelial cell function and that epithelium is not just a passive barrier but it also actively regulates the immune responses and prevents the penetration of luminal microbes.

Enteroendocrine cells

Gut enteroendocrine cells can secrete various bioactive products in response to the luminal bacterial signals (Garrett et al. 2010) (Fig. 2). These products among others include serotonin, somatostatin, gastrin, motilin, neurotensin, and cholecystokinin (Rindi et al. 2004). Hormones of enteroendocrine cells can be secreted into the blood stream to manifest their effects on the distant tissues or they can locally regulate the functions of other types of intestinal epithelial cells (Garrett et al. 2010; Rindi et al. 2004). The gut hormones of enteroendrocine cells regulate absorption, secretion of mucus and AMPs, motility, cell proliferation and differentiation (Garrett et al. 2010; Rindi et al. 2004). Defects in the functions of enteroendocrine cells, especially in the serotonin production, are correlated to intestinal diseases such as irritable bowel syndrome (IBS) (Coates et al. 2004; Spiller et al. 2000; Spiller 2003). IBS is the most common intestinal disorder and it is found that these patients have alterations in the microflora (Malinen et al. 2005; Posserud et al. 2007). Altered microflora possibly alters the serotonin production by enteroendocrine cells, therefore contributing to the pathology of IBS. In the line with this, some probiotics are shown to be efficient in the treatment of IBS symptoms (Brenner et al. 2009).

M-cells

In contrast to enterocytes, M-cells lack the villus structures on the apical surface (Nagler-Anderson 2001) (Fig. 2). These specialized cells play an important role in the intestinal immune system. They are localized over the PP and ILF in the small intestine and due to expression of various PRRs they efficiently sense the microbes and their metabolites (Garrett et al. 2010) (Fig. 1). Using the transcellular vesicular transport, M-cells carry the luminal antigens and present them to APC of underlying inductive sites of the intestinal immune system (Nagler-Anderson 2001). Therefore, M-cells have fewer lysosomes, ensuring low intracellular antigen degradation and delivery of immunogenic form of antigens to APC (Miron and Cristea 2012). The existence of the M-cells in intestinal epithelium far away from the PP and any lymphoid tissue is also reported. These PP-independent M-cells are called villous M-cells and have the same function as the normal M-cells (Scholtens et al. 2012) (Fig. 1). Existence of these specialized M-cells and villous M-cells ensure the interaction between intestinal microflora and the immune system is localized and regulated and this is essential for the host's defence against pathogens.

Innate Lymphoide Cells

ILCs have been recently described and represent a novel cell family of haematopoietic origin. These cells play a role in the effector responses of innate immunity, in the formation of lymphoid tissue and in post-damage tissue remodelling (Spits and Di Santo 2011). Despite the differences in functions these cells constitute a single family since they are developmentally related and require the transcriptional repressor Id2 and signals through common γ-chain of IL-2 receptor for their development (Spits and Di Santo 2011). This cell family include natural killer (NK) cells with a subtype of NKT cells, lymphoid tissue inducer (LTi) cells, ILC22 and ICL17 cells (Spits and Di Santo 2011). ILCs rapidly respond to the pathogen assault and are crucial for the development of the later response of the host to the intestinal infection.

NK and NKT cells

The typical innate effector cells are NK and NKT cells. Using cytotoxic mechanisms to eliminate infected or damaged cells and IFN-γ production to direct and prime immune responses, these cells efficiently fight the pathogens (Sanos et al. 2009; Yadav et al. 2011). These cells kill their targets in a cell contact-dependent manner either by exocytosis of granulas containing granzimes and perforines or by engagement of death receptors (Yadav et al. 2011). NK cells express specific receptors that recognize cell surface pathogen-induced ligands (Spits and Di Santo 2011). On the other hand, NKT cells have an invariant T cell receptor (TCR) and recognize lipid antigens presented by CD1d molecule and they can secrete IL-4 and IL-13 (Hooper et al. 2012; Olszak et al. 2012). Microflora represents an important regulator of NK and NKT cell functions in the intestine. The number of NK cells in the GF intestine is not determined (Table 1), but colonization of immunodeficient mice with intestinal bacteria induces the production of IFN-γ by intraepithelial NK cells. This effect is not observed in immunocompetent mice indicating that an adaptive immune system regulates the NK cell responses to intestinal microbes (Keilbaugh et al. 2005). GF mice have an increased numbers of NKT cells in the colonic LP and these mice are more sensitive in the NKT cell-dependent colitis model (Table 1). Accumulation of NKT cells in the colon of GF mice was shown to be dependent on CXCL16 chemokine expression and the microbial colonization in neonatal, but not adult period protected these animals from colitis (Olszak et al. 2012). Furthermore, the expression of CXCL16 is differentially regulated in patients with CD (Diegelmann et al. 2010) pointing to the important role of NKT cells in intestinal homeostasis.

LTi cells

The most important haematopoietic cells in the formation of organized lymphoid tissue are LTi cells (Spits and Di Santo 2011). These cells require the transcriptional factor retinoic acid, receptor-related orphan receptor (ROR) γt and IL-7-dependndent signals for their development (Eberl et al. 2004; Yoshida et al. 2002). In the fetal period, LTi cells interact with stromal cells to induce the formation of lymphoid tissue and then secrete many chemokines that attract B and T lymphocytes and DC (Spits and Di Santo 2011). Some of the fetal LTi cells persist in the adulthood, presumably having the role in the repair of damaged lymphoid tissue and in the microflora-dependent formation of ILF in the intestine (Spits and Di Santo 2011). However, the number of LTi cells in the GF intestine is the same as CV (Table 1). LTi cells can produce IL-17 and IL-22 cytokines, indicating their role in immune defence, but the signals that start the production of these cytokines by LTi cells are not yet known (Ouyang et al. 2008).

ILC22 cells

Sharing the characteristics of LTi and NK cells, ICL22 cells are defined by the expression of RORγt transcriptional regulator and surface marker NKp46 (Kinnebrew and Pamer 2012). However these cells do not participate in the lymphoid tissue formation nor do they show the cytotoxic activity and IFN-γ production (Kinnebrew and Pamer 2012; Spits and Di Santo 2011). These cells are mainly located at mucosal sites like the intestine and are characterized by the constitutive production of IL-22 (Sawa et al. 2011; Spits and Di Santo 2011). As already discussed, IL-22 induce the AMPs production by Paneth cells. ILC22 cells are found in GF mice but evidences exist that microbiota induce the differentiation of these cells and their IL-22 production (Table 1). However, one study reported that microbita repress IL-22 production by ILC22 cells (Sawa et al. 2011). The function of ILC22 is indirectly regulated by intestinal pathogens. It has been shown that CX_3CR1^+ mononuclear phagocytes induce IL-22 production by ILC22 cells after the infection with the mouse pathogen *Citrobacter rodentium* (Manta et al. 2012). Further studies are needed to fully describe the role and function of this novel cell type.

ILC17 cells

In contrast to ILC22, ILC17 cells do not express NKp46 surface marker, but express CD90 that can be found on both NK and ILC22 cells (Spits and Di Santo 2011). They also require the transcriptional regulator RORγt for development and functioning (Kinnebrew and Pamer 2012; Spits and Di Santo 2011). Ashe effector molecule ILC17 produce IL-17, the potent pro-inflammatory cytokine (Spits and Di Santo 2011). This is why ILC17 cells accumulate in the intestine during inflammation (Spits and Di Santo 2011), but the possible interactions of these cells with intestinal microflora are not yet known (Table 1).

Myeloide cells

Myeloid cells are non-lymphocyte haematopoietic cells that develop from a common myeloid precursor. The most important myeloid cells in the intestine are monocyte-derived macrophages and DC. These cells are mononuclear, express numerous PRRs and posses the ability of phagocytosis. They densely populate the intestinal LP, forming a complex network (Kinnebrew and Pamer 2012). Both macrophages and DC serve as microbe-sensing and effector cells of the innate immunity, but also have an important role in directing adoptive immune responses.

Macrophages

The most abound phagocytes in the intestine are macrophages (Denning et al. 2007). Located in the LP, these cells sense the bacteria breaking the mucosal barrier and rapidly destroy them by autophagy (Hooper et al. 2012). Macrophages are able to effectively kill intracellular and extracellular bacteria by the fusion of the phagosome with the lysosomal compartment (Amer et al. 2005). A large number of macrophages accumulate in the intestinal mucosa during infection (Kinnebrew and Pamer 2012). These cells produce large amounts of pro-inflammatory mediators like tumour necrosis factor (TNF)-α, IL-1, IL-6, nitric-oxide and chemokines (Mowat

2003). Macrophages effectively regulate the translocation of intestinal microflora across the epithelium, so mice deficient in enzymes involved in the synthesis of reactive oxygen species suffer from wide spread bacteria abscess formation (Shiloh et al. 1999). The presence of microflora activates the intestinal macrophage since neonatal macrophages, after stimulation with lipopolysaccharide (LPS) showed reduced TNF-α and IL-1 production (Levy 2007; Philbin and Levy 2009). However, the number of these cells is the same in the intestine of GF and CV mice (Table 1).

Dendritic cells

The population of professional APC that coordinate both innate and adaptive immune responses are called DC. These cells are located at the intestinal LP and at the inductive sites. They express various PRRs and after encountering the antigen they are able to effectively process it and present it to the lymphocytes in an immunogenic active form at the cell surface coupled with MHC molecules. Antigen coupled with the MHC molecule is the first signal required for the full activation of lymphocytes and the second signal is provided by co-stimulatory molecules expressed also at the DC surface (Abbas et al. 2007). Furthermore, antigen activated DC secrete various cytokines and chemokines that regulate the immune responses in the intestine. The expression of CD11c surface marker is often used for the characterization of DC, but some well defined macrophages also express CD11c (Mowat and Bain 2011). In the murine intestine, $CD11c^{+}$ DC can be divided into CD103 of CX_3CR1 (fractalkine receptor) expressing populations (Mowat and Bain 2011; Niess and Adler 2010). After antigen encounter in LP, $CD103^{+}$ DC migrate to the MLN where they prime the naïve lymphocyte and imprint the gut-homing markers on them to make sure the primed cells will migrate to the same location where the antigen was found (Mowat and Bain 2011; Pabst et al. 2007). Studies showed that intestinal $CD103^{+}$ DC are especially efficient in the induction of regulatory T cells (Tregs) and this ability was attributed to their secretion of retinoic acid (RA) and transforming growth factor (TGF)-β (du Pre and Samsom 2011; Pabst et al. 2007). On the other side, some of the CX_3CR1^{+} $CD11c^{+}$ cells in murine intestine also express the typical macrophage markers (F4/80, CD68) (Mowat and Bain 2011; Niess and Adler 2010; Rescigno et al. 2001). These cells can extend their cellular processes in between the intestinal epithelial cells, directly sampling the luminal antigens (Niess et al. 2005), but the significance of this is still not clear. However, the intestinal microbiota promote the formation of the transepithelial cellular processes by CX_3CR1^{+} cells (Niess and Adler 2010) indicating that they are important for the host-microbiota interactions. There is no evidence that CX_3CR1^{+} DC migrate to MLN (Schulz et al. 2009), so they probably prime the lymphocytes directly in LP. GF mice have the same number of total $CD11c^+$ and $CD103^+$ DC subtype in the intestine, but the number of CX_3CR1^+ phagocytes is reduced compared to CV mice (Table 1). However it is clearly shown that microflora regulate the functions of intestinal DC since in the neonatal period the number of these cells is decreased and they show the immature phenotype with reduced MHC and co-stimulatory molecules expression, as well as reduced pro-inflammatory cytokine production after *in vitro* PRRs stimulation (Levy 2007; Philbin and Levy 2009). *Lactobacillus* spp. induce the maturation of intestinal DC through class II MHC (MHC-II) molecule and co-stimulatory CD86 molecule expression (Christensen et al. 2002). However different *Lactobacillus* strains induce different cytokine responses by DC, some promote IL-12 and TNF-α production while the others promote IL-10 and IL-6 (Christensen et al. 2002).

The Adaptive Immune System in the Intestine

The cells of the adaptive immune system, lymphocytes, are able to start very effective antigen-specific immune responses activating the immune mechanisms that will most efficiently fight the given antigen. In contrast to the cells of the innate immunity, lymphocyte create the appropriate immune memory for the case of repeated infection with the same antigen (Abbas et al. 2007). Memory lymphocytes react more rapidly and robustly, ensuring the fastest possible clearance of the repeated infection. As already discussed, for full activation lymphocytes need help of the APC, most often DC but macrophages and enterocytes can also function as APC in the intestine (Mowat 2003; Mowat and Bain 2011; Niess and Adler 2010). Normally, microflora-derived antigens do not induce pro-inflammatory lymphocyte responses, but it is believed that these antigens are constantly processed by intestinal DC and presented to the lymphocytes to induce the state of normal physiologic inflammation in the intestine. The disruption of this delicate balance in the intestine leads to exaggerated immune responses to the intestinal microbiota and to the pathology. There are two types of adaptive immune responses: cellular and humoral.

Cellular Immunity

Cellular immune responses are mediated by T lymphocytes. They express TCR that recognizes protein antigens coupled with MHC-II for CD4 T lymphocytes and MHC-I for CD8 T lymphocytes. The effector mechanism of CD4 T lymphocytes is mainly the secretion of various cytokines and chemokines that direct the innate and adaptive immune responses. That is why these cells are calls T helper (Th) lymphocytes. Depending of the type of antigen, DC induce the differentiation of different types of Th cells (Th1, Th2, Th17 or Tregs) ensuring the most efficient response to the antigen. CD8 T lymphocytes possess cytotoxic activity that is similar to NK cells with the difference that CD8 T lymphocytes recognize only protein antigens presented by MHC-I molecules on the surface of target cells. These cells are also called cytotoxic lymphocytes (CTL). $CD8^+$ CTL are dominant T lymphocyte type in the intraepithelial intestinal compartment while $CD4^+$ Th cells with the memory phenotype are dominant in LP (Smith and Garrett 2011).

Th1 cells

In response to the intracellular pathogens, DC drive the generation of Th1 CD4 T cells from naïve lymphocytes. Beside the TCR-mediated signals, these cells need IL-12, IFN type I or type II that will activate the signal transducer and activator of transcription (Stat)1, Stat4 and T-box transcription factor (T-bet) necessary for their differentiation (Szabo et al. 2000). Th1 cells produce IFN-γ, TNF-α and IL-12, cytokines that induce the cytotoxic activity of NK cells and CTL lymphocytes, recruit more of these cells on the site of infection and by positive feed-back loop, ensure further maturation of DC and generation of Th1 cells. These events lead to the efficient clearance of the intracellular pathogen that started the response.

Th2 cells

Helmints and parasites induce Th2 CD4 T cells responses by promoting IL-4, IL-5 and IL-13 production by DC. These cytokines together with TCR signals activate the GATA-3 transcriptional factor that governs the Th2 differential pathway (Zheng and Flavell 1997). Th2 cells secrete IL-4, IL-5, IL-10 and IL-13 that induce further generation of Th2 cells and immunoglobulin (Ig)E production by B lymphocytes. IgE antibodies bind to helmints and parasites, ensuring their elimination. Exacerbated Th2 responses are often connected to the pathology in lungs where they can cause asthma, but also in the intestine since some ulcerative colitis (UC) patients have increased levels of IL-13, the Th2 specific cytokine (Strober et al. 2007; van Wijk and Cheroutre 2010).

Th17 cells

The newest pro-inflammatory T cell type typically induced by extracellular bacteria is named Th17. These cells are characterized by the expression of transcriptional factors RORγt, interferon regulatory factor (IRF)4 and Stat3

(Ivanov et al. 2006). Th17 cells require IL-6 or IL-21 together with TGF-β for their differentiation, while IL-23 is needed for maintenance and expansion of these cells (Korn et al. 2007; Zhou et al. 2007). As the effector cytokines Th17 cells produce IL-17A, IL-17F, IL-21 and IL-22 that further direct the immune responses. Studies showed that Th17 cells are necessary for the protection from *Citrobacter rodentium, Klebsiella pneumoniae* (Happel et al. 2005) and *Salmonella typhimurium* (Godinez et al. 2009; Schulz et al. 2008). However, IL-17 producing lymphocytes seem to be involved in IBD pathogenesis, but the exact role of Th17 cells in this process is still to be determined.

Tregs cells

Active suppression of the immune responses by Treg cells is particularly important in the intestine to prevent the excessive responses to numerous microflora-derived antigens. This is why Treg cells are numerous in intestinal LP and several types of them are recognized. The most studied are forkhead box protein (Foxp)3$^+$ Treg cells. Foxp3 is the transcriptional factor required for both differentiation and functioning of these cells. TGF-β and RA can induce Foxp3 expression and therefore Treg cells generation. Foxp3 Tregs suppress the immune responses through the cell contact-dependent mechanism using the membrane bound form of TGF-β, but these cells can also produce the secretory form of TGF-β that suppress the immune responses without cell contact. These cells are shown to suppress the proliferation of CD4 T cells and the differentiation of Th1 and Th2 lymphocytes *in vitro* (Chen et al. 2003; du Pre and Samsom 2011; Fantini et al. 2006; Gorelik and Flavell 2002). Studies showed that Foxp3$^+$ Treg cells are protective *in vivo* in murine models of colitis (Haribhai et al. 2009; Murai et al. 2009), arthritis (Kelchtermans et al. 2005), autoimmune diseases (Huter et al. 2008; Zhang et al. 2010), and even tumor treatments in humans (Clarke et al. 2006; Kobayashi et al. 2007). Mice with genetic deletion of Foxp3 gene develop lymphopreliferative autoimmune syndrome (Brunkow et al. 2001) and humans with inactive Foxp3 gene develop immunodysregulation-polyendocrinopathy-entheropathy-X-linked (IPEX) syndrome (Bennett et al. 2001). TGF-β-producing Th3 cells induced in the murine models of oral tolerance are also considered Tregs since they showed the suppressive properties (du Pre and Samsom 2011; Faria and Weiner 2006; Li et al. 2006; Mowat 2003). Besides TGF-β, IL-10 is recognized as anti-inflammatory cytokine since it can inhibit the secretion of pro-inflammatory cytokines (Garside et al. 1999). IL-10 producing T regulatory type 1 (Tr1) cells can be differentiated *in vitro* in the presence of antigen, IL-10 or type I IFN (Mowat 2003). These cells *in vivo* inhibited a number of disease models in mice including colitis in IL-10-dependent manner (Izcue et al. 2009).

Microflora regulate the CD4 T cell responses in intestine

The number of CD4 T cells in LP of CV neonatal and adult GF mice is greatly reduced (Hapfelmeier et al. 2010; Macpherson and Harris 2004; Niess et al. 2008; Sawa et al. 2011; Steege et al. 1997). GF mice are also characterized by lack of expansion of CD4 T cells, but this can be reversed by colonization with *Bacteroides fragilis* or treatment with purified polysaccharide A (PSA) of this bacterial specie (Sekirov et al. 2010). CD4 T cells of gut mucosa are normally in an activated state due to the conditions of physiological inflammation, but in GF mice these cells do not express the typical surface markers of activated cells CD45RB^{low} and CD69 (Cebra 1999; Radulovic et al. 2012). Th17 cells are completely missing from the gut mucosa of GF mice (Table 1) and the systemic CD4 T cell response is skewed toward Th2 type (Kaplan et al. 2011; Mazmanian et al. 2005; Sekirov et al. 2010). Therefore, the number of Th1 cells is reduced, but Th2 increased in the intestine of GF mice (Table 1). The fetal immune system is also Th2-based, presumably to avoid the pro-inflammatory Th1 and Th17 responses that can lead to maternal rejection of the fetus or pre-term delivery (Dimmitt et al. 2010; Philbin and Levy 2009). In neonates, upon microflora colonization, CD4 T responses switch to Th1 and Th17 to facilitate the pathogen defence (Dimmitt et al. 2010). *Bacteroides fragilis*-derived PSA can correct the Th17 deficiency and aberrant Th1/Th2 ratio in GF mice (Mazmanian et al. 2005; Round et al. 2011; Sekirov et al. 2010). Also, monocolonization of GF mice with spore-forming segmented filamentous bacteria (SFB), *Escherichia coli* or *Escherichia faecalis* induce Th17 and Th1 responses (Chinen and Rudensky 2012). SFB were shown to be the most efficient in the Th17 cell induction since they adhere to the intestinal epithelium and induce the expression of cytokines necessary for Th17 cell differentiation (Arboleya et al. 2012). The frequency and number of Tregs in MLN, but not intestinal LP, of GF animals is reduced and these cells showed less suppressive activity *in vitro* compared to CV Treg cells (Ishikawa et al. 2008). *Clostridium* species are shown to be the best inducers of Foxp3 Treg cells and Tr1 cells leading to the down-regulation of pro-inflammatory responses and induction of tolerance (Atarashi et al. 2011). Hence, intestinal microflora are able to prompt the generation of different CD4 T cell subsets in the different conditions, being therefore crucial in the pathogenesis of the intestinal diseases.

CD8 T cells

Residing mainly in the intestinal intraepithelial compartment, CD8 CTL together with innate immune cells represents the first line of the defence against intestinal pathogens. CD8 T responses efficiently eliminate the infection with intracellular parasite *Listeria monocytogenes* (Huang et al.

2011) and viruses (Isakov et al. 2011). However, some studies showed that in contrast to systemic CTL, intestinal CD8 T cells also have a regulatory role and promote oral tolerance induction (Li et al. 1995; Metwali et al. 2006). Since these cells are in close contact with the luminal content, it is possible that they are important for the tolerogenic reactions to microflora. Murine neonatal have reduced number of intraepithelial lymphocytes (Steege et al. 1997) and adult GF intestine has reduced numbers of both LP and intraepithelial CD8 T lymphocytes (Table 1). Adult GF mice have reduced numbers of both LP and Intraepithelial CD8 T cells (O'Hara and Shanahan 2006; Strachan 1989). Microbial colonization induces the clonal expansion of CD8 T cells locally in the gut (Helgeland et al. 2004). The most important microflora members that induce CD8 T cell proliferation and maturation are SFB (Cebra 1999).

Humoral Immunity

The effector cells of the humoral immunity are B lymphocytes. These cells are numerous among intestinal LP lymphocytes. When primed in the B cell folicles of inductive sites by APC, B lymphocytes mature into the Ig-producing plasma cells. Depending on the type of recognized antigen and also on the cell location, plasma cells can produce different isotypes of Ig antibodies. Every B cell developed with the ability to produce IgD antibodies and after priming the class-switching of Ig genes, can change the Ig isotype to IgM, IgG, IgE or IgA. Normally, IgA is produced exclusively in mucosal surfaces such as intestine, lungs and urogenital tract. About three quarters of all Ig production in a mammalian body is IgA and the daily production of it by the intestinal plasma cells is about 40 mg/kg body weight (Macpherson et al. 2012; Ouwehand et al. 2002). DC loaded with live bacteria that they encounter in LP prime the B cells in intestinal inductive sites and probably in a TGF-β-dependent manner, promote the class-switching to IgA (Coffman et al. 1989; Macpherson et al. 2012). IgA induction is shown to be very effective with the help of CD4 T cells and their cytokines (T cell-dependent pathway of IgA induction), but it is shown that DC alone can induce IgA without the presence of CD4 T cells (T cell-independent IgA induction) (Macpherson et al. 2012). Produced in LP, IgA is actively transported to the intestinal lumen by enterocytes. During this transport, IgA retains the part of the receptor molecule bound to it and this form of antibody that can be found in intestinal lumen is called secretory IgA (SIgA). SIgA stays in the mucus layer where it can bind microflora- and dietary-derived antigens and preventing them from reaching the epithelial layer and starting the immune responses (Cerutti and Rescigno 2008; Lindner et al. 2012; Peterson et al. 2007). Furthermore, IgA can bind the microbes that broke the mucosal barrier and reached the intestinal LP and tranlocate them back into the lumen through the same transport system on enterocytes that make SIgA (Pasquier et al. 2005; Phalipon et al. 2002). Therefore IgA and the plasma cells producing it are very important for maintaining the host-microbota symbiotic relationships and in protection from pathogens. Mice impaired in IgA production show the expansion of anaerobic bacteria in the gut with increased susceptibility to intestinal pathogens (Chinen and Rudensky 2012). Commensal bacteria drive the production of IgA and in CV animals the amount of IgA is greatly increased compared to GF (Table 1). Gram-negative *Bacteroides* were more effective in IgA induction than Gram-positive *Lactobacilli* (Sekirov et al. 2010). However, early colonization with SFB strongly induced the generation of IgA-producing plasma cells in intestinal LP (Cebra 1999). After efficient colonization of GF mice, bacteria stayed in the intestinal lumen for over a year continuously coated with IgA (Cebra 1999). Intestinal colonization of GF mice deficient in B cells led to the increased translocation of bacteria to MLN confirming that IgA production does prevent the penetration of intestinal microflora into the host body (Lindner et al. 2012). Microflora also increased the diversity of IgA by promoting the somatic mutations in the IgA genes of intestinal B cells by yet unknown mechanism (Lindner et al. 2012).

Unresponsiveness to Intestinal Microflora

In healthy individuals, inflammatory immune responses against intestinal microflora are not observed. If such immune responses do occur they lead to the intestinal pathology and IBD development. Being in constant contact with microflora, the intestinal immune system developed effective mechanisms to suppress the inflammatory responses to microflora-derived antigens. On the other hand, different hypotheses try to explain how the systemic immune system deals with intestinal microflora presence.

The Intestinal Immune System Actively Tolerates the Microflora

Intestinal immune system recognizes microflora-derived antigens as self antigens and favours the induction of Tregs and the production of IgA in the response to them. Specific Treg-inducing CD103$^+$ DC can be found only at mucosal surfaces. Intestinal Treg cells can strongly bind the antigens of heat-killed luminal content of CV mice, but not GF mice (Littman and Pamer 2011). The generation of both Foxp3 Tregs and Th3 cells is observed in the murine models of oral tolerance induction. As already discussed, IgA prevents the penetration of luminal antigens and exclude already penetrated ones from intestinal LP and it is demonstrated that after the monocolonization of GF mice, intestinal plasma cells produce IgA directed against specific bacterial specie used for monocolonization (Macpherson et al. 2012). Furthermore, IgA is shown to be the tolerant Ig isotype since it is not efficient in the initiation of pro-inflammatory responses (Nagler-Anderson 2001). The regulatory cytokines IL-10 and TGF-ß abound in the intestine and their signals are necessary for tolerance of microflora since mice deficient in IL-10, IL-10R, TGF-β1 or TGF-βRII develop spontaneous colitis which can be treated with antibiotics (Kulkarni et al. 1993; Madsen 2001; Shull et al. 1992). Furthermore, IL-10 knockout mice showed the signs of disease only when housed in CV, but not when housed in GF conditions (Sellon et al. 1998).

Enterocytes are shown to have a very important role in microflora tolerance. Expression of PRRs allows enterocytes to recognize and respond to microbial antigens. However, symbiotic and pathogenic microbes express the same PAMPS and the question is how the host can distinguish between harmless and harmful luminal microbes? The answer to this question is not yet clear but recent studies have offered the convincing explanation. Namely, pathogenic bacteria behave differently from commensal ones as the pathogens try to break the mucosal barrier while commensals stay peacefully in the lumen. Therefore, pathogens would preferably interact with PRRs on the basolateral sides of entrocytes, while commensals with the PRRs on the apical side (Fava and Danese 2011). PRRs are not expressed in high levels on the apical surface of enterocytes and studies of TLRs showed that their activation on apical surface triggers different intracellular responses compared to the activation on basolateral surface (Chinen and Rudensky 2012; Kinnebrew and Pamer 2012). Basolateral TLR9 signals activate NFkB while apical TLR9 signals actually prevent NFkB activation (Lee et al. 2006). TLR5 ligand flagelin induce the pro-inflammatory response of enterocytes only when introduced at the basolateral side of the cells (Rhee et al. 2005). Enterocytes express TLR4 only at intracellular Golgi apparatus ensuring its activation only if the ligand, lipopolysaccharide (LPS) penetrate into the cell (Espevik et al. 2003). This polarization of PRR responses in enterocytes enables the tolerance of microbes that stay in the intestinal lumen. Furthermore, enterocytes are responsible for the high amount of regulatory cytokines in the intestine. Enterocytes produce TGF-β and thymic stromal lymphopoietin (TSLP) which induce IL-10 production by $CD103^+$ DC (Miron and Cristea 2012).

Tolerance or Ignorance of Microflora by the Systemic Immune System

According to the first hypothesis, the systemic immune system also actively sense and tolerate microflora-derived antigens. This hypothesis is supported by findings that oral administration of antigen induces systemic unresponsiveness to substantial subcutaneous administration of the same antigen (Scurlock et al. 2010). This is how experimental oral tolerance models are designed. Study of this phenomenon showed that MLN and CD4 T cells are necessary for oral tolerance induction and systemic CD4 T cells can transfer oral tolerance *in vivo* (Chen et al. 1994; Garside et al. 1995; Hirahara et al. 1995; Worbs et al. 2006). Furthermore, oral tolerance is transferable through serum (Strobel et al. 1983) and it is discovered that supra-molecular, exosome-like structures called tolerozomes mediate this effect (Karlsson et al. 2001). Tolerozomes originate from enterocytes that produce them as MHC-II-associated peptides packed in exosomal form and they detach from the basal surface (Karlsson et al. 2001; Miron and Cristea 2012; Ostman et al. 2005). Protective effects of antigen pre-feeding and oral tolerance induction are shown in murine models of autoimmune and inflammatory diseases such as experimental autoimmune encephalomyelitis (Chen et al. 1995; Chen et al. 1994), arthritis (Zhang et al. 2001), colitis (Neurath et al. 1996), diabetes (Bergerot et al. 1999) and asthma (Almqvist et al. 2008).

According to the second hypothesis, the systemic immune system does not even sense the microflora-derived antigens due to anatomical separation of the intestinal immune system from the systemic one (Macpherson et al. 2012). Instead of tolerating intestinal microflora, the systemic immune system ignores it. Supporting this hypothesis, studies showed that there is no systemic Ig response to microflora and orally delivered antigens induce only local IgA response in the intestine (Konrad et al. 2006; Macpherson et al. 2000). Absence of TLR-signaling in MyD88/TRIF double knockout mice led to the systemic IgG response to intestinal microbes due to the defective elimination of small number of commensals that break the microbial barrier (Macpherson et al. 2012). Penetration of orally delivered commensal bacteria in GF animals is limited only to MLN and these bacteria do not reach the systemic circulation, but if mice lack MLN then bacteria can be found in the spleen (Macpherson and Uhr 2004). Normally, DC that encounter commensals in LP carry live bacteria to the inductive sites of intestinal immune system but do not reach further than MLN (Hooper et al. 2012; Macpherson and Uhr 2004). In this way the mucosal compartmentalization minimizes the exposure of intestinal microflora to the systemic immune system and promotes the ignorance.

Manipulation of the Microflora for the Benefit of the Host

Understanding of the host-microflora interactions in the intestine led to the development of therapeutic strategies based on microflora manipulation. As already discussed, some bacterial species promote inflammatory while the others promote regulatory immune responses. The latest are being studied intensively as candidates for the treatment of intestinal inflammatory disorders. These species are named probiotic and defined as "live microorganisms that are beneficial to the host beyond nutritional value" (Kaplan et al. 2011).

Some *Lactobacillus* species showed convincing probiotic effects in animal models of IBD (Fabia et al. 1993; Mao et al. 1996; Whary et al. 2011) but the clinical studies in IBD patients so far have not shown significant benefits (Gupta et al. 2000; Rolfe et al. 2006; Van Gossum et al. 2007). *Bacteroides thetanotaomicron* showed strong anti-inflammatory effects in *in vitro* experimental models (Kelly et al. 2004) but in animal models this bacterial strain promoted the disease (Hansen et al. 2012). Certain beneficial effects in patients with IBD are reported after treatment with non-pathogenic *Escherichia coli Nissle* (Kruis et al. 2004; Matthes et al. 2010), *Saccharomyces boulardii* (Guslandi et al. 2003; Guslandi et al. 2000) and some *Bifidobacteria* strains (Cui et al. 2004; Ishikawa et al. 2003). However, *Bifidobacteria* strains differ drastically in their anti-inflammatory effects and recent studies in mice pointed out to *Bifidobacterium bifidum S17* strain as the best candidate for IBD treatment (Philippe et al. 2011; Preising et al. 2010; Riedel et al. 2006). *Clostridium* species that can induce Treg cells are also candidates for probiotc studies.

Future therapeutic strategies will consider even the genetically manipulated bacteria as treatment or vaccine for inflammatory disorders. In this area some studies on animal models have shown remarkable success (Foligne et al. 2007; Steidler et al. 2000), but more details regarding the safety of this therapy is needed. However, probiotics are a promising therapeutic approach for the treatment of intestinal disorders.

Conclusion

There is a rising number of data showing the great importance of microflora for the development and functioning of the intestinal immune system. The presence of microflora changes the characteristic and functions of immune cells and different microbial commensal species influence different cell subtypes. The molecular mechanisms and pathways of the host-microbiota interactions are being discovered, introducing the exciting and promising possibility for use of specific species of commensal bacteria or the products of their metabolism for the therapeutic treatment of inflammatory diseases affecting intestinal tissues.

Acknowledgement

This work was supported by DFG grants Ni575/6-2 and Ni575/7-1 to J.H.N.

References Cited

Abbas, A.K., A.H. Lichtman and P. Shiv. 2007. Cellular and molecular immunology. SAUNDERS Elsevier. USA.

Abraham, C. and J.H. Cho. 2009. Inflammatory bowel disease. N. Engl. J. Med. 361: 2066–2078.

Adlerberth, I., D.P. Strachan, P.M. Matricardi, S. Ahrne, L. Orfei, N. Aberg, M.R. Perkin, S. Tripodi, B. Hesselmar, R. Saalman, A.R. Coates, C.L. Bonanno, V. Panetta and A.E. Wold. 2007. Gut microbiota and development of atopic eczema in 3 European birth cohorts. J. Allergy. Clin. Immunol. 120: 343–350.

Akira, S. and K. Takeda. 2004. Toll-like receptor signalling. Nat. Rev. Immunol. 4: 499–511.

Almqvist, N., A. Lonnqvist, S. Hultkrantz, C. Rask and E. Telemo. 2008. Serum-derived exosomes from antigen-fed mice prevent allergic sensitization in a model of allergic asthma. Immunology. 125: 21–27.

Amer, A.O., B.G. Byrne and M.S. Swanson. 2005. Macrophages rapidly transfer pathogens from lipid raft vacuoles to autophagosomes. Autophagy. 1: 53–58.

Arboleya, S., A. Binetti, N. Salazar, N. Fernandez, G. Solis, A. Hernandez-Barranco, A. Margolles, C.G. de Los Reyes-Gavilan and M. Gueimonde. 2012. Establishment and development of intestinal microbiota in preterm neonates. FEMS Microbiol. Ecol. 79: 763–772.

Atarashi, K., T. Tanoue, T. Shima, A. Imaoka, T. Kuwahara, Y. Momose, G. Cheng, S. Yamasaki, T. Saito, Y. Ohba, T. Taniguchi, K. Takeda, S. Hori, Ivanov, II, Y. Umesaki, K. Itoh and K. Honda. 2011. Induction of colonic regulatory T cells by indigenous Clostridium species. Science. 331: 337–341.

Atuma, C., V. Strugala, A. Allen and L. Holm. 2001. The adherent gastrointestinal mucus gel layer: thickness and physical state *in vivo*. Am. J. Physiol. Gastrointest. Liver. Physiol. 280: G922–929.

Backhed, F., R.E. Ley, J.L. Sonnenburg, D.A. Peterson and J.I. Gordon. 2005. Host-bacterial mutualism in the human intestine. Science. 307: 1915–1920.

Beaugerie, L. and J.C. Petit. 2004. Microbial-gut interactions in health and disease. Antibiotic-associated diarrhoea. Best. Pract. Res. Clin. Gastroenterol. 18: 337–352.

Bennett, C.L., J. Christie, F. Ramsdell, M.E. Brunkow, P.J. Ferguson, L. Whitesell, T.E. Kelly, F.T. Saulsbury, P.F. Chance and H.D. Ochs. 2001. The immune dysregulation, polyendocrinopathy, enteropathy, X-linked syndrome (IPEX) is caused by mutations of FOXP3. Nat. Genet. 27: 20–21.

Bergerot, I., G.A. Arreaza, M.J. Cameron, M.D. Burdick, R.M. Strieter, S.W. Chensue, S. Chakrabarti and T.L. Delovitch. 1999. Insulin B-chain reactive CD4+ regulatory T-cells induced by oral insulin treatment protect from type 1 diabetes by blocking the cytokine secretion and pancreatic infiltration of diabetogenic effector T-cells. Diabetes. 48: 1720–1729.

Black, J.G. 1996. Microbiology Principles and Applications. Prantice hall, New Jersey. USA.

Bouskra, D., C. Brezillon, M. Berard, C. Werts, R. Varona, I.G. Boneca and G. Eberl. 2008. Lymphoid tissue genesis induced by commensals through NOD1 regulates intestinal homeostasis. Nature. 456: 507–510.

Brenner, D.M., M.J. Moeller, W.D. Chey and P.S. Schoenfeld. 2009. The utility of probiotics in the treatment of irritable bowel syndrome: a systematic review. Am. J. Gastroenterol. 104: 1033–1049; quiz 1050.

Brunkow, M.E., E.W. Jeffery, K.A. Hjerrild, B. Paeper, L.B. Clark, S.A. Yasayko, J.E. Wilkinson, D. Galas, S.F. Ziegler and F. Ramsdell. 2001. Disruption of a new forkhead/winged-helix protein, scurfin, results in the fatal lymphoproliferative disorder of the scurfy mouse. Nat. Genet. 27: 68–73.

Cash, H.L., C.V. Whitham, C.L. Behrendt and L.V. Hooper. 2006. Symbiotic bacteria direct expression of an intestinal bactericidal lectin. Science. 313: 1126–1130.

Cebra, J.J., S.B. Periwal, G. Lee, F. Lee and K.E. Shroff. 1998. Development and maintenance of the gut-associated lymphoid tissue (GALT): the roles of enteric bacteria and viruses. Dev. Immunol. 6: 13–18.

Cebra, J.J. 1999. Influences of microbiota on intestinal immune system development. Am. J. Clin. Nutr. 69: 1046S–1051S.

Cerutti, A. and M. Rescigno. 2008. The biology of intestinal immunoglobulin A responses. Immunity. 28: 740–750.

Chen, W., W. Jin, N. Hardegen, K.J. Lei, L. Li, N. Marinos, G. McGrady and S.M. Wahl. 2003. Conversion of peripheral CD4+CD25- naive T cells to CD4+CD25+ regulatory T cells by TGF-beta induction of transcription factor Foxp3. J. Exp. Med. 198: 1875–1886.

Chen, Y., J. Inobe and H.L. Weiner. 1995. Induction of oral tolerance to myelin basic protein in CD8-depleted mice: both CD4+ and CD8+ cells mediate active suppression. J. Immunol. 155: 910–916.

Chen, Y., V.K. Kuchroo, J. Inobe, D.A. Hafler and H.L. Weiner. 1994. Regulatory T cell clones induced by oral tolerance: suppression of autoimmune encephalomyelitis. Science. 265: 1237–1240.

Chinen, T. and A.Y. Rudensky. 2012. The effects of commensal microbiota on immune cell subsets and inflammatory responses. Immunol. Rev. 245: 45–55.

Christensen, H.R., H. Frokiaer and J.J. Pestka. 2002. Lactobacilli differentially modulate expression of cytokines and maturation surface markers in murine dendritic cells. J. Immunol. 168: 171–178.

Clarke, S.L., G.J. Betts, A. Plant, K.L. Wright, T.M. El-Shanawany, R. Harrop, J. Torkington, B.I. Rees, G.T. Williams, A.M. Gallimore and A.J. Godkin. 2006. CD4+CD25+FOXP3+ regulatory T cells suppress anti-tumor immune responses in patients with colorectal cancer. PLoS One. 1: e129.

Coates, M.D., C.R. Mahoney, D.R. Linden, J.E. Sampson, J. Chen, H. Blaszyk, M.D. Crowell, K.A. Sharkey, M.D. Gershon, G.M. Mawe and P.L. Moses. 2004. Molecular defects in mucosal serotonin content and decreased serotonin reuptake transporter in ulcerative colitis and irritable bowel syndrome. Gastroenterology. 126: 1657–1664.

Coffman, R.L., D.A. Lebman and B. Shrader. 1989. Transforming growth factor beta specifically enhances IgA production by lipopolysaccharide-stimulated murine B lymphocytes. J. Exp. Med. 170: 1039–1044.

Costello, E.K., K. Stagaman, L. Dethlefsen, B.J. Bohannan and D.A. Relman. 2012. The application of ecological theory toward an understanding of the human microbiome. Science. 336: 1255–1262.

Cui, H.H., C.L. Chen, J.D. Wang, Y.J. Yang, Y. Cun, J.B. Wu, Y.H. Liu, H.L. Dan, Y.T. Jian and X.Q. Chen. 2004. Effects of probiotic on intestinal mucosa of patients with ulcerative colitis. World. J. Gastroenterol. 10: 1521–1525.

Davis, B.K., H. Wen and J.P. Ting. 2011. The inflammasome NLRs in immunity, inflammation, and associated diseases. Annu. Rev. Immunol. 29: 707–735.

Denning, T.L., Y.C. Wang, S.R. Patel, I.R. Williams and B. Pulendran. 2007. Lamina propria macrophages and dendritic cells differentially induce regulatory and interleukin 17-producing T cell responses. Nat. Immunol. 8: 1086–1094.

Diegelmann, J., J. Seiderer, J.H. Niess, D. Haller, B. Goke, H.C. Reinecker and S. Brand. 2010. Expression and regulation of the chemokine CXCL16 in Crohn's disease and models of intestinal inflammation. Inflamm. Bowel. Dis. 16: 1871–1881.

Dimmitt, R.A., E.M. Staley, G. Chuang, S.M. Tanner, T.D. Soltau and R.G. Lorenz. 2010. Role of postnatal acquisition of the intestinal microbiome in the early development of immune function. J. Pediatr. Gastroenterol. Nutr. 51: 262–273.

Douwes, J., S. Cheng, N. Travier, C. Cohet, A. Niesink, J. McKenzie, C. Cunningham, G. Le Gros, E. von Mutius and N. Pearce. 2008. Farm exposure in utero may protect against asthma, hay fever and eczema. Eur. Respir. J. 32: 603–611.

du Pre, M.F. and J.N. Samsom. 2011. Adaptive T-cell responses regulating oral tolerance to protein antigen. Allergy. 66: 478–490.

Eberl, G., S. Marmon, M.J. Sunshine, P.D. Rennert, Y. Choi and D.R. Littman 2004. An essential function for the nuclear receptor RORgamma(t) in the generation of fetal lymphoid tissue inducer cells. Nat. Immunol. 5: 64–73.

Ege, M.J., M. Mayer, A.C. Normand, J. Genuneit, W.O. Cookson, C. Braun-Fahrlander, D. Heederik, R. Piarroux, E. von Mutius and G.T.S. Group. 2011. Exposure to environmental microorganisms and childhood asthma. N. Engl. J. Med. 364: 701–709.

Elinav, E., T. Strowig, J. Henao-Mejia and R.A. Flavell. 2011. Regulation of the antimicrobial response by NLR proteins. Immunity. 34: 665–679.

Espevik, T., E. Latz, E. Lien, B. Monks and D.T. Golenbock. 2003. Cell distributions and functions of Toll-like receptor 4 studied by fluorescent gene constructs. Scand. J. Infect. Dis. 35: 660–664.

Fabia, R., A. Ar'Rajab, M.L. Johansson, R. Willen, R. Andersson, G. Molin and S. Bengmark. 1993. The effect of exogenous administration of Lactobacillus reuteri R2LC and oat fiber on acetic acid-induced colitis in the rat. Scand. J. Gastroenterol. 28: 155–162.

Falth-Magnusson, K. and N.I. Kjellman. 1992. Allergy prevention by maternal elimination diet during late pregnancy--a 5-year follow-up of a randomized study. J. Allergy. Clin. Immunol. 89: 709–713.

Fanaro, S., R. Chierici, P. Guerrini and V. Vigi. 2003. Intestinal microflora in early infancy: composition and development. Acta. Paediatr. Suppl. 91: 48–55.

Fantini, M.C., C. Becker, I. Tubbe, A. Nikolaev, H.A. Lehr, P. Galle and M.F. Neurath. 2006. Transforming growth factor beta induced FoxP3+ regulatory T cells suppress Th1 mediated experimental colitis. Gut. 55: 671–680.

Faria, A.M. and H.L. Weiner 2006. Oral tolerance: therapeutic implications for autoimmune diseases. Clin. Dev. Immunol. 13: 143–157.

Faust, K., J.F. Sathirapongsasuti, J. Izard, N. Segata, D. Gevers, J. Raes and C. Huttenhower. 2012. Microbial Co-occurrence Relationships in the Human Microbiome. PLoS. Comput. Biol. 8: e1002606.

Faust, K. and J. Raes. 2012. Microbial interactions: from networks to models. Nat. Rev. Microbiol. 10: 538–550.

Fava, F. and S. Danese. 2011. Intestinal microbiota in inflammatory bowel disease: friend of foe? World. J. Gastroenterol. 17: 557–566.

Foligne, B., R. Dessein, M. Marceau, S. Poiret, M. Chamaillard, B. Pot, M. Simonet and C. Daniel. 2007. Prevention and treatment of colitis with Lactococcus lactis secreting the immunomodulatory Yersinia LcrV protein. Gastroenterology. 133: 862–874.

Fouhy, F., R.P. Ross, G.F. Fitzgerald, C. Stanton and P.D. Cotter. 2012. Composition of the early intestinal microbiota: knowledge, knowledge gaps and the use of high-throughput sequencing to address these gaps. Gut. Microbes. 3: 203–220.

Foureau, D.M., D.W. Mielcarz, L.C. Menard, J. Schulthess, C. Werts, V. Vasseur, B. Ryffel, L.H. Kasper and D. Buzoni-Gatel. 2010. TLR9-dependent induction of intestinal alpha-defensins by Toxoplasma gondii. J. Immunol. 184: 7022–7029.

Franchi, L., R. Munoz-Planillo and G. Nunez. 2012. Sensing and reacting to microbes through the inflammasomes. Nat. Immunol. 13: 325–332.

Frank, D.N. and N.R. Pace. 2008. Gastrointestinal microbiology enters the metagenomics era. Curr. Opin. Gastroenterol. 24: 4–10.

Garrett, W.S., J.I. Gordon and L.H. Glimcher. 2010. Homeostasis and inflammation in the intestine. Cell. 140: 859–870.

Garside, P., M. Steel, F.Y. Liew and A.M. Mowat. 1995. CD4+ but not CD8+ T cells are required for the induction of oral tolerance. Int. Immunol. 7: 501–504.

Garside, P., A.M. Mowat and A. Khoruts 1999. Oral tolerance in disease. Gut. 44: 137–142.

Gill, N., M. Wlodarska and B.B. Finlay. 2011. Roadblocks in the gut: barriers to enteric infection. Cell. Microbiol. 13: 660–669.

Godinez, I., M. Raffatellu, H. Chu, T.A. Paixao, T. Haneda, R.L. Santos, C.L. Bevins, R.M. Tsolis and A.J. Baumler. 2009. Interleukin-23 orchestrates mucosal responses to Salmonella enterica serotype Typhimurium in the intestine. Infect. Immun. 77: 387–398.

Gordon, H.A. and L. Pesti. 1971. The gnotobiotic animal as a tool in the study of host microbial relationships. Bacteriol. Rev. 35: 390–429.

Gorelik, L. and R.A. Flavell. 2002. Transforming growth factor-beta in T-cell biology. Nat. Rev. Immunol. 2: 46–53.

Gupta, P., H. Andrew, B.S. Kirschner and S. Guandalini. 2000. Is lactobacillus GG helpful in children with Crohn's disease? Results of a preliminary, open-label study. J. Pediatr. Gastroenterol. Nutr. 31: 453–457.

Guslandi, M., P. Giollo and P.A. Testoni 2003. A pilot trial of *Saccharomyces boulardii* in ulcerative colitis. Eur. J. Gastroenterol. Hepatol. 15: 697–698.

Guslandi, M., G. Mezzi, M. Sorghi and P.A. Testoni. 2000. *Saccharomyces boulardii* in maintenance treatment of Crohn's disease. Dig. Dis. Sci. 45: 1462–1464.

Gustafsson, B.E., T. Midtvedt and K. Strandberg. 1970. Effects of microbial contamination on the cecum enlargement of germfree rats. Scand. J. Gastroenterol. 5: 309–314.

Hansen, J.J., Y. Huang, D.A. Peterson, L. Goeser, T.J. Fan, E.B. Chang and R.B. Sartor. 2012. The Colitis-Associated Transcriptional Profile of Commensal Bacteroides thetaiotaomicron Enhances Adaptive Immune Responses to a Bacterial Antigen. PLoS. One. 7: e42645.

Hapfelmeier, S., M.A. Lawson, E. Slack, J.K. Kirundi, M. Stoel, M. Heikenwalder, J. Cahenzli, Y. Velykoredko, M.L. Balmer, K. Endt, M.B. Geuking, R. Curtiss, 3rd, K.D. McCoy and A.J. Macpherson. 2010. Reversible microbial colonization of germ-free mice reveals the dynamics of IgA immune responses. Science. 328: 1705–1709.

Happel, K.I., P.J. Dubin, M. Zheng, N. Ghilardi, C. Lockhart, L.J. Quinton, A.R. Odden, J.E. Shellito, G.J. Bagby, S. Nelson and J.K. Kolls. 2005. Divergent roles of IL-23 and IL-12 in host defense against Klebsiella pneumoniae. J. Exp. Med. 202: 761–769.

Haribhai, D., W. Lin, B. Edwards, J. Ziegelbauer, N.H. Salzman, M.R. Carlson, S.H. Li, P.M. Simpson, T.A. Chatila and C.B. Williams. 2009. A central role for induced regulatory T cells in tolerance induction in experimental colitis. J. Immunol. 182: 3461–3468.

Helgeland, L., E. Dissen, K.Z. Dai, T. Midtvedt, P. Brandtzaeg and J.T. Vaage. 2004. Microbial colonization induces oligoclonal expansions of intraepithelial CD8 T cells in the gut. Eur. J. Immunol. 34: 3389–3400.

Henao-Mejia, J., E. Elinav, T. Strowig and R.A. Flavell. 2012. Inflammasomes: far beyond inflammation. Nat. Immunol. 13: 321–324.

Hill, M.J. 1997. Intestinal flora and endogenous vitamin synthesis. Eur. J. Cancer. Prev. 6 Suppl. 1: S43-45.

Hirahara, K., T. Hisatsune, K. Nishijima, H. Kato, O. Shiho and S. Kaminogawa. 1995. CD4+ T cells anergized by high dose feeding establish oral tolerance to antibody responses when transferred in SCID and nude mice. J. Immunol. 154: 6238–6245.

Hooper, L.V., T.S. Stappenbeck, C.V. Hong and J.I. Gordon. 2003. Angiogenins: a new class of microbicidal proteins involved in innate immunity. Nat. Immunol. 4: 269–273.

Hooper, L.V., D.R. Littman and A.J. Macpherson. 2012. Interactions between the microbiota and the immune system. Science. 336: 1268–1273.

Huang, Y., Y. Park, Y. Wang-Zhu, A. Larange, R. Arens, I. Bernardo, D. Olivares-Villagomez, D. Herndler-Brandstetter, N. Abraham, B. Grubeck-Loebenstein, S.P. Schoenberger, L. Van Kaer, M. Kronenberg, M.A. Teitell and H. Cheroutre. 2011. Mucosal memory CD8(+) T cells are selected in the periphery by an MHC class I molecule. Nat. Immunol. 12: 1086–1095.

Huter, E.N., G.H. Stummvoll, R.J. DiPaolo, D.D. Glass and E.M. Shevach. 2008. Cutting edge: antigen-specific TGF beta-induced regulatory T cells suppress Th17-mediated autoimmune disease. J. Immunol. 181: 8209–8213.

Isakov, D., A. Dzutsev, J.A. Berzofsky and I.M. Belyakov. 2011. Lack of IL-7 and IL-15 signaling affects interferon-gamma production by, more than survival of, small intestinal intraepithelial memory CD8+ T cells. Eur. J. Immunol. 41: 3513–3528.

Ishikawa, H., K. Tanaka, Y. Maeda, Y. Aiba, A. Hata, N.M. Tsuji, Y. Koga and T. Matsumoto. 2008. Effect of intestinal microbiota on the induction of regulatory CD25+ CD4+ T cells. Clin. Exp. Immunol. 153: 127–135.

Ishikawa, H., I. Akedo, Y. Umesaki, R. Tanaka, A. Imaoka and T. Otani. 2003. Randomized controlled trial of the effect of bifidobacteria-fermented milk on ulcerative colitis. J. Am. Coll. Nutr. 22: 56–63.

Ivanov, II, B.S. McKenzie, L. Zhou, C.E. Tadokoro, A. Lepelley, J.J. Lafaille, D.J. Cua and D.R. Littman. 2006. The orphan nuclear receptor RORgammat directs the differentiation program of proinflammatory IL-17+ T helper cells. Cell. 126: 1121–1133.

Izcue, A., J.L. Coombes and F. Powrie. 2009. Regulatory lymphocytes and intestinal inflammation. Annu. Rev. Immunol. 27: 313–338.

Juhr, N.C. and M. Ladeburg. 1986. Intestinal accumulation of urea in germ-free animals—a factor in caecal enlargement. Lab. Anim. 20: 238–241.

Kandori, H., K. Hirayama, M. Takeda and K. Doi. 1996. Histochemical, lectin-histochemical and morphometrical characteristics of intestinal goblet cells of germfree and conventional mice. Exp. Anim. 45: 155–160.

Kaplan, J.L., H.N. Shi and W.A. Walker. 2011. The role of microbes in developmental immunologic programming. Pediatr. Res. 69: 465–472.

Karlsson, M., S. Lundin, U. Dahlgren, H. Kahu, I. Pettersson and E. Telemo. 2001. "Tolerosomes" are produced by intestinal epithelial cells. Eur. J. Immunol. 31: 2892–2900.

Keilbaugh, S.A., M.E. Shin, R.F. Banchereau, L.D. McVay, N. Boyko, D. Artis, J.J. Cebra and G.D. Wu. 2005. Activation of RegIIIbeta/gamma and interferon gamma expression in the intestinal tract of SCID mice: an innate response to bacterial colonisation of the gut. Gut. 54: 623–629.

Kelchtermans, H., B. De Klerck, T. Mitera, M. Van Balen, D. Bullens, A. Billiau, G. Leclercq and P. Matthys. 2005. Defective CD4+CD25+ regulatory T cell functioning in collagen-induced arthritis: an important factor in pathogenesis, counter-regulated by endogenous IFN-gamma. Arthritis. Res. Ther. 7: R402–415.

Kelly, D., J.I. Campbell, T.P. King, G. Grant, E.A. Jansson, A.G. Coutts, S. Pettersson and S. Conway. 2004. Commensal anaerobic gut bacteria attenuate inflammation by regulating nuclear-cytoplasmic shuttling of PPAR-gamma and RelA. Nat. Immunol. 5: 104–12.

Kinnebrew, M.A. and E.G. Pamer. 2012. Innate immune signaling in defense against intestinal microbes. Immunol. Rev. 245: 113–131.

Kobayashi, K.S., M. Chamaillard, Y. Ogura, O. Henegariu, N. Inohara, G. Nunez and R.A. Flavell. 2005. Nod2-dependent regulation of innate and adaptive immunity in the intestinal tract. Science. 307: 731–734.

Kobayashi, N., N. Hiraoka, W. Yamagami, H. Ojima, Y. Kanai, T. Kosuge, A. Nakajima and S. Hirohashi. 2007. FOXP3+ regulatory T cells affect the development and progression of hepatocarcinogenesis. Clin. Cancer. Res. 13: 902–911.

Koboziev, I., F. Karlsson and M.B. Grisham. 2010. Gut-associated lymphoid tissue, T cell trafficking, and chronic intestinal inflammation. Ann. N. Y. Acad. Sci. 1207 Suppl. 1: E86–93.

Konrad, A., Y. Cong, W. Duck, R. Borlaza and C.O. Elson. 2006. Tight mucosal compartmentation of the murine immune response to antigens of the enteric microbiota. Gastroenterology. 130: 2050–2059.

Korn, T., E. Bettelli, W. Gao, A. Awasthi, A. Jager, T.B. Strom, M. Oukka and V.K. Kuchroo. 2007. IL-21 initiates an alternative pathway to induce proinflammatory T(H)17 cells. Nature. 448: 484–487.

Kruis, W., P. Fric, J. Pokrotnieks, M. Lukas, B. Fixa, M. Kascak, M.A. Kamm, J. Weismueller, C. Beglinger, M. Stolte, C. Wolff and J. Schulze. 2004. Maintaining remission of ulcerative colitis with the probiotic *Escherichia coli* Nissle 1917 is as effective as with standard mesalazine. Gut. 53: 1617–1623.

Kulkarni, A.B., C.G. Huh, D. Becker, A. Geiser, M. Lyght, K.C. Flanders, A.B. Roberts, M.B. Sporn, J.M. Ward and S. Karlsson. 1993. Transforming growth factor beta 1 null mutation in mice causes excessive inflammatory response and early death. Proc. Natl. Acad. Sci. USA 90: 770–774.

Kuwahara, T., Y. Ogura, K. Oshima, K. Kurokawa, T. Ooka, H. Hirakawa, T. Itoh, H. Nakayama-Imaohji, M. Ichimura, K. Itoh, C. Ishifune, Y. Maekawa, K. Yasutomo, M. Hattori and T. Hayashi. 2011. The lifestyle of the segmented filamentous bacterium: a non-culturable gut-associated immunostimulating microbe inferred by whole-genome sequencing. DNA Res. 18: 291–303.

Lee, J., J.H. Mo, K. Katakura, I. Alkalay, A.N. Rucker, Y.T. Liu, H.K. Lee, C. Shen, G. Cojocaru, S. Shenouda, M. Kagnoff, L. Eckmann, Y. Ben-Neriah and E. Raz. 2006. Maintenance of colonic homeostasis by distinctive apical TLR9 signalling in intestinal epithelial cells. Nat. Cell. Biol. 8: 1327–1336.

Levy, O. 2007. Innate immunity of the newborn: basic mechanisms and clinical correlates. Nat. Rev. Immunol. 7: 379–390.

Li, M.O., Y.Y. Wan, S. Sanjabi, A.K. Robertson and R.A. Flavell. 2006. Transforming growth factor-beta regulation of immune responses. Annu. Rev. Immunol. 24: 99–146.

Li, Y., X.Y. Yio and L. Mayer. 1995. Human intestinal epithelial cell-induced CD8+ T cell activation is mediated through CD8 and the activation of CD8-associated p56lck. J. Exp. Med. 182: 1079–1088.

Lindner, C., B. Wahl, L. Fohse, S. Suerbaum, A.J. Macpherson, I. Prinz and O. Pabst. 2012. Age, microbiota, and T cells shape diverse individual IgA repertoires in the intestine. J. Exp. Med. 209: 365–377.

Littman, D.R. and E.G. Pamer. 2011. Role of the commensal microbiota in normal and pathogenic host immune responses. Cell. Host. Microbe. 10: 311–323.

Macpherson, A.J., M.B. Geuking, E. Slack, S. Hapfelmeier and K.D. McCoy. 2012. The habitat, double life, citizenship, and forgetfulness of IgA. Immunol. Rev. 245: 132–146.

Macpherson, A.J. and T. Uhr. 2004. Induction of protective IgA by intestinal dendritic cells carrying commensal bacteria. Science. 303: 1662–1665.

Macpherson, A.J. and N.L. Harris. 2004. Interactions between commensal intestinal bacteria and the immune system. Nat. Rev. Immunol. 4: 478–485.

Macpherson, A.J., D. Gatto, E. Sainsbury, G.R. Harriman, H. Hengartner and R.M. Zinkernagel. 2000. A primitive T cell-independent mechanism of intestinal mucosal IgA responses to commensal bacteria. Science. 288: 2222–2226.

Madara, J.L. 1998. Regulation of the movement of solutes across tight junctions. Annu. Rev. Physiol. 60: 143–159.

Madsen, K.L. 2001. Inflammatory bowel disease: lessons from the IL-10 gene-deficient mouse. Clin. Invest. Med. 24: 250–257.

Makita, S., T. Kanai, Y. Nemoto, T. Totsuka, R. Okamoto, K. Tsuchiya, M. Yamamoto, H. Kiyono and M. Watanabe. 2007. Intestinal lamina propria retaining CD4+CD25+ regulatory T cells is a suppressive site of intestinal inflammation. J. Immunol. 178: 4937–4946.

Malinen, E., T. Rinttila, K. Kajander, J. Matto, A. Kassinen, L. Krogius, M. Saarela, R. Korpela and A. Palva. 2005. Analysis of the fecal microbiota of irritable bowel syndrome patients and healthy controls with real-time PCR. Am. J. Gastroenterol. 100: 373–382.

Manta, C., E. Heupel, K. Radulovic, V. Rossini, N. Garbi, C.U. Riedel and J.H. Niess. 2012. CX(3)CR1(+) macrophages support IL-22 production by innate lymphoid cells during infection with Citrobacter rodentium. Mucosal. Immunol.

Mao, Y., S. Nobaek, B. Kasravi, D. Adawi, U. Stenram, G. Molin and B. Jeppsson. 1996. The effects of Lactobacillus strains and oat fiber on methotrexate-induced enterocolitis in rats. Gastroenterology. 111: 334–344.

Matthes, H., T. Krummenerl, M. Giensch, C. Wolff and J. Schulze. 2010. Clinical trial: probiotic treatment of acute distal ulcerative colitis with rectally administered *Escherichia coli* Nissle 1917 (EcN). BMC Complement. Altern. Med. 10: 13.

Mazmanian, S.K., C.H. Liu, A.O. Tzianabos and D.L. Kasper. 2005. An immunomodulatory molecule of symbiotic bacteria directs maturation of the host immune system. Cell. 122: 107–118.

McDole, J.R., L.W. Wheeler, K.G. McDonald, B. Wang, V. Konjufca, K.A. Knoop, R.D. Newberry and M.J. Miller. 2012. Goblet cells deliver luminal antigen to CD103+ dendritic cells in the small intestine. Nature. 483: 345–349.

McGuckin, M.A., S.K. Linden, P. Sutton and T.H. Florin. 2011. Mucin dynamics and enteric pathogens. Nat. Rev. Microbiol. 9: 265–278.

Metges, C.C. 2000. Contribution of microbial amino acids to amino acid homeostasis of the host. J. Nutr. 130: 1857S–1864S.

Metwali, A., T. Setiawan, A.M. Blum, J. Urban, D.E. Elliott, L. Hang and J.V. Weinstock. 2006. Induction of CD8+ regulatory T cells in the intestine by Heligmosomoides polygyrus infection. Am. J. Physiol. Gastrointest. Liver. Physiol. 291: G253–259.

Miron, N. and V. Cristea. 2012. Enterocytes: active cells in tolerance to food and microbial antigens in the gut. Clin. Exp. Immunol. 167: 405–412.

Mowat, A.M. 2003. Anatomical basis of tolerance and immunity to intestinal antigens. Nat. Rev. Immunol. 3: 331–341.

Mowat, A.M. and C.C. Bain. 2011. Mucosal macrophages in intestinal homeostasis and inflammation. J. Innate. Immun. 3: 550–564.

Murai, M., O. Turovskaya, G. Kim, R. Madan, C.L. Karp, H. Cheroutre and M. Kronenberg. 2009. Interleukin 10 acts on regulatory T cells to maintain expression of the transcription factor Foxp3 and suppressive function in mice with colitis. Nat. Immunol. 10: 1178–1184.

Nagler-Anderson, C. 2001. Man the barrier! Strategic defences in the intestinal mucosa. Nat. Rev. Immunol. 1: 59–67.

Neurath, M.F., I. Fuss, B.L. Kelsall, D.H. Presky, W. Waegell and W. Strober. 1996. Experimental granulomatous colitis in mice is abrogated by induction of TGF-beta-mediated oral tolerance. J. Exp. Med. 183: 2605–2616.

Niess, J.H., F. Leithauser, G. Adler and J. Reimann. 2008. Commensal gut flora drives the expansion of proinflammatory CD4 T cells in the colonic lamina propria under normal and inflammatory conditions. J. Immunol. 180: 559–568.

Niess, J.H., S. Brand, X. Gu, L. Landsman, S. Jung, B.A. McCormick, J.M. Vyas, M. Boes, H.L. Ploegh, J.G. Fox, D.R. Littman and H.C. Reinecker. 2005. CX3CR1-mediated dendritic cell access to the intestinal lumen and bacterial clearance. Science. 307: 254–258.

Niess, J.H. and G. Adler. 2010. Enteric flora expands gut lamina propria CX3CR1+ dendritic cells supporting inflammatory immune responses under normal and inflammatory conditions. J. Immunol. 184: 2026–2037.

O'Hara, A.M. and F. Shanahan. 2006. The gut flora as a forgotten organ. EMBO Rep. 7: 688–693.

Olszak, T., D. An, S. Zeissig, M.P. Vera, J. Richter, A. Franke, J.N. Glickman, R. Siebert, R.M. Baron, D.L. Kasper and R.S. Blumberg. 2012. Microbial exposure during early life has persistent effects on natural killer T cell function. Science. 336: 489–493.

Osorio, F. and C. Reis e Sousa. 2011. Myeloid C-type lectin receptors in pathogen recognition and host defense. Immunity. 34: 651–664.

Ostman, S., M. Taube and E. Telemo. 2005. Tolerosome-induced oral tolerance is MHC dependent. Immunology. 116: 464–476.

Ouwehand, A., E. Isolauri and S. Salminen. 2002. The role of the intestinal microflora for the development of the immune system in early childhood. Eur. J. Nutr. 41 Suppl. 1: I32–37.

Ouyang, W., J.K. Kolls and Y. Zheng. 2008. The biological functions of T helper 17 cell effector cytokines in inflammation. Immunity. 28: 454–467.

Pabst, O., G. Bernhardt and R. Forster. 2007. The impact of cell-bound antigen transport on mucosal tolerance induction. J. Leukoc. Biol. 82: 795–800.

Park, H.K., S.S. Shim, S.Y. Kim, J.H. Park, S.E. Park, H.J. Kim, B.C. Kang and C.M. Kim. 2005. Molecular analysis of colonized bacteria in a human newborn infant gut. J. Microbiol. 43: 345–353.

Pasquier, B., P. Launay, Y. Kanamaru, I.C. Moura, S. Pfirsch, C. Ruffie, D. Henin, M. Benhamou, M. Pretolani, U. Blank and R.C. Monteiro. 2005. Identification of FcalphaRI as an inhibitory receptor that controls inflammation: dual role of FcRgamma ITAM. Immunity. 22: 31–42.

Penders, J., E.E. Stobberingh, P.A. van den Brandt and C. Thijs. 2007. The role of the intestinal microbiota in the development of atopic disorders. Allergy. 62: 1223–1236.

Peterson, D.A., N.P. McNulty, J.L. Guruge and J.I. Gordon. 2007. IgA response to symbiotic bacteria as a mediator of gut homeostasis. Cell. Host. Microbe. 2: 328–339.

Phalipon, A., A. Cardona, J.P. Kraehenbuhl, L. Edelman, P.J. Sansonetti and B. Corthesy. 2002. Secretory component: a new role in secretory IgA-mediated immune exclusion *in vivo*. Immunity. 17: 107–115.

Philbin, V.J. and O. Levy. 2009. Developmental biology of the innate immune response: implications for neonatal and infant vaccine development. Pediatr. Res. 65: 98R–105R.

Philippe, D., E. Heupel, S. Blum-Sperisen and C.U. Riedel. 2011. Treatment with Bifidobacterium bifidum 17 partially protects mice from Th1-driven inflammation in a chemically induced model of colitis. Int. J. Food. Microbiol. 149: 45–49.

Posserud, I., P.O. Stotzer, E.S. Bjornsson, H. Abrahamsson and M. Simren. 2007. Small intestinal bacterial overgrowth in patients with irritable bowel syndrome. Gut. 56: 802–808.

Preising, J., D. Philippe, M. Gleinser, H. Wei, S. Blum, B.J. Eikmanns, J.H. Niess and C.U. Riedel. 2010. Selection of bifidobacteria based on adhesion and anti-inflammatory capacity *in vitro* for amelioration of murine colitis. Appl. Environ. Microbiol. 76: 3048–3051.

Qin, J., R. Li, J. Raes, M. Arumugam, K.S. Burgdorf, C. Manichanh, T. Nielsen, N. Pons, F. Levenez, T. Yamada, D.R. Mende, J. Li, J. Xu, S. Li, D. Li, J. Cao, B. Wang, H. Liang, H. Zheng, Y. Xie, J. Tap, P. Lepage, M. Bertalan, J.M. Batto, T. Hansen, D. Le Paslier, A. Linneberg, H.B. Nielsen, E. Pelletier, P. Renault, T. Sicheritz-Ponten, K. Turner, H. Zhu, C. Yu, M. Jian, Y. Zhou, Y. Li, X. Zhang, N. Qin, H. Yang, J. Wang, S. Brunak, J. Dore, F. Guarner, K. Kristiansen, O. Pedersen, J. Parkhill, J. Weissenbach, P. Bork and S.D. Ehrlich. 2010. A human gut microbial gene catalogue established by metagenomic sequencing. Nature. 464: 59–65.

Radulovic, K., C. Manta, V. Rossini, K. Holzmann, H.A. Kestler, U.M. Wegenka, T. Nakayama and J.H. Niess. 2012. CD69 regulates type I IFN-induced tolerogenic signals to mucosal CD4 T cells that attenuate their colitogenic potential. J. Immunol. 188: 2001–2013.

Rakoff-Nahoum, S. and R. Medzhitov. 2007. Regulation of spontaneous intestinal tumorigenesis through the adaptor protein MyD88. Science. 317: 124–127.

Reddy, B.S., J.R. Pleasants, D.R. Zimmerman and B.S. Wostmann. 1965a. Iron and copper utilization in rabbits as affected by diet and germfree status. J. Nutr. 87: 189–196.

Reddy, B.S., B.S. Wostmann and J.R. Pleasants. 1965b. Iron, Copper, and Manganese in Germfree and Conventional Rats. J. Nutr. 86: 159–168.

Rescigno, M., M. Urbano, B. Valzasina, M. Francolini, G. Rotta, R. Bonasio, F. Granucci, J.P. Kraehenbuhl and P. Ricciardi-Castagnoli. 2001. Dendritic cells express tight junction proteins and penetrate gut epithelial monolayers to sample bacteria. Nat. Immunol. 2: 361–367.

Rhee, S.H., E. Im, M. Riegler, E. Kokkotou, M. O'Brien and C. Pothoulakis. 2005. Pathophysiological role of Toll-like receptor 5 engagement by bacterial flagellin in colonic inflammation. Proc. Natl. Acad. Sci. USA. 102: 13610–13615.

Riedel, C.U., F. Foata, D. Philippe, O. Adolfsson, B.J. Eikmanns and S. Blum. 2006. Anti-inflammatory effects of bifidobacteria by inhibition of LPS-induced NF-kappaB activation. World. J. Gastroenterol. 12: 3729–3735.

Rindi, G., A.B. Leiter, A.S. Kopin, C. Bordi and E. Solcia. 2004. The "normal" endocrine cell of the gut: changing concepts and new evidences. Ann. N. Y. Acad. Sci. 1014: 1–12.

Rolfe, V.E., P.J. Fortun, C.J. Hawkey and F. Bath-Hextall. 2006. Probiotics for maintenance of remission in Crohn's disease. Cochrane. Database. Syst. Rev. CD004826.

Rossi, M., A. Amaretti and S. Raimondi. 2011. Folate production by probiotic bacteria. Nutrients. 3: 118–134.

Round, J.L., S.M. Lee, J. Li, G. Tran, B. Jabri, T.A. Chatila and S.K. Mazmanian. 2011. The Toll-like receptor 2 pathway establishes colonization by a commensal of the human microbiota. Science. 332: 974–977.

Sanos, S.L., V.L. Bui, A. Mortha, K. Oberle, C. Heners, C. Johner and A. Diefenbach. 2009. RORgammat and commensal microflora are required for the differentiation of mucosal interleukin 22-producing NKp46+ cells. Nat. Immunol. 10: 83–91.

Satoh-Takayama, N., C.A. Vosshenrich, S. Lesjean-Pottier, S. Sawa, M. Lochner, F. Rattis, J.J. Mention, K. Thiam, N. Cerf-Bensussan, O. Mandelboim, G. Eberl and J.P. Di Santo. 2008. Microbial flora drives interleukin 22 production in intestinal NKp46+ cells that provide innate mucosal immune defense. Immunity. 29: 958–970.

Sawa, S., M. Lochner, N. Satoh-Takayama, S. Dulauroy, M. Berard, M. Kleinschek, D. Cua, J.P. Di Santo and G. Eberl. 2011. RORgammat+ innate lymphoid cells regulate intestinal homeostasis by integrating negative signals from the symbiotic microbiota. Nat. Immunol. 12: 320–326.

Schmitz, F., A. Heit, S. Guggemoos, A. Krug, J. Mages, M. Schiemann, H. Adler, I. Drexler, T. Haas, R. Lang and H. Wagner. 2007. Interferon-regulatory-factor 1 controls Toll-like receptor 9-mediated IFN-beta production in myeloid dendritic cells. Eur. J. Immunol. 37: 315–327.

Scholtens, P.A., R. Oozeer, R. Martin, K.B. Amor and J. Knol. 2012. The early settlers: intestinal microbiology in early life. Annu. Rev. Food. Sci. Technol. 3: 425–447.

Schulz, O., E. Jaensson, E.K. Persson, X. Liu, T. Worbs, W.W. Agace and O. Pabst. 2009. Intestinal CD103+, but not CX3CR1+, antigen sampling cells migrate in lymph and serve classical dendritic cell functions. J. Exp. Med. 206: 3101–3114.

Schulz, S.M., G. Kohler, C. Holscher, Y. Iwakura and G. Alber. 2008. IL-17A is produced by Th17, gammadelta T cells and other CD4- lymphocytes during infection with Salmonella enterica serovar Enteritidis and has a mild effect in bacterial clearance. Int. Immunol. 20: 1129–1138.

Scurlock, A.M., B.P. Vickery, J.O. Hourihane and A.W. Burks. 2010. Pediatric food allergy and mucosal tolerance. Mucosal. Immunol. 3: 345–354.

Sekirov, I., S.L. Russell, L.C. Antunes and B.B. Finlay. 2010. Gut microbiota in health and disease. Physiol. Rev. 90: 859–904.

Sellon, R.K., S. Tonkonogy, M. Schultz, L.A. Dieleman, W. Grenther, E. Balish, D.M. Rennick and R.B. Sartor. 1998. Resident enteric bacteria are necessary for development of spontaneous colitis and immune system activation in interleukin-10-deficient mice. Infect. Immun. 66: 5224–5231.

Shiloh, M.U., J.D. MacMicking, S. Nicholson, J.E. Brause, S. Potter, M. Marino, F. Fang, M. Dinauer and C. Nathan. 1999. Phenotype of mice and macrophages deficient in both phagocyte oxidase and inducible nitric oxide synthase. Immunity. 10: 29–38.

Shull, M.M., I. Ormsby, A.B. Kier, S. Pawlowski, R.J. Diebold, M. Yin, R. Allen, C. Sidman, G. Proetzel, D. Calvin et al. 1992. Targeted disruption of the mouse transforming growth factor-beta 1 gene results in multifocal inflammatory disease. Nature. 359: 693–699.

Smith, P.M. and W.S. Garrett. 2011. The gut microbiota and mucosal T cells. Front. Microbiol. 2: 111.

Spiller, R.C., D. Jenkins, J.P. Thornley, J.M. Hebden, T. Wright, M. Skinner and K.R. Neal. 2000. Increased rectal mucosal enteroendocrine cells, T lymphocytes, and increased gut permeability following acute Campylobacter enteritis and in post-dysenteric irritable bowel syndrome. Gut. 47: 804–811.

Spiller, R.C. 2003. Postinfectious irritable bowel syndrome. Gastroenterology. 124: 1662–1671.

Spits, H. and J.P. Di Santo. 2011. The expanding family of innate lymphoid cells: regulators and effectors of immunity and tissue remodeling. Nat. Immunol. 12: 21–27.

Stappenbeck, T.S. 2009. Paneth cell development, differentiation, and function: new molecular cues. Gastroenterology. 137: 30–33.

Steege, J.C., W.A. Buurman and P.P. Forget. 1997. The neonatal development of intraepithelial and lamina propria lymphocytes in the murine small intestine. Dev. Immunol. 5: 121–128.

Steidler, L., W. Hans, L. Schotte, S. Neirynck, F. Obermeier, W. Falk, W. Fiers and E. Remaut. 2000. Treatment of murine colitis by Lactococcus lactis secreting interleukin-10. Science. 289: 1352–1355.

Stephani, J., K. Radulovic and J.H. Niess. 2011. Gut microbiota, probiotics and inflammatory bowel disease. Arch. Immunol. Ther. Exp. (Warsz) 59: 161–177.

Strachan, D.P. and C.H. Sanders. 1989. Damp housing and childhood asthma; respiratory effects of indoor air temperature and relative humidity. J. Epidemiol. Community. Health. 43: 7–14.

Strachan, D.P. 1989. Hay fever, hygiene, and household size. BMJ. 299: 1259–1260.

Strobel, S., A.M. Mowat, H.E. Drummond, M.G. Pickering and A. Ferguson. 1983. Immunological responses to fed protein antigens in mice. II. Oral tolerance for CMI is due to activation of cyclophosphamide-sensitive cells by gut-processed antigen. Immunology. 49: 451–456.

Strober, W., I. Fuss and P. Mannon. 2007. The fundamental basis of inflammatory bowel disease. J. Clin. Invest. 117: 514–521.

Strowig, T., J. Henao-Mejia, E. Elinav and R. Flavell. 2012. Inflammasomes in health and disease. Nature. 481: 278–286.

Szabo, S.J., S.T. Kim, G.L. Costa, X. Zhang, C.G. Fathman and L.H. Glimcher. 2000. A novel transcription factor, T-bet, directs Th1 lineage commitment. Cell. 100: 655–669.

Takeuchi, O. and S. Akira 2010. Pattern recognition receptors and inflammation. Cell. 140: 805–820.

Tap, J., S. Mondot, F. Levenez, E. Pelletier, C. Caron, J.P. Furet, E. Ugarte, R. Munoz-Tamayo, D.L. Paslier, R. Nalin, J. Dore and M. Leclerc. 2009. Towards the human intestinal microbiota phylogenetic core. Environ. Microbiol. 11: 2574–2584.

The human microbiome project consortium. 2012a. A framework for human microbiome research. Nature. 486: 215–221.

The human microbiome project consortium. 2012b. Structure, function and diversity of the healthy human microbiome. Nature. 486: 207–214.

Thim, L., F. Madsen and S.S. Poulsen. 2002. Effect of trefoil factors on the viscoelastic properties of mucus gels. Eur. J. Clin. Invest. 32: 519–527.

Vaishnava, S., M. Yamamoto, K.M. Severson, K.A. Ruhn, X. Yu, O. Koren, R. Ley, E.K. Wakeland and L.V. Hooper. 2011. The antibacterial lectin RegIIIgamma promotes the spatial segregation of microbiota and host in the intestine. Science. 334: 255–258.

Vaishnava, S., C.L. Behrendt, A.S. Ismail, L. Eckmann and L.V. Hooper. 2008. Paneth cells directly sense gut commensals and maintain homeostasis at the intestinal host-microbial interface. Proc. Natl. Acad. Sci. USA. 105: 20858–20863.

Van Gossum, A., O. Dewit, E. Louis, G. de Hertogh, F. Baert, F. Fontaine, M. DeVos, M. Enslen, M. Paintin and D. Franchimont. 2007. Multicenter randomized-controlled clinical trial of probiotics (*Lactobacillus johnsonii*, LA1) on early endoscopic recurrence of Crohn's disease after lleo-caecal resection. Inflamm. Bowel. Dis. 13: 135–142.

van Wijk, F. and H. Cheroutre. 2010. Mucosal T cells in gut homeostasis and inflammation. Expert. Rev. Clin. Immunol. 6: 559–566.

Vanhoutvin, S.A., F.J. Troost, H.M. Hamer, P.J. Lindsey, G.H. Koek, D.M. Jonkers, A. Kodde, K. Venema and R.J. Brummer. 2009. Butyrate-induced transcriptional changes in human colonic mucosa. PLoS One. 4: e6759.

Vedantam, G. and D.W. Hecht. 2003. Antibiotics and anaerobes of gut origin. Curr. Opin. Microbiol. 6: 457–461.
Whary, M.T., N.S. Taylor, Y. Feng, Z. Ge, S. Muthupalani, J. Versalovic and J.G. Fox. 2011. Lactobacillus reuteri promotes Helicobacter hepaticus-associated typhlocolitis in gnotobiotic B6.129P2-IL-10(tm1Cgn) (IL-10(–/–)) mice. Immunology. 133: 165–178.
Williams, A.M., C.S. Probert, R. Stepankova, H. Tlaskalova-Hogenova, A. Phillips and P.W. Bland. 2006. Effects of microflora on the neonatal development of gut mucosal T cells and myeloid cells in the mouse. Immunology. 119: 470–478.
Woodcock, A., L.A. Lowe, C.S. Murray, B.M. Simpson, S.D. Pipis, P. Kissen, A. Simpson and A. Custovic. 2004. Early life environmental control: effect on symptoms, sensitization, and lung function at age 3 years. Am. J. Respir. Crit. Care. Med. 170: 433–439.
Worbs, T., U. Bode, S. Yan, M.W. Hoffmann, G. Hintzen, G. Bernhardt, R. Forster and O. Pabst. 2006. Oral tolerance originates in the intestinal immune system and relies on antigen carriage by dendritic cells. J. Exp. Med. 203: 519–527.
Wostmann, B.S., C. Larkin, A. Moriarty and E. Bruckner-Kardoss. 1983. Dietary intake, energy metabolism, and excretory losses of adult male germfree Wistar rats. Lab. Anim. Sci. 33: 46–50.
Xavier, R.J. and D.K. Podolsky. 2007. Unravelling the pathogenesis of inflammatory bowel disease. Nature. 448: 427–434.
Yadav, P.K., C. Chen and Z. Liu. 2011. Potential role of NK cells in the pathogenesis of inflammatory bowel disease. J. Biomed. Biotechnol. 2011: 348530.
Yamamoto, M., S. Sato, K. Mori, K. Hoshino, O. Takeuchi, K. Takeda and S. Akira. 2002. Cutting edge: a novel Toll/IL-1 receptor domain-containing adapter that preferentially activates the IFN-beta promoter in the Toll-like receptor signaling. J. Immunol. 169: 6668–6672.
Yoshida, H., A. Naito, J. Inoue, M. Satoh, S.M. Santee-Cooper, C.F. Ware, A. Togawa and S. Nishikawa. 2002. Different cytokines induce surface lymphotoxin-alphabeta on IL-7 receptor-alpha cells that differentially engender lymph nodes and Peyer's patches. Immunity. 17: 823–833.
Zachar, Z. and D.C. Savage. 1979. Microbial interference and colonization of the murine gastrointestinal tract by Listeria monocytogenes. Infect. Immun. 23: 168–174.
Zhang, H., J.R. Podojil, J. Chang, X. Luo and S.D. Miller 2010. TGF-beta-induced myelin peptide-specific regulatory T cells mediate antigen-specific suppression of induction of experimental autoimmune encephalomyelitis. J. Immunol. 184: 6629–6636.
Zhang, X., L. Izikson, L. Liu and H.L. Weiner. 2001. Activation of CD25(+)CD4(+) regulatory T cells by oral antigen administration. J. Immunol. 167: 4245–4253.
Zheng, W. and R.A. Flavell. 1997. The transcription factor GATA-3 is necessary and sufficient for Th2 cytokine gene expression in CD4 T cells. Cell. 89: 587–596.
Zhou, L., Ivanov, II, R. Spolski, R. Min, K. Shenderov, T. Egawa, D.E. Levy, W.J. Leonard and D.R. Littman 2007. IL-6 programs T(H)-17 cell differentiation by promoting sequential engagement of the IL-21 and IL-23 pathways. Nat. Immunol. 8: 967–974.

2

Presentation of Microbial Signals via Maternal Cells

An Evolutionary Advantage of Mammals

Pablo F. Perez,[1,*] *Anne Donnet-Hughes*[2] and *Eduardo J. Schiffrin*[3]

Introduction

The evolutionary lineage leading to mammals began over 300 million years ago when mammal's ancestors diverged from those of reptiles and birds (Lefèvre et al. 2010). As one of the main challenges of offspring survival was moisture loss, early mammals developed skin secretions to prevent dehydration. These secretions later evolved to milk (Oftedal 2012), whose early form was composed primarily of antimicrobial compounds but later evolved to a complex mixture of components that maximized offspring survival.

[1] Centro de Investigación y Desarrollo en Criotecnología de Alimentos (CIDCA-CCT CONICET), and Departamento de Ciencias Biológicas, Facultad de Ciencias Exactas, Cátedra de, Microbiología, UNLP, 1900, La Plata, Argentina.
E-mail: pfp@biol.unlp.edu.ar
[2] Nestlé Research Center, Nestec Ltd., Vers-chez-les-Blanc, 1000 Lausanne 26.
E-mail: anne.donnet@rdls.nestle.com
[3] Clinical Development Unit, Nestec Ltd., Vers-chez-les-Blanc, 1000 Lausanne 26.
E-mail : eduardo.schiffrin@rdls.nestle.com
* Corresponding author

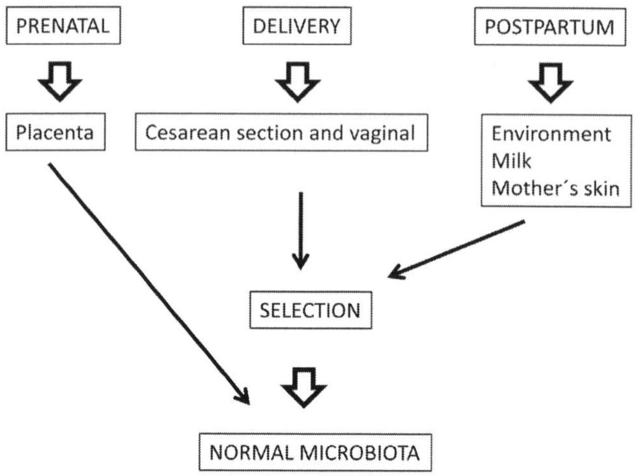

Figure 1. Sources of microorganisms of the intestinal microbiota.

Microorganisms were the earliest forms of life. They are considered to have played a major role in determining the Earth's environment and in supporting larger and more complex life forms, including mammals. While the classical concept of microbial-mammalian interaction is a confrontational relationship that leads to different outcomes for each (Ulvestad 2009), it is now widely accepted that the mammalian host has developed an intimate and mutually beneficial relationship with very complex microbial communities.

Until the advent of molecular techniques, the study of the human microbiota was accomplished using culture-based methods whose principles were established in the 19th century. The use of DNA-based analysis has widened our knowledge of the microbial communities that inhabit the human body and has provided a greater appreciation of how culturable and non-culturable microbes interact with human genes and tissues. The term "microbiome" has been adopted to reflect this new understanding of host-microbe interactions (Pflughoeft and Versalovic 2012).

The use of molecular tools has provided accumulating evidence of a human milk microbiome which may impact neonatal intestinal colonization. During the birthing process, fetal tissues are rapidly exposed to a vast array of microbes in the maternal genital tract and feces. Remarkably, they allow specific microorganisms to colonize in the absence of any adverse immune response. This suggests that some form of immune education occurs during fetal life. Differences between breast-fed and formula-fed infants in intestinal bacterial colonization and in susceptibility to intestinal disease or sepsis suggest that further education is achieved via breast milk.

Protection and education of offspring are integral parts of evolutionary success. This chapter will introduce how certain constituents of breast milk, together with microbes, might contribute to this success.

Colonization of the Neonatal Intestine

Establishment of the microbiota of the gastrointestinal tract partially depends on the intake of microorganisms from the external environment and it is driven by a complex cross-talk between the incoming microorganisms and host cells. In humans, the intestinal microbiota is more dynamic in the early years of life, becoming more stable with age (Zoetendal et al. 1998; Tap et al. 2009; Claesson et al. 2011). Similar observations have been made in the nematode model *Caenorhabditis elegans* (Portal-Celhay and Blaser 2012). Furthermore, it has been demonstrated that colonization in this model system depends on nematode genotype and the innate immune response in the intestine. The genotype is also an important determining factor for lifespan (Portal-Celhay et al. 2012).

Genotype also influences the composition of the human intestinal microbiota. A higher degree of similarity is seen in the fecal bacterial DNA patterns of monozyotic twins than in those of dizygotic twins (Stewart et al. 2005). Moreover, several host genes have been linked to the composition of the microbiota. Most of these genes code for factors involved in immune response (e.g., MEFV, MyD88, NOD2, TLR, defensins IgA, HLA, IFN-gamma) but a few are involvedin metabolism (e.g., APOA1, leptin) (Spor et al. 2011; De Palma et al. 2012).

Several studies examining the fecal microbiota of infants using culture-based methods, have shown that in infants delivered vaginally, gut colonization by *Bacteroides*, lactobacilli, staphylococci, enterococci, bifidobacteria and enterobacteria follows different kinetics depending on whether infants are breast-fed or formula-fed. Furthermore, higher numbers of bifidobacteria and lower levels of enterobacteria and bacteroides are found in breast-fed infants (Yoshioka et al. 1983). In contrast, two interesting studies, (Bezirtzoglou and Romond 1990, 1991), assessing the effect of feeding practices on intestinal colonization in infants delivered by cesarean section, have demonstrated that while bifidobacteria are favored by breast feeding, *Clostridium perfringens* levels are higher in infants receiving artificial feeding. The origin of the *C. perfringens* inoculum was clearly environmentally but no hypotheses on the possible routes of colonization by bifidobacteria were given in the study. Nevertheless, regular isolation of viable microorganisms from meconium has been considered indicative of selective colonization of the newborn digestive tract by environmentally derived bacterial strains (Patte et al. 1979).

Selection of colonizing microorganisms certainly involves interplay between host- and microorganism- derived signals. In the perinatal period, both the immune status of the mother and the infant favor the establishment of specific microorganisms. While the mechanisms underlying the selection of specific genera, species and strains remain largely unknown, there are nevertheless some clues as to how this might be achieved.

It has been demonstrated that communication between bacteria is mediated by chemical signals involving small molecules such as acyl-homoserine lactones and short peptides. Interestingly, some of these molecules are structurally similar to human signaling molecules and as such, they may also permit communication with the host (Pacheco and Sperandio 2009). The host can interfere with microbial signals in a way that dramatically modifies colonization. Many mechanisms are involved in this interference but a role for lactone-hydrolyzing enzymes has been highlighted. Paraoxonases, which hydrolyze paraoxon and other organophosphates, also possess lactonase activity (Pezzulo et al. 2012). Of these, human paraoxonase 1 (PON1) is the most extensively studied. Expression of human PON1 in *Drosophila melanogaster* leads to dramatic changes in the antimicrobial activity of host cells as well as in gut colonization (Pezzulo et al. 2012). The mechanisms underlying these effects are multifactorial and involve altered expression of genes associated with oxidative stress, as well as the subsequent decrease in superoxide levels in the fly's gut. It is also noteworthy that expression of PON1 modifies the proportion of different bacterial populations and leads to higher levels of Firmicutes (Bacillales and Lactobacillales) as assessed by high throughput analysis of the V2 region of the bacterial 16S rRNA. In addition, culture-dependent methods have demonstrated that although the total bacterial load is not altered by the expression of PON1, the number of *Lactobacillus* spp., mainly *L. plantarum*, is significantly increased (Pezzulo et al. 2012).

Differential interactions between microorganisms and key components of the innate human immune system have also been described. While *E. coli*, *Bacillus cereus* and *Lactobacillus delbrueckii* subspecies *bulgaricus* (strain CIDCA 331) are sensitive to cationic antimicrobial extracts derived from the human intestinal epithelial cell line Caco-2, a selected strain of *Lactobacillus delbrueckii* subspecies *lactis* (strain CIDCA 133) is resistant (Hugo et al. 2010). The resistant strain is also able to grow in the presence of purified human beta-defensins (Hugo et al. 2010). Later studies showed that during exposure to defensins, this strain retains envelope integrity and that liposomes prepared from the bacteria show lower release of 5(6)-carboxyfluorescein compared to liposomes derived from strain CIDCA 331. These findings in strain CIDCA 133 correlated with the higher glycoplipid: phospholipid ratio which modifies surface charge and thereby, decreases interaction with antimicrobial peptides (Hugo et al. 2012). Clearly, such interactions influence the fate of microorganisms and may therefore affect host immune responses. Expression of exopolysaccharide (EPS) on bacterial cell surfaces has also been shown to influence bacterial persistence and the capacity to modulate immune responses. While an EPS-producing strain of *Bifidobacterium breve* is resistant to acid and bile and evades adaptive responses of the host, an isogenic EPS-deficient strain elicits a strong adaptive immune response (Fanning et al. 2012).

Colonization of the newborn infant gut occurs in a series of steps that, until recently, were considered to be well-conserved. Previously, culture-based methods determined that the early inhabitants of the human gut are *E. coli*, enterococci (*E. faecalis* and *E. faecium*) and staphylococci. It was shown that the metabolic activity of these organisms leads to a low redox potential that subsequently favors the establishment of strict anaerobes such as *Bifidobacterium*, *Clostridium* and *Bacteroides* species. This sequential process was the accepted mode of intestinal colonization until the 80's. However, the use of more sensitive molecular techniques suggests that the timing of colonization and the switch from facultative to strict anaerobes may occur earlier than previously thought. Indeed, a recent study using multiple analytical approaches like culture, qPCR, Sanger sequencing and pyrosequencing to assess bacterial colonization in vaginally delivered, breast fed infants, has challenged the dogma that enterobacteria are always more predominant than strict anaerobes (*Bacteroides* and *Bifidobacterium* species) at 3–6 days postpartum (Jost et al. 2012). While the ratio of enterobacteria to strict anaerobes in some infants was in agreement with previously accepted values, other infants had inverse proportions of these communities. Albeit the small number of infants included in the study, inter-individual variation was clearly demonstrated.

Besides the traditional sources of colonizing microorganisms such as the environment, the vagina during delivery, maternal fecal microbiota, maternal skin and milk, and the nursing personnel, additional new sources have recently been proposed (Perez et al. 2007; Jiménez et al. 2008; Jeurink et al. 2013) (Fig. 1). Irrespective of the origin of the bacterial inoculum, there is a specific window for selecting those microorganisms which ultimately colonize and predominate in the gut. However, other extrinsic factors, such as the use of antibiotics, may impact the final composition of the neonatal microbiota (Keski-Nisula et al. 2013).

Greater certainty about the origin of microbes in the neonatal intestine is now possible using molecular tools to compare isolates from different sources. These tools offer greater sensitivity and the possibility of detecting non-culturable bacteria. Thus, a more complete spectrum of bacterial signals can be catalogued in biological tissues and fluids. Examination of the fecal microbiota of mother-infant pairs using such tools has provided support for vertical transmission of microorganisms from the mother to the infant.

Nevertheless, a recent study has demonstrated that, although the same *Bacteroides* species were detected in paired mother-infant fecal samples, *Bifidobacterium* species in the feces of mothers and their breast-fed babies were distinct (Jost et al. 2012). While *B. adolescentis* and *B. catenulatum* were isolated from the maternal feces, *B. breve*, *B. longum* and *B. pseudocatenulatum* were isolated from their infant's stools. The same authors also found that the levels of *Bacteroides* species in infant feces were inversely correlated to those of *Bifidobacterium* (Jost et al. 2012).

Breast Milk Regulation of Microbial-Host Immune System Cross Talk

The host immune system samples microbes in the intestinal lumen via specialized epithelial cells, called M cells, in the Peyer's patches. These M cells then interact with immune cells, including dendritic cells (DC), in the underlying tissues. However, intestinal DC also sample the luminal contents directly by sending dendrites across the epithelium (Rescigno et al. 2001) Mucosal DC then play a decisive role in determining the nature and amplitude of the immune response which follows. In the healthy intestine, this results in tolerance to the normal microbiota and protection against pathogenic attack.

The immune system of the human newborn is anatomically intact at birth but functionally immature and yet, it is already capable of discriminating the diverse microbial challenge during the birthing process and permits assembly of specific bacterial communities in mucosal tissues during the postpartum period. Breast milk factors play a major role in facilitating this process and in avoiding untoward reactions to the organisms encountered. It has traditionally been accepted that protection of mammalian offspring by breast feeding is a passive process by virtue of which the newborn receives "ready-to-use" protective factors. This paradigm has been challenged by evidence showing that the protection afforded by breast feeding requires the active participation of the newborn in order to orchestrate the future immune repertoire.

Breast milk, nevertheless, plays a principal role in shaping the composition of the neonatal microbiota and immune defense mechanisms, and in simultaneously providing protection against harmful microorganisms. Indeed, this dichotomy is the foundation of the long-lasting protective effects of breast milk.

There is a plethora of soluble breast milk factors which modulate how neonatal cells perceive microbes and/or react to microbial contact. They are too numerous to mention here but examples are oligosaccharides, glycoproteins and glycolipids that, through immune exclusion, limit specific microbes from reaching and interacting with cells in host tissues (Newburg 2009; Choi et al. 2012) as well as antimicrobial peptides such as defensins (Jia et al. 2001), lysozyme (Ibrahim et al. 2011) and the iron-sequestering molecule, lactoferrin (Håversen et al. 2010) which affect the viability of microorganisms, influence the composition of the microbes in the intestinal lumen and also modulate immune responses through multiple mechanisms. Other factors such as milk trefoil factors (TFF) contribute to the reparation of the epithelium, down-regulate production of pro-inflammatory cytokines and promote beta-defensin production (Vestergaard et al. 2008).

The most recent research now suggests that breast milk is permissive, and indeed encourages, interaction between particular microbial motifs and the neonatal immune system but avoids an energetic innate reaction (Donnet-Hughes et al. 2008). In so doing, breast milk promotes a healthy relationship between the neonatal immune system and the various bacterial communities that constitute the permanent microbiota of the gut (Donnet-Hughes et al. 2008).

Human milk oligosaccharides (HMOs) have long been considered as prebiotic compounds that stimulate the growth of beneficial intestinal microorganisms such as bifidobacteria (Yu et al. 2013). They also antagonize interaction of pathogens with enterocytes through molecular mimicry (Newburg et al. 2005). Furthermore, they modulate the activity of intestinal epithelial and immune cells (Bode 2012) by enhancing bifidobacteria-mediated expression of tight junction proteins and anti-inflammatory mediators, as well as increasing lymphocyte maturation and activation (Eiweggeret al. 2004).

Milk fat globules are surrounded by a membrane, the milk fat globule membrane, which contains the glycoproteins mucin (MUC)-1 and lactadherin. These glycoproteins, by virtue of their sialic acid residues (Yolken et al. 1992), and many other milk glycoconjugates in milk, such as the gangliosie GM1, globosyl ceramide and glucosaminoglycans, are also able to interfere with the adhesion of bacteria, toxins and virus to target cells (Dai et al. 2000).

Immediately after birth, there is a period of insufficient antibody production that results from delayed B cell activation and maturation. This potentially harmful status is compensated by the maternal antibodies provided by breast milk. During lactation, mammary gland and milk lymphoblasts originating from the maternal respiratory and gastrointestinal tracts, confer protection against infectious agents to which the mother has been exposed (Bourges et al. 2008; Brandtzaeg 2003). The main immunoglobulin in breast milk, secretory IgA (sIgA), inhibits interaction of bacteria, parasites, virus and toxins with neonatal cells and in so doing, avoids inflammatory tissue damage (Brandtzaeg 2007, 2003; Corthésy 2013). Although sIgA binds to many intestinal bacteria, it is not eliminated from the intestine (van der Waaij et al. 1996). As milk sIgA facilitates sampling of antigens by M cells overlying the Peyer's patches, it is plausible that it guides the neonatal immune responses to the flourishing intestinal microbiota (Kadaoui and Corthésy 2007).

Other soluble factors in breast milk modulate the interaction of bacteria and their components with pattern recognition receptors (PRR) on host cells. These PRR recognize evolutionarily conserved microbial motifs and in so doing, empower host cells to recognize and respond appropriately to a number of microbial components. Indeed, human breast milk contains high levels of soluble PRR including CD14 (sCD14) and TLR 2 (Labéta et al. 2000; Filipp et al. 2001; LeBouder et al. 2003) thus modulating responses to microbial lipopolysaccharide (LPS) and lipoteicoic acid (LTA) respectively. The cytokines IL-10 and TGF-beta can further modify cell responses by respectively increasing or decreasing PRR expression on monocytes and macrophages (Imai et al. 2000).

Interplay between multiple milk and neonatal factors leads to a controlled immune response. The antimicrobial milk lactoferrin downregulates LPS-induced production of proinflammatory cytokines and inhibits sCD14-LPS- induced expression of IL-8, E-selectin and ICAM-1 by endothelial cells (Baveye et al. 2000). Interestingly, while LPS stimulates proinflammatory cellular responses in the presence of milk through interaction with sCD14, LTA in the presence of milk, does not trigger proinflammatory cytokine production. However, LTA in milk inhibits the responsiveness of monocytes and intestinal epithelial cells to milk sCD14-mediated LPS effects (Vidal et al. 2002).

Taken together, the above studies clear demonstrate a dual role for breast milk. It favors a limited innate response to microbial danger signals but avoids an uncontrolled inflammatory response. Anti-inflammatory milk factors such as TGF-beta, lactoferrin and lysozyme, possibly contributing to the latter by inhibiting NF-kB signal transduction (Lönnerdal and Iyer 1995). Still other soluble milk factors, such as osteoprotogerin, IL-1 receptor antagonist (Vidal et al. 2004; Buescher and Malinowska 1996) and soluble TNF-alpha receptor (Buescher and Malinowska 1996), acting as decoy receptors, may provide additional control of inflammatory responses.

The Breast Milk Microbiome as an Inoculum for the Neonatal Intestine

Human breast milk provides a source of important nutrients and a wide range of bioactive factors that are widely recognized to nourish and protect the developing infant. Traditionally considered to be sterile, there is increasing evidence that human milk is also a source of microorganisms for the suckling infant (Martín et al. 2003; Perez et al. 2007; Collado et al. 2009; Martín et al. 2009; Jeurink et al. 2013; Jost et al. 2013). These studies have used both culture-dependent and independent methods to examine breast milk. They have identified a range of species including *Streptococcus, Staphylococcus, Lactobacillus, Bifidobacterium, Propionibacterium, Enterococcus, Peptostreptococcus, Corynebacterium, Bacteroides, Clostridium, Eubacterium, Faecalibacterium and/or Roseburia.*

To date, the precise role of milk-borne bacteria in the breast-fed infant is unknown but it has been suggested that such microbes form part of the bacterial inoculum that colonizes the infant's gut and that they may modulate the neonatal immune system- a specific form of immune modulation being achieved by microbes that are delivered to the gut via maternal cells (Perez et al. 2007; Martín et al. 2003). How such microbes contribute to the protective effects of breast milk against specific immune diseases such as allergy, type 1 diabetes mellitus and celiac disease (Verhasselt 2010; Ludvigsson and Fsano 2012) has yet to be investigated.

For breast milk to contain a natural bacterial inoculum that includes obligate anaerobes that clearly do not arise from the mother's skin, the infant's mouth or from the environment, microbes must originate from an endogenous source without compromising maternal health.

Human breast milk contains numerous cells, including leukocytes, epithelial cells and stem cells, whose total concentration differs greatly between individuals (10^4–10^7 cells per ml). These variations are related to stage of lactation, breast fullness, health status of the mother/infant pair, permeability of the basement membrane and development of the breast epithelium (Hassiotou et al. 2013). The leukocyte population arises from cells that have migrated from the maternal mucosa-associated lymphoid tissue to the lactating mammary gland via the lymphatics and the circulation (Roux et al 1977). It has been postulated that this cellular circuitry is used to traffic maternal gut microbes to the breast during late pregnancy and lactation (Martín et al. 2003; Perez et al. 2007) but how might this occur?

Studies by the group of Andrew MacPherson were a milestone in our understanding of the events underlying intestinal microbial colonization (Macpherson and Uhr 2004). This group demonstrated that a limited number of commensal bacteria can translocate from the healthy gut lumen to the mesenteric lymph nodes (MLN) where they trigger an IgA-mediated immune response. This "controlled translocation" did not extend beyond the MLN and was associated with transport of bacteria within DC exiting the Peyer's patches. Interestingly, these migrating DCs readily phagocytose intestinal bacteria but are relatively ineffective at killing internalized organisms (Nagl et al. 2002). In contrast, invading pathogens such as *Salmonella* were efficiently phagocytosed and killed by intestinal macrophages. We considered that the natural, low-level of microbial translocation seen in healthy conditions, might be amplified in specific physiological conditions such as pregnancy and lactation. Certainly, transport of factors across the intestinal epithelium and migration of immune cells towards the mammary gland are enhanced by lactogenic hormones (Berotto, et al. 1991; Charoenphandhu et al. 2009; Santos et al. 2009). However, the altered immune status and the nature of the intestinal microbiota during pregnancy and lactation would also be important in determining the quantity and type of organisms which translocate.

Changes in the immune status of the pregnant woman are reflected by a shift in the composition of the intestinal microbiota. In a recent study (Koren et al. 2012), culture-independent methods were used to analyze the fecal microbiota of woman at two time points during pregnancy (weeks 14 and 34) and at 1 month postpartum. The fecal microbiota of women during the first trimester of pregnancy was found to be similar to that of non- pregnant women. Thereafter, greater phylogenetic (beta) diversity was observed with increasing gestational age and by the third trimester the composition of the microbiota was dramatically different, with greater abundances of Proteobacteria and Actinobacteria (Koren et al. 2012). The high diversity persisted until at least 1 month postpartum. In addition, the microbiota of infants aged 1 month and 6 months had elevated beta-diversity but by 4 years of age, children had decreased levels of beta-diversity that were similar to those of mothers during the first trimester. Taken together, these results show that the physiological changes occurring during pregnancy have a profound effect on both the mother and the infant's relationship with microbes.

It is clear that the transport of live bacteria within cells migrating from the intestine to the mammary gland will depend on the killing capacity of the cells. Early studies showed that breast milk phagocytes do not kill ingested bacteria efficiently (Pickering et al. 1980; Buescher and McIlheran 1993). It has been postulated that components such as milk lipid may interfere with intracellular killing mechanisms. However, it is also noteworthy that the pregnancy hormone, progesterone interferes with phagosome maturation via suppression of Toll-like receptor triggered signals (Blander and Medzhitov 2004; Sun et al. 2012 and that it suppresses production of the inflammatory cytokines TNF-alpha and IL-1 beta but not the phagocytic activity of DC (Butts et al. 2007).

Taken together, the above studies support the concept that cells destined for the mammary gland during pregnancy and lactation, may indeed be able to transport live microbes from the intestine and ultimately to the breast-fed infant. Studies conducted in humans and mice lend further credence to this hypothesis. More specifically, molecular analysis of milk cells, peripheral blood mononuclear cells and feces from the mother and corresponding infant fecal samples during lactation revealed that certain bacterial ribosomal DNA sequences were common to all samples (Perez et al 2007). Species such as *Bacteroides, Bifidobacterium,* Clostridia, *Eubacterium* and Enterobacteria, among a total of up to 15 genera typical of ileal and colonic microbiota, were identified. More recent studies using other molecular tools to analyse maternal and infant fecal samples and/or breast milk samples have provided further evidence that bacteria present in breast milk do indeed originate from the mother's intestine (Makino et al. 2011; Martín et al. 2012; Jost et al. 2013). Paired breast milk and infant fecal samples were found to share specific bacterial strains of *Staphylococcus, Lactobacillus and Bifidobacterium* (Martín et al. 2012). It has also been shown that *Lactobacillus* species given orally to mothers can be recovered in their breast milk (Jiménez et al. 2008b; Abrahamsson et al. 2009).

The presence of bacteria in different tissues of pregnant and lactating mice has also been studied (Perez et al. 2007). An increased translocation of bacteria beyond the MLN was observed pre- and postpartum. This physiological translocation of microorganisms may explain the presence of viable bacteria within selected immune cells of breast milk (Fig. 2). In accordance with this, Peyer's patches are more prominent in pregnant and lactating animals and a noticeable efflux of mononuclear cells through lymphatic vessels can be seen (Perez et al. 2007). A role for DC in the transport of the bacteria is suggested by the reduced numbers of circulating maternal DC precursors during lactation (Perez et al. 2007) and the presence of partially differentiated DC in breast milk (Ichikawa et al. 2003; Perez et al. 2007). Of relevance, reduced numbers of circulating DC subsets have also been observed in late pregnancy (Ueda et al. 2003). It is feasible that decreased numbers of DC-like cells during pregnancy and lactation may reflect cellular homing to the breast.

Increased intestinal translocation of bacteria or their components during pregnancy and lactation may impact maternal immune status. Physiological activation of innate immune responses, with cell activation and increased production of IL-12 and TNF-α, is certainly observed during pregnancy (Sacks et al. 2003; Naccasha et al. 2001; Kupferminc et al. 1994). In keeping with this, CpG oligodeoxinucleotides have been shown to stimulate innate immunity in pregnant mice, improve maternal survival and prevent pathogen transmission to the fetus (Ito et al. 2004). Perhaps dysregulation of the quantity and/or type of translocating microbes explains the association between specific intrauterine infections and the risk of premature birth or abortion.

Recent observations in human have shown that the number of microorganisms comprising the human mammary microbiome increases during the last trimester of pregnancy, remains stable throughout lactation

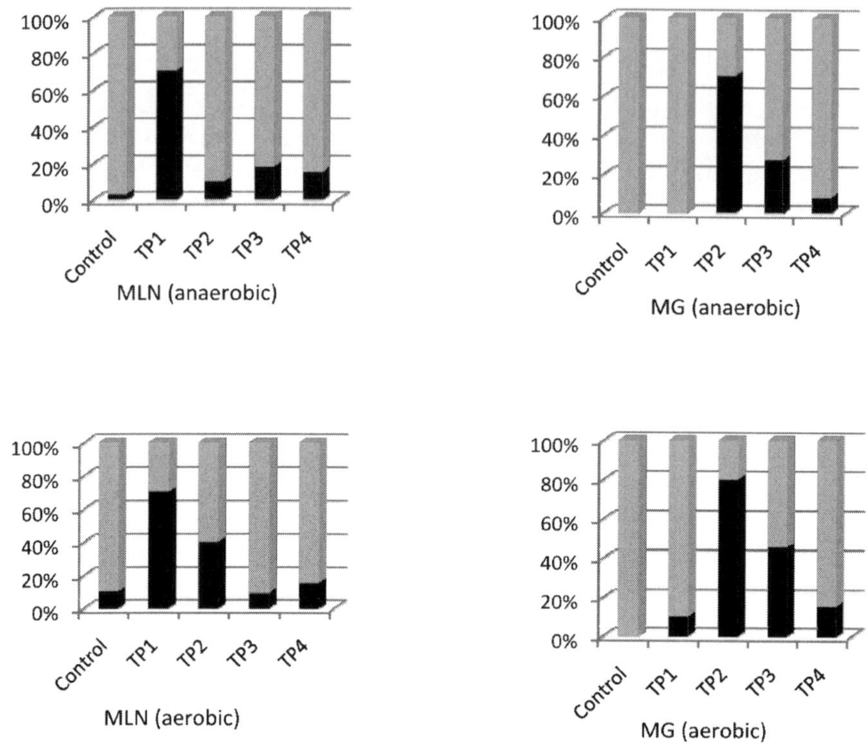

Figure 2. Microbiological analysis of mesenteric lymph nodes (MLN) and mammary glands (MG) of non-pregnantmice (control), pregnant mice (TP1) and lactating mice at 0–1 day (TP2); 3–4 days (TP3) and 14–15 days (TP4) after delivery. Grey bars represent percentage of negative animals and black bars represent percentage of positive animals.

Color image of this figure appears in the color plate section at the end of the book.

but decreases at weaning (Fernández et al. 2013). These results support an earlier study in humans showing that the levels of bacteria in breast milk at 1 day postpartumar significantly higher than those found in milk at 3 months postpartum (Solís et al. 2010) and a study of lactating mouse mammary tissue which found that translocation to the mammary gland begins late in pregnancy but subsides with increasing lactation (Perez et al. 2007).

It is possible that the immune status of the fetus and the newborn facilitate acquisition of the maternal microbiota. To avoid a detrimental immune effect, the fetus downregulates immune responses during intrauterine life. This evolutive strategy involves an increased ratio of regulatory T-cells, a decreased proliferation of antigen-specific T-cells and a suppressed response via specific metabolites and FasL-induced apoptosis of activated T-cells. The functions of neutrophils, antigen presenting cells and B cells are also modified in the fetus. The isolation of commensal bacteria in the meconium (Jiménez et al. 2008a), in the umbilical cord blood of infants born by caesarian section and in the amniotic fluid of pregnant mice (Jiménez et al. 2005) suggests that colonization by maternal organisms may already begin *in utero*. Interestingly, *Enterococcus faecium* strain orally administered to pregnant mice can be recovered in the meconium of offspring delivered by cesarean section (Jiménez et al. 2008a). The high permeability of the intestinal epithelial layer, decreased secretion of antimicrobial compounds and differences in the glycosylation of mucins that are typical features of the neonatal gastrointestinal tract (M'Rabet et al. 2008) suggest that colonization by maternally derived microbes may be favored further in the postnatal period. Taken together, these studies suggest that microorganisms from the mother's gut not only home to the mammary gland via an endogenous route, they also target the fetal gut. The isolation of microbial DNA from human breast milk cells and peripheral blood cells (Perez et al. 2007) and the observation that there is transplacental passage of circulating maternal cells to fetal tissues (Zhou et al. 2000) support a role for maternal immune cells in the trafficking of bacteria from the maternal gut to both sites.

In spite of all the precautions taken to ensure aseptic collection of breast milk samples in various studies, it is still possible that some milk isolates arise from contamination. Nevertheless, the increasing number of studies which consistently identify bacteria in breast milk, the sharing of certain sequences with microbes in other maternal sampling sites and in infant feces, and the presence of obligate anaerobic species which do not persist outside the body, together suggest that breast milk harbors a natural microbial community that is transferred to the suckling infant.

It is feasible that environmentally derived bacteria in milk may be biologically relevant to the breast-fed infant. Indeed, *Staphylococcus* species are common constituents of the early neonatal microbiota (Jimenez-Truque et al. 2012). However, microbes delivered to the neonatal intestine within a milk cellular compartment may play a different role in the infant than "free" microorganisms which colonize the neonatal intestine. Cell-associated organisms may eventually colonize neonatal tissues, but we have considered that they educate the neonatal immune system in the recognition of commensal-associated molecular patterns of bacteria.

It has been previously shown that human milk contains soluble PRR for bacterial molecular motifs and that these may mediate a differential response to Gram negative and Gram positive organisms (Labéta et al. 2000; Vidal et al. 2002). It is also worth noting that breast milk cells are able to enter the infant's systemic circulation by diapedesis through the intestinal mucosa, and can be identified in neonatal tissues (Weiler et al. 1983; Hughes et al. 1988; Zhou et al. 2000; Jain et al. 1989; Schnorr and Pearson 1984). Translocation of maternal cells begins *in utero* (Zhou et al. 2000; Barinaga 2002) and continues via breast feeding. Transplantation studies demonstrate that the immune system of breast-fed infants is tolerized to the maternal antigens and cells delivered in the breast milk (Hanson 2000; Zhang et al. 1991).

Exploitation of maternal immune cells to transport maternal gut microbes and their components to the mammary gland, and thereafter to the breastfeeding infant, may offer several advantages over a non-cell associated homing and delivery of microbes. Firstly, it may avoid potentially dangerous effects on maternal health during endogenous microbial homing to the breast. Secondly, it may protect viable maternal microbes during passage through the neonatal gastrointestinal tract and allow them to become established in the sparsely populated neonatal intestine. Additionally, it may allow intracellular microbes and their components to be released at sites other than the neonatal intestinal lumen, wherein they induce particular immune responses. While the contribution of the milk cellular component to breast milk's protective effects has yet to be fully appreciated, we favor the hypothesis that prolonged penetration of inconspicuous bacterial molecular patterns via maternal immune cells educates the neonatal immune system to initiate appropriate differential responses to commensals, opportunistic pathogens and true pathogens and in so doing, contributes to the long-term protective effects of breast milk.

References Cited

Abrahamsson, T.R., G. Sinkiewicz, T. Jakobsson, M. Fredrikson and B. Björkstén. 2009. Probiotic lactobacilli in breast milk and infant stool in relation to oral intake during the first year of life. J. Pediatr. Gastroenterol. Nutr. 49: 349–354.

Adlerberth, I. and A.E. Wold. 2009. Establishment of the gut microbiota in Western infants. Acta. Pædiatr. 98: 229–238.

Barinaga, M. 2002. Cells Exchanged During Pregnancy Live On. Science. 296: 2169–2172.

Baveye, S., E. Elass, D.G. Fernig, C. Blanquart, J. Mazurier and D. Legrand. 2000. Human lactoferrin interacts with soluble CD14 and inhibits expression of endothelial adhesion molecules, E-Selectin and ICAM-1, induced by the CD14-lipopolysaccharide complex. Infect. Immun. 68: 6519–6525.

Berotto, A., R. Gerli, G. Castelluci, F. Scalise and R. Vaccaro. 1991. Human milk lymphocytes bearing the gamma/delta T-cell receptor are mostly delta TCS1-positive cells. Immunol. 74: 360–361.

Bezirtzoglou, E. and C. Romond. 1990. Effect of the feeding practices on the establishment of bacterial interactions in the intestine of the newborn delivered by cesarian section. J. Perinat. Med. 18: 139–143.

Bezirtzoglou, E. and C. Romond. 1991. Behavior of *Clostridium perfringens* in the intestine of newborn infants born by cesarean section. Ann. Pediatr. (Paris) 38: 331–335.

Blander, J.M. and R. Medzhitov. 2004. Regulation of phagosome maturation by signals from Toll-Like Receptors. Science. 304: 1014–1018.

Bode, L. 2012. Human milk oligosaccharides: Every baby needs a sugar mama. Glycobiology. 22: 1147–1162.

Bourges, D., F. Meurens, M. Berri, C. Chevaleyre, G. Zanello, B. Levast, S. Melo, V. Gerdts and H. Salmon. 2008. New insights into the dual recruitment of IgA+ B cells in the developing mammary gland. Mol. Immunol. 45: 3354–3362.

Brandtzaeg, P. 2003. Role of secretory antibodies in the defence against infections. Int. J. Med. Microbiol. 293: 3–15.

Brandtzaeg, P. 2007. Induction of secretory immunity and memory at mucosal surfaces. Vaccine. 25: 5467–5484.

Buescher, E.S. and S.M. McIlheran. 1993. Polymorphonuclear leukocytes and human colostrum: effects of *in vivo* and *in vitro* exposure. J. Pediatr. Gastroenterol. Nutr. 17: 424–433.

Buescher, E.S. and I. Malinowska. 1996. Soluble receptors and cytokine antagonists in human milk. Pediatr. Res. 40: 839–844.

Butts, C.L., S.A. Shukair, K.M. Duncan, E. Bowers, C. Horn, E. Belyavskaya, L. Tonelli and E.M. Sternberg. 2007. Progesterone inhibits mature rat dendritic cells in a receptor-mediated fashion. Int. Immunol. 19: 287–296.

Charoenphandhu, N., L.-i. Nakkrasae, K. Kraidith, J. Teerapornpuntakit, K. Thongchote, N. Thongon and N. Krishnamra. 2009. Two-step stimulation of intestinal Ca2+ absorption during lactation by long-term prolactin exposure and suckling-induced prolactin surge. Am. J. Physiol. Endocrinol. Metab. 297: E609–E619.

Choi, K.Y., L.N.Y. Chow and N. Mookherjee. 2012. Cationic host defence peptides: multifaceted role in immune modulation and inflammation. J. Inn. Immun. 4: 361–370.

Claesson, M.J., S. Cusack, O. O'Sullivan, R. Greene-Diniz, H. de Weerd, E. Flannery, J.R. Marchesi, D. Falush, T. Dinan, G. Fitzgerald, F. Shanahan, C. Twomey, R. Hill and P.W. O'Toole. 2011. Composition, variability, and temporal stability of the intestinal microbiota of the elderly. Proc. Natl. Acad. Sci. USA. 108: 4586–4591.

Collado, M.C., S. Delgado, A. Maldonado and J.M. Rodríguez. 2009. Assessment of the bacterial diversity of breast milk of healthy women by quantitative real-time PCR. Lett. Appl. Microbiol. 48: 523–528.

Corthésy, B. 2013. Role of secretory IgA in infection and maintenance of homeostasis. Autoimmun. Rev. 12: 661–665.

Dai, D., N.N. Nanthkumar, D.S. Newburg, and W.A. Walker. 2000. Role of oligosaccharides and glycoconjugates in intestinal host defense. J. Pediatr. Gastroenterol. Nutr. 30: S23–S33.

De Palma, G., J. Kamanova, J. Cinova, M. Olivares, H. Drasarova, L. Tuckova and Y. Sanz. 2012. Modulation of phenotypic and functional maturation of dendritic cells by intestinal bacteria and gliadin: relevance for celiac disease. J. Leukoc. Biol. 92: 1043–1054.

Donnet-Hughes, A., E.J. Schiffrin and W.A. Walker. 2008. Protective factors in human milk. Invited chapter in book: Nutrition in pediatrics: Basic science and clinical applications. 4th edition. Eds. C. Duggan, J. Watkins and W.A. Walker. Publisher: BC Decker Inc., Hamilton, Ontario, Canada.

Eiwegger, T., B. Stahl, J. Schmitt, G. Boehm, M. Gerstmayr, J. Pichler, E. Dehlink, C. Loibichler, R. Urbanek and Z. Szepfalusi. 2004. Human milk-derived oligosaccharides and plant-derived oligosaccharides stimulate cytokine production of cord blood T-cells *in vitro*. Pediatr. Res. 56: 536–540.

Fanning, S., Hall, L.J. and van Sinderen, D. 2012. *Bifidobacterium breve* UCC2003 surface exopolysaccharide production is a beneficial trait mediating commensal-host interaction through immune modulation and pathogen protection. Gut. Microbes. 3: 420–425.

Fernández, L., S. Langa, V. Martín, A. Maldonado, E. Jiménez, R. Martín and J.M. Rodríguez. 2013. The human milk microbiota: Origin and potential roles in health and disease. Pharmacol. Res. 69: 1–10.

Filipp, D., K. Alizadeh-Khiavi, C. Richardson, A. Palma, N. Paredes, O. Takeuchi, S. Akira and M. Julius. 2001. Soluble CD14 enriched in colostrum and milk induces B cell growth and differentiation. Proc. Natl. Acad. Sci. USA 98: 603–608.

Hanson, L. 2000. The mother-offspring dyad and the immune system. Acta. Paediatr. 89: 252–258.

Håversen, L., N. Kondori, L. Baltzer, L.Å. Hanson, G.T. Dolphin, K. Dunér and I. Mattsby-Baltzer. 2010. Structure-microbicidal activity relationship of synthetic fragments derived from the antibacterial α-helix of human lactoferrin. Antimicrob. Agents. Chemother. 54: 418–425.

Hassiotou, F., D.T. Geddes and P.E. Hartmann. 2013. Cells in human milk: State of the science. J. Hum. Lact. 29: 171–182.

Ho, P.C. and J.W. Lawton. 1978. Human colostral cells: phagocytosis and killing of *E. coli* and *C. albicans*. J. Pediatr. 93: 910–915.

Hughes, A., J.H. Brock, D.M. Parrott and F. Cockburn. 1988. The interaction of infant formula with macrophages: effect on phagocytic activity, relationship to expression of class II MHC antigen and survival of orally administered macrophages in the neonatal gut. Immunol. 64: 213–218.

Hugo, A.A., G.L. De Antoni and P.F. Pérez. 2010. *Lactobacillus delbrueckii* subsp. lactis (strain CIDCA 133) resists the antimicrobial activity triggered by molecules derived from enterocyte-like Caco-2 cells. Lett. Appl. Microbiol. 50: 335–340.

Hugo, A.A., E.E. Tymczyszyn, A. Gómez-Zavaglia and P.F. Pérez. 2012. Effect of human defensins on lactobacilli and liposomes. J. Appl. Microbiol. 113: 1491–1497.

Ibrahim, H.R., K. Imazato and H. Ono. 2011. Human lysozyme possesses novel antimicrobial peptides within Its N-terminal domain that target bacterial respiration. J. Agric. Food. Chem. 59: 10336–10345.

Ichikawa, M., M. Sugita, M. Takahashi, S. Satomi, T. Takeshita, T. Araki and H. Takahashi. 2003. Breast milk macrophages spontaneously produce granulocyte–macrophage colony-stimulating factor and differentiate into dendritic cells in the presence of exogenous interleukin-4 alone. Immunol. 108: 189–195.

Imai, K., A. Takeshita and S. Hanazawa. 2000. Transforming growth factor-β inhibits lipopolysaccharide-stimulated expression of inflammatory cytokines in mouse macrophages through downregulation of activation protein 1 and CD14 receptor expression. Infect. Immun. 68: 2418–2423.

Ito, S.-i., K.J. Ishii, H. Shirota and D.M. Klinman. 2004. CpG Oligodeoxynucleotides improve the survival of pregnant and fetal mice following *Listeria monocytogenes* Infection. Infect. Immun. 72: 3543–3548.

Jain, L., D. Vidyasagar, M. Xanthou, V. Ghai, S. Shimada and M. Blend. 1989. *In vivo* distribution of human milk leucocytes after ingestion by newborn baboons. Arch. Dis. Child. 64: 930–933.

Jeurink, P.V., J. van Bergenhenegouwen, E. Jiménez, L.M.J. Knippels, L. Fernández, J. Garssen, J. Knol, J.M. Rodríguez and R. Martín. 2013. Human milk: a source of more life than we imagine. Beneficial. Microbes. 4: 17–30.

Jia, H.P., T. Starner, M. Ackerman, P. Kirby, B.F. Tack and P.B. McCray. 2001. Abundant human β-defensin-1 expression in milk and mammary gland epithelium. J. Pediatr. 138: 109–112.

Jiménez, E., L. Fernández, M. Marín, R. Martín, J. Odriozola, C. Nueno-Palop, A. Narbad, M. Olivares, J. Xaus and J. Rodríguez. 2005. Isolation of commensal bacteria from umbilical cord blood of healthy neonates born by cesarean section. Curr. Microbiol. 51: 270–274.

Jiménez, E., M.L. Marín, R. Martín, J.M. Odriozola, M. Olivares, J. Xaus, L. Fernández and J.M. Rodríguez. 2008a. Is meconium from healthy newborns actually sterile? Res. Microbiol. 159: 187–193.

Jiménez, E., L. Fernandez, A. Maldonado, R. Martin, M. Olivares, J. Xaus and J.M. Rodriguez. 2008b. Oral administration of *Lactobacillus* strains isolated from breast milk as an alternative for the treatment of infectious mastitis during lactation. Appl. Environ. Microbiol. 74: 4650–4655.

Jimenez-Truque, N., S. Tedeschi, E.J. Saye, B.D. McKenna, W. Langdon, J.P. Wright, A. Alsentzer, S. Arnold, B.P. Saville, W. Wang et al. 2012. Relationship between maternal and neonatal *Staphylococcus aureus* colonization. Pediatr. 129: e1252–e1259.

Jost, T., C. Lacroix, C.P. Braegger and C. Chassard. 2012. New insights in gut microbiota establishment in healthy breast fed eonates. PLoS. ONE. 7: e44595.

Jost, T., C. Lacroix, C. Braegger and C. Chassard. 2013. Assessment of bacterial diversity in breast milk using culture-dependent and culture-independent approaches. Br. J. Nutr. 14: 1–10.

Kadaoui, K.A. and B. Corthésy. 2007. Secretory IgA mediates bacterial translocation to dendritic cells in mouse Peyer's Patches with restriction to mucosal compartment. J. Immunol. 179: 7751–7757.

Keski-Nisula, L., H.-R. Kyynäräinen, U. Kärkkäinen, J. Karhukorpi, S. Heinonen and J. Pekkanen. 2013. Maternal intrapartum antibiotics and decreased vertical transmission of *Lactobacillus* to neonates during birth. Acta. Paediatr. 102: 480–485.

Koren, O., Goodrich, Julia K., Cullender, Tyler C., Spor, A., Laitinen, K., Kling Bäckhed, H., Gonzalez, A., Werner, Jeffrey J., Angenent, Largus T., Knight, R. et al. 2012. Host remodeling of the gut microbiome and metabolic changes during pregnancy. Cell. 150: 470–480.

Kupferminc, M.J., A.M. Peaceman, T.R. Wigton, R.K. Tamura, K.A. Rehnberg and M.L. Socol. 1994. Immunoreactive tumor necrosis factor-alpha is elevated in maternal plasma but undetected in amniotic fluid in the second trimester. Am. J. Obstet. Gynecol. 171: 976–979.

Labéta, M.O., K. Vidal, J.E.R. Nores, M. Arias, N. Vita, B.P. Morgan, J.C. Guillemot, D. Loyaux, P. Ferrara, D. Schmid, M. Affolter, L.K. Borysiewicz, A. Donnet-Hughes and E.J. Schiffrin. 2000. Innate recognition of bacteria in human milk is mediated by a milk-derived highly expressed pattern recognition receptor, soluble Cd14. J. Exp. Med. 191: 1807–1812.

LeBouder, E., J.E. Rey-Nores, N.K. Rushmere, M. Grigorov, S.D. Lawn, M. Affolter, G.E. Griffin, P. Ferrara, E.J. Schiffrin, B.P. Morgan and M.O. Labéta. 2003. Soluble forms of toll-like receptor (TLR)2 capable of modulating TLR2 signaling are present in human plasma and breast milk. J. Immunol. 171: 6680–6689.

Lefèvre, C.M., J.A. Sharp and K.R. Nicholas. 2010. Evolution of lactation: Ancient origin and extreme adaptations of the lactation system. Annu. Rev. Genom. Human. Genet. 11: 219–238.

Lönnerdal, B. and S. Iyer. 1995. Lactoferrin: Molecular structure and biological function. Annu. Rev. Nutr. 15: 93–110.

Ludvigsson, J.F. and A. Fasano. 2012. Timing of introduction of gluten and celiac disease risk. Ann. Nutr. Metab. 60: 22–29.

Macpherson, A.J. and T. Uhr. 2004. Induction of protective IgA by intestinal dendritic cells carrying commensal bacteria. Science. 303: 1662–1665.

Macpherson, A.J. and T. Uhr, T. 2004. Compartmentalization of the mucosal immune responses to commensal intestinal bacteria. Ann. N. Y. Acad. Sci. 1029: 36–43.

Makino, H., A. Kushiro, E. Ishikawa, D. Muylaert, H. Kubota, T.Sakai, K. Oishi, R. Martin, K. Ben Amor, R. Oozeer et al. 2011. Transmission of intestinal *Bifidobacterium longum* subsp. longum strains from mother to infant, determined by multilocus sequencing typing and amplified fragment length polymorphism. Appl. Environ. Microbiol. 77: 6788–6793.

Martín, R., S. Langa, C. Reviriego, E. Jimínez, M.L. Marín, J. Xaus, L. Fernández and J.M. Rodríguez. 2003. Human milk is a source of lactic acid bacteria for the infant gut. J. Pediatr. 143: 754–758.

Martín, R., E. Jiménez, H. Heilig, L. Fernández, M.L. Marín, E.G. Zoetendal and J.M. Rodríguez. 2009. Isolation of *Bifidobacteria* from breast milk and assessment of the bifidobacterial population by PCR-denaturing gradient gel electrophoresis and quantitative real-time PCR. Appl. Environ. Microbiol. 75: 965–969.

Martín, V., A. Maldonado-Barragán, L. Moles, M. Rodriguez-Baños, R.d. Campo, L. Fernández, J.M. Rodríguez and E. Jiménez. 2012. Sharing of bacterial strains between breast milk and infant feces. J. Human. Lact. 28: 36–44.

M'Rabet, L., A.P. Vos, G. Boehm and J. Garssen. 2008. Breast-feeding and its role in early development of the immune system in infants: consequences for health later in life. J. Nutr. 138: 1782S–1790S.

Naccasha, N., M.-T. Gervasi, T. Chaiworapongsa, S. Berman, B.H. Yoon, E. Maymon and R. Romero. 2001. Phenotypic and metabolic characteristics of monocytes and granulocytes in normal pregnancy and maternal infection. Am. J. Obstet. Gynecol. 185: 1118–1123.

Nagl, M., L. Kacani, B. Müllauer, E.-M. Lemberger, H. Stoiber, G.M. Sprinzl, H. Schennach and M.P. Dierich. 2002. Phagocytosis and killing of bacteria by professional phagocytes and dendritic cells. Clin. Diagn. Lab. Immunol. 9: 1165–1168.

Newburg, D.S. 2009. Neonatal protection by an innate immune system of human milk consisting of oligosaccharides and glycans. J. Anim. Sci. 87: 26–34.

Newburg, D.S., G.M. Ruiz-Palacios and A.L. Morrow. 2005. Human milk glycans protect infants against enteric pathogens. Ann. Rev. Nutr. 25: 37–58.

Oftedal, O.T. 2012. The evolution of milk secretion and its ancient origin. Animal. 6: 355–368.

Pacheco, A.R. and V. Sperandio. 2009. Inter-kingdom signaling: chemical language between bacteria and host. Curr. Opin. Microbiol. 12: 192–198.

Patte, C., C. Tancrède, P. Raibaud and R. Ducluzeau.1979. First steps in the bacterial colonization of the digestive tract of neonates. Ann. Microbiol. (Paris) 130: 69–84.

Perez, P.F., J. Doré, M. Leclerc, F. Levenez, J. Benyacoub, P. Serrant, I. Segura-Roggero, E.J. Schiffrin and A. Donnet-Hughes. 2007. Bacterial imprinting of the neonatal immune system: lessons from maternal cells? Pediatrics. 119: e724–e732.

Pezzulo, A.A., E.E. Hornick, M.V. Rector, M. Estin, A.C. Reisetter, P.J. Taft, S.C. Butcher, A.B. Carter, J.R. Manak, D.A. Stoltz and J. Tabner. 2012. Expression of human paraoxonase 1 decreases superoxide levels and alters bacterial colonization in the gut of *Drosophila melanogaster*. PLoS. ONE. 7: e43777.

Pflughoeft and Versalovic. 2012. Human microbiome in health and disease. Annu. Rev. Pathol. Mech. Dis. 7: 99–112.

Pickering, L.K., T.G. Cleary, S. Kohl and S. Get. 1980. Polymorphonuclear leukocytes of human colostrum. I. Oxidative metabolism and kinetics of killing of radiolabeled *Staphylococcus aureus*. J. Infect. Dis. 142: 685–693.

Portal-Celhay and M.J. Blaser. 2012. Competition and resilience between founder and introduced bacteria in the *Caenorhabditis elegans* gut Infect. Immun. 80: 1288–1299.

Portal-Celhay, C., E. Bradley and M. Blaser. 2012. Control of intestinal bacterial proliferation in regulation of lifespan in *Caenorhabditis elegans*. BMC Microbiol. 12: 49.

Rescigno, M., M. Urbano, B. Valzasina, M. Francolini, G. Rotta, R. Bonasio, F. Granucci, J.P. Kraehenbuhl and P. Ricciardi-Castagnoli. 2001. Dendritic cells express tight junction proteins and penetrate gut epithelial monolayers to sample bacteria. Nat. Immunol. 2: 361–367.

Roux, M.E., M. McWilliams, J.M. Phillips-Quagliata, P. Weisz-Carrington and M.E. Lamm. 1977. Origin of IgA secretory plasma cells in the mammary gland. J. Exp. Med. 146: 1311–1322.

Sacks, G.P., C.W.G. Redman and I.L. Sargent. 2003. Monocytes are primed to produce the Th1 type cytokine IL-12 in normal human pregnancy: an intracellular flow cytometric analysis of peripheral blood mononuclear cells. Clin. Exp. Immunol. 131: 490–497.

Santos, S.J., M.D. Aupperlee, J. Xie, S. Durairaj, R. Miksicek, S.E. Conrad, J.R. Leipprandt, Y.S. Tan, R.C. Schwartz and S.Z. Haslam. 2009. Progesterone receptor A-regulated gene expression in mammary organoid cultures. J. Steroid. Biochem. Mol. Biol. 115: 161–172.
Schnorr, K.L. and L.D. Pearson. 1984. Intestinal absorption of maternal leucocytes by newborn lambs. J. Reprod. Immunol. 6: 329–337.
Solís, G., C.G. de los Reyes-Gavilan, N. Fernández, A. Margolles and M. Gueimonde. 2010. Establishment and development of lactic acid bacteria and bifidobacteria microbiota in breast-milk and the infant gut. Anaerobe. 16: 307–310.
Spor, A., O. Koren and R. Ley. 2011. Unravelling the effects of the environment and host genotype on the gut microbiome. Nat. Rev. Micro. 9: 279–290.
Stewart, J.A., V.S. Chadwick and A. Murray. 2005. Investigations into the influence of host genetics on the predominant eubacteria in the faecal microflora of children. J. Med. Microbiol. 54: 1239–1242.
Sun, Y., J. Cai, F. Ma, P. Lö, H. Huang and J. Zhou. 2012. Mir-155 mediates suppressive effect of progesterone on TLR3, TLR4-triggered immune response. Immunol. Lett. 146: 25–30.
Tap, J., S. Mondot, F. Levenez, E. Pelletier, C. Caron, J.-P. Furet, E. Ugarte, R. Muñoz-Tamayo, D.L.E. Paslier, R. Nalin, J. Doré and M. Leclerc. 2009. Towards the human intestinal microbiota phylogenetic core. Environ. Microbiol. 11: 2574–2584.
Ueda, Y., M. Hagihara, A. Okamoto, A. Higuchi, A. Tanabe, K. Hirabayashi, S. Izumi, T. Makino, S. Kato and T. Hotta. 2003. Frequencies of dendritic cells (myeloid DC and plasmacytoid DC) and their ratio reduced in pregnant women: comparison with umbilical cord blood and normal healthy adults. Hum. Immunol. 64: 1144–1151.
Ulvestad, E. 2009. Cooperation and conflict in host–microbe relations. APMIS. 117: 311–322.
van der Waaij, L.A., L.P.C., Mesander G. and van der Waaij D. 1996. *In vivo* IgA coating of anaerobic bacteria in human faeces. Gut. 38: 348–354.
Verhasselt, V. 2010. Neonatal tolerance under breastfeeding influence: The presence of allergen and transforming growth factor-β in breast milk protects the progeny from allergic asthma. J. Pediatr. 156: S16–S20.
Vestergaard, E.M., E. Nexo, A. Wendt and F. Guthmann. 2008. Trefoil factors in human milk. Early Hum. Dev. 84: 631–635.
Vidal, K., A. Donnet-Hughes and D. Granato. 2002. Lipoteichoic acids from *Lactobacillus johnsonii* strain La1 and *Lactobacillus acidophilus* strain La10 antagonize the responsiveness of human intestinal epithelial HT29 cells to lipopolysaccharide and gram-negative bacteria. Infect. Immun. 70: 2057–2064.
Vidal, K., P. Serrant, B. Schlosser, P. van den Broek, F. Lorget and A. Donnet-Hughes, A. 2004. Osteoprotegerin production by human intestinal epithelial cells: A potential regulator of mucosal immune responses. Am. J. Physiol. Gastrointest. Liver. Physiol. 287: G836–G844.
Weiler, I.J., W. Hickler and R. Sprenger. 1983. Demonstration that milk cells invade the suckling neonatal mouse. Am. J. Reprod. Immunol. 4: 95–98.
Yolken, R.H., J.A. Peterson, S.L. Vonderfecht, E.T. Fouts, K. Midthun and D.S. Newburg. 1992. Human milk mucin inhibits rotavirus replication and prevents experimental gastroenteritis. J. Clin. Invest. 90: 1984–1991.
Yoshioka, H., K. Iseki and K. Fujita. 1983. Development and differences of intestinal flora in the neonatal period in breast-fed and bottle-fed infants. Pediatrics. 72: 317–321
Yu, Z.-T., C. Chen, D.E. Kling, B. Liu, J.M. McCoy, M. Merighi, M. Heidtman and D.S. Newburg. 2013. The principal fucosylated oligosaccharides of human milk exhibit prebiotic properties on cultured infant microbiota. Glycobiology. 23: 169–177.
Zhang, L., S. van Bree, J.J. van Rood and F.H. Claas. 1991. Influence of breast feeding on the cytotoxic T cell allorepertoire in man. Transplantation. 52: 914–916.

Zhou, L., Y. Yoshimura, Y.Y. Huang, R. Suzuki, M. Yokoyama, M. Okabe and M. Shimamura. 2000. Two independent pathways of maternal cell transmission to offspring: through placenta during pregnancy and by breast-feeding after birth. Immunol. 101: 570–580.
Zoetendal, E.G., A.D.L. Akkermans and W.M. De Vos. 1998. Temperature gradient gel electrophoresis analysis of 16S rRNA from human fecal samples reveals stable and host-specific communities of active bacteria. App. Environ. Microbiol. 64: 3854–3859.

3

Interactions of the Intestinal Microbiota with Mucosal Epithelial Cells

Cherbuy Claire, Tomas Julie, Thomas Muriel and *Langella Philippe**

Introduction

The mammalian intestine is in constant contact with a vast array of commensal bacteria and faces the challenge of having to maintain its capacity to respond to intestinal pathogens while ensuring that the inflammatory response to commensal organisms is minimal (Garrett et al. 2010; Hooper and Macpherson 2010). The intestinal epithelium is at the crossroads between the gut microbiome and the host, and the intestinal epithelial cells (IECs) play a major role in tolerance to commensal microbes (Goto and Kiyono 2012). The multiple specialized intestinal epithelial lineages are called upon to develop a rich array of essential strategies for sustaining the complex and dynamic microbial communities of the intestine. Preservation of the functions of the intestinal epithelium is important, as shown by genome-wide association studies linking loci implicated in epithelial barrier function with inflammatory bowel diseases (IBD)

Commensal and Probiotics-Host Interactions Team, INRA, UMR 1319 MICALIS, Domaine de Vilvert, F-78350 Jouy-en-Josas, France.
E-mails: claire.cherbuy@jouy.inra.fr; julie.tomas@jouy.inra.fr; muriel.thomas@jouy.inra.fr
* Corresponding author: philippe.langella@jouy.inra.fr

(Cho and Brant 2011). A recent study in B cell-deficient mice showed that failure of the adaptive immune system led to commensal bacteria forcing IECs to upregulate genes controlling immunity at the expense of those regulating metabolism, revealing an additional level of complexity in the crosstalk between microbiota and the intestinal epithelium (Shulzhenko et al. 2011).

In this chapter, we provide an overview of the interactions responsible for preserving the symbiotic nature of the relationships between the gut microbiota and the intestinal epithelium and illustrate how complex, rich and attractive these interactions are. We will first briefly describe the intestinal epithelium and present evidence to suggest that despite the interaction between commensal bacteria and IECs, direct contact is minimized, mostly by the mucus layer and antimicrobial peptides, which keep the bacteria at some distance from the epithelium. However, some commensal bacteria escape these mechanisms and are found in close contact with the epithelium. We will also focus on the reliable cohesion of the epithelium as a wall preventing mucosal immune cells from encountering the heavy bacterial load. We will also illustrate the increasing body of data indicating that commensal bacteria and probiotics help to preserve this optimal epithelial barrier. We will also discuss the roles of pattern recognition receptors and intestinal alkaline phosphatase in managing this vast consortium of commensal bacteria. Conversely, we will present evidence that the gut microbial communities have helped to shape the development and physiology of the intestinal epithelium. The function of microfold cells (M cells) and secreted IgA are touched upon only briefly in this chapter, as these elements are essentially beyond the scope of this chapter.

We show here that understanding crosstalk at the mucosal surface is a major issue in deciphering the colonization mechanisms contributing to the stability of the ecosystem. Health benefits might be obtained by making use of these positive interactions, particularly for the development and design of promising track, such as probiotics.

The Intestinal Epithelium: A Dynamic and Multifunctional Tissue

The intestinal epithelium is one of the most dynamic tissues in the body as it continually undergoes rapid homeostatic renewal over a cycle lasting a few days. Within the small intestine, the epithelium is folded into villi, which project into the lumen, increasing the surface area available for the absorption of nutrients (Fig. 1A). The villi are aligned with the crypts—invaginations into the underlying mesenchyme. The colon displays a similar

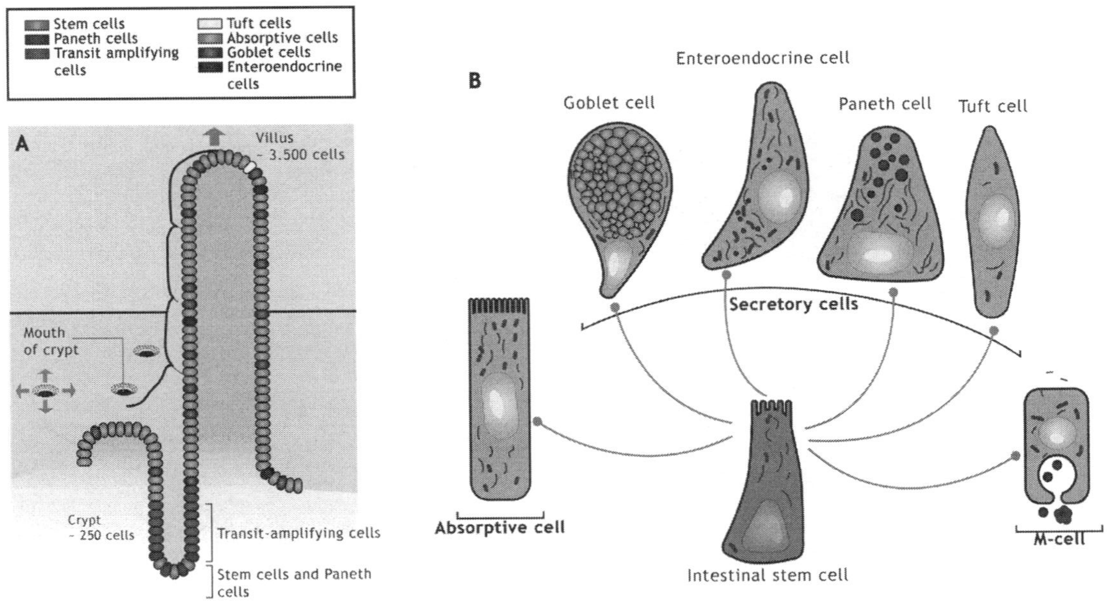

Figure 1. Structure of the Epithelia of the Small Intestine and Colon. The intestinal epithelium is spatially structured into invaginations called crypts, which extend into finger-like protrusions called villi in the small intestine and into the flat surface of the colon (A). Stem cells are found close to the base of the crypt. They generate highly proliferative cells that move upwards and subsequently differentiate into the various specialized lineages of the intestinal epithelium: absorptive cells (enterocytes or colonocytes), goblet cells, enteroendocrine cells, tuft cells, Paneth cells and the microfold (M) cells (B). The renewal process ends with the exfoliation of terminally differentiated cells at the tips of villi or from the surface of the colon epithelium. Paneth cells migrate towards the bottom of the crypt, where they remain for 6–8 weeks, supporting stem cell function. Cells expressing Paneth cell markers have been detected at the base of the crypts in the colon. These processes occur continuously and are highly coordinated, to allow tissue homeostasis.

Color image of this figure appears in the color plate section at the end of the book.

folding into crypts, but the epithelium at the surface is flat, with no villi. This modular organization into crypts and villi results in morphologically and functionally different compartments in which the proliferative and differentiated cell populations are spatially separated (van der Flier and Clevers 2009).

The perpetual self-renewal of the intestinal epithelium is fuelled by stem cells residing at the bottom of crypts in the small intestine and colon. Stem cells proliferate continually, generating highly proliferative, transit-amplifying daughter cells that move out of the crypt towards the tips of the villi or the surface of the epithelium in the colon (Fig. 1A). Proliferation ceases when the cells migrate, and the cells undergo terminal differentiation to generate the various lineages of IECs. The renewal process ends with the exfoliation of cells arriving at the tips of the villi or the surface of the epithelium in the colon (van der Flier and Clevers 2009). Cells can differentiate into several different lineages (Fig. 1B): (in descending order of frequency) the absorptive enterocytes or colonocytes responsible for nutrient, ion and water absorption; the goblet or caliciform cells scattered throughout the epithelium and responsible for producing the protective mucus layer; the enteroendocrine cells producing various peptide hormones, the Paneth cells restricted to the small intestine and playing an antimicrobial role (Karam 1999) and the tuft cells, which produce β-endorphin (Gerbe et al. 2012; Gerbe et al. 2011). Another specialized type of IECs is the M cells located at the luminal surface of the dome structure of Peyer's patches, which transfer microbiota-derived antigens to antigen-presenting cells. *In vivo* imaging has recently shown that goblet cells are involved in the delivery of small-molecular weight antigens from the intestinal lumen to the underlying immune cells (McDole et al. 2012).

Leucine-rich repeat-containing G-protein-coupled receptor 5 (Lgr5) has been identified as a marker for the small, cycling, intestinal stem cells residing at the base of the crypts (Barker et al. 2007). Lineage tracing experiment have indicated that $Lgr5^+$ cells fuel daily self-renewal of the intestine and generate the types of differentiated cells, including M cells (Barker et al. 2007; de Lau et al. 2012). Unlike the other differentiated cells, Paneth cells migrate downward and reside at the bottom of the crypts for two months. Paneth cells, which are located between the $Lgr5^+$ stem cells, support these stem cells by secreting essential signals for stem-cell maintenance (Sato et al. 2011). The genetic elimination of Paneth cells results in a loss of stem cells (Sato et al. 2011). In the colon, cells resembling Paneth cells have been found in the vicinity of stem cells (Sato et al. 2011).

The proliferation and acquisition of a single cell fate are tightly coordinated by a small number of highly conserved signaling pathways. In particular, the Wingless-related MMTV integration site (Wnt), bone morphogenetic protein, epidermal growth factor and Notch pathways have been shown to play major roles in the regulation of these processes (Schuijers and Clevers 2012). Indeed, inactivation of the Wnt/β-catenin/Tcf4 signaling pathway leads to the loss of crypt epithelial progenitors (Korinek et al. 1998), whereas genetic deletions of effectors from the Notch signaling pathway drive the undifferentiated cells toward a secretory fate, thereby modifying the numbers of goblet, enteroendocrine and Paneth cells (van der Flier and Clevers 2009).

Good Fences Make Good Neighbors: The Role of the Mucus Layer

There is crosstalk between the gut microbiota and the intestinal epithelium, but direct interactions are limited. Recent data indicate that most of the commensal microbes are physically separated from the epithelium. This maintenance of the bacterial population at some distance from the active epithelium probably contributes to peaceful coexistence, as patients with IBD have larger numbers of bacteria close to the epithelium (Swidsinski et al. 2005). Like most mucosal surfaces, the digestive epithelium is covered with a protective mucus layer that has the consistency of a viscous and permeable gel (Fig. 2). This layer plays a role in defense against enteric bacterial infections and expels pathogens from the mucosal surface (Bergstrom et al. 2010). The mucus layer is composed largely of secreted mucins, large glycoproteins linked to diverse *O*-linked oligosaccharides. Mucin-type *O*-linked oligosaccharides (*O*-glycans) account for more than 70% of the weight of the mucin molecule. MUC2 is the most abundant secreted mucin in the small and large intestines (Gum et al. 1989). MUC2 is synthesized, assembled and secreted by intestinal goblet cells, in a multistep processes. Once secreted into the lumen, mucins expand, increasing in volume by a factor of 100 to 1000, due to their ability to take up large quantities of water (Johansson et al. 2011; Kim and Ho 2010).

The commensal bacteria and the two mucus layers

In the distal part of the digestive tract in mice, the mucus layer extends 150 µm above the epithelial cells and is organized into two stratified layers (Fig. 2). The inner mucus layer closest to the epithelium adheres strongly to the epithelial cells, has a high density and is resistant to bacterial penetration. By contrast, the outer layer is loose and easy to remove (Atuma et al. 2001; Johansson et al. 2008). Resident bacteria display differential associations with these two mucus layers (Fig. 2). They are excluded from the dense, compact, inner mucus layer, resulting in the maintenance of a zone containing very few bacteria next to the surface of the IECs. Microbes

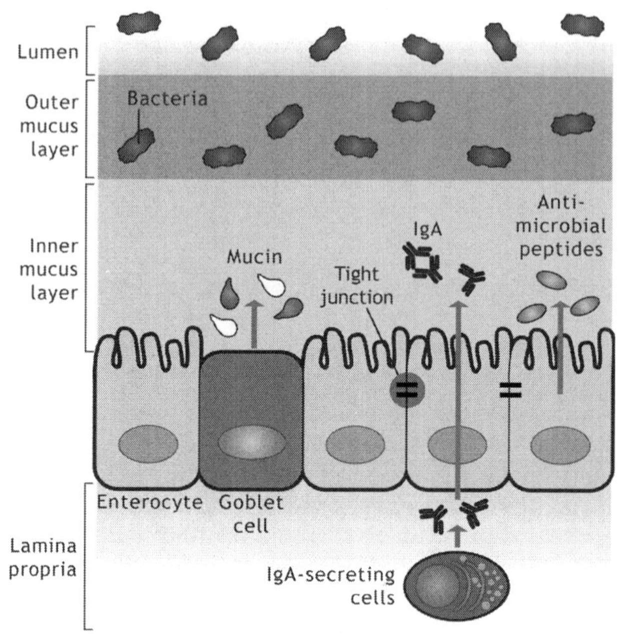

Figure 2. Mechanisms Minimizing Direct Contact Between Intestinal Bacteria and Epithelial Cells and Preventing Bacterial Penetration. Several mechanisms act in concert to limit direct interactions between the dense luminal microbial community and the intestinal epithelial cell surface. Goblet cells secrete gel-forming mucins, which assemble into a bilayer of intestinal mucus. The firm inner layer is dense and resistant to bacterial penetration, whereas the loose outer layer is a rich and reliable source of nutrients for the intestinal bacteria. The mucus layer concentrates antimicrobial peptides (AMPs), which are secreted principally by Paneth cells, but also by enterocytes. Some AMPs are produced constitutively, whereas others—such as RegIIIγ, a member of the C-type lectin family—are induced by commensal microorganisms. The epithelial cells are bound tightly together by intercellular junctional complexes, which regulate paracellular permeability. Another mechanism involve the transcytosis of IgA secreted by the underlying immune cells, which helps to limit the number of bacteria close to the epithelium and to prevent the penetration of host tissues by bacteria. This latter aspect is not considered in detail in this chapter.

Color image of this figure appears in the color plate section at the end of the book.

associate preferentially with the outer, loose mucus layer, and use the glycans present in the mucus as major source of energy (Johansson et al. 2011; Johansson et al. 2008).

The mucus gel acts as a mesh, concentrating other protective effectors secreted by goblet cells and vital for host innate defense. These factors include trefoil factor peptides (TFF3), resistin-like molecule β (RELMβ) and Fc-γ binding protein (Fcgbp). These molecules stabilize the mucin network in the mucus layer or have healing and immunoregulatory roles (Kim and Ho 2010). In addition, secretory IgA, which is transported across

IECs by the polymeric immunoglobulin receptor, also resides in the mucus layer (Fig. 2). The mucus covering the epithelium of the small intestine is less dense than that in the colon, probably in order to facilitate nutrient absorption (Atuma et al. 2001). As discussed below, the mucus layer in the small intestine concentrates antimicrobial peptides close to the epithelial cell surface, keeping the bacteria away from the epithelium (Meyer-Hoffert et al. 2008). These molecules thus help to limit the penetration of bacteria into the zone adjacent to the epithelium.

Studies of genetically modified mice have revealed the important role in host protection played by the mucus layer, by limiting the total number of commensal bacteria associated with the surface of the colon. The absence or mutation of *muc2*, impairing the biosynthesis or assembly of the protein, results in spontaneous inflammation and high susceptibility to chemically induced inflammation (Heazlewood et al. 2008; Van der Sluis et al. 2006). Similar results have been obtained in mice with impaired intestinal MUC2 *O*-glycosylation (An et al. 2007; Fu et al. 2011) or palmitoylation (Wei et al. 2012). The pattern of mucin glycosylation also affects susceptibility to chemically induced colitis (Tobisawa et al. 2010). It has been shown that changes to these protective layers bring bacteria into close contact with epithelial cells, allowing them to penetrate deeper into the crypts and enter epithelial cells (Bergstrom et al. 2010; Johansson et al. 2008; Wei et al. 2012). Such close contact between bacteria and the epithelium may account for the inflammatory phenotype observed in the absence of *muc2*. Indeed, antibiotic treatment attenuates the inflammation, suggesting that resident bacteria are involved in the disease (Wei et al. 2012). Interestingly, some of the traits described in the mouse model resemble severe human ulcerative colitis, with goblet cell loss and a decrease in the thickness of the mucus layer (Strugala et al. 2008).

The mucus layer: an attractive niche for the greedy commensal bacteria

The *O*-glycans on the mucins provide a rich reliable source of nutrients for members of gut microbiota. In addition, the mucus layer provides numerous microbial attachment sites, allowing commensal microbes to avoid being washed out from the gut and to establish themselves in close proximity for the sharing of nutrients (Koropatkin et al. 2012). Some commensal bacterial species from the genera *Bifidobacterium*, *Bacteroides* and *Ruminococcus*, and a representative of the phylum Verrucomicrobia, *Akkermansia muciniphila*, degrade mucins (Ouwehand et al. 2005). An ability to metabolize *O*-linked oligosaccharides is a key factor for the stability of the gut ecosystem, as these compounds can be used as a source of nutrients in the absence of polysaccharides. This has been shown for the glycophilic gut symbiont

Bacteroides thetaiotaomicron (Sonnenburg et al. 2005), which then distributes its digestion products to other microbiota less able to degrade glycans (Mahowald et al. 2009). An ability to forage host glycans may also help a species to colonize the gut during changes in dietary composition, such as the transition from suckling to weaning, for example (Marcobal et al. 2011). However, excessive consumption of host glycans may be detrimental to the host, as it may lead to the mucus barrier being breached. The substantial metabolic cost of glycan degradation and the complex enzymatic equipment required to break mucins which is harbored by only a few commensal bacteria, probably prevent excessive degradation of endogenous *O*-linked glycans *in vivo* (Hoskins and Boulding 1981).

Antimicrobial Peptides to Tame the Gut Microbial Ecosystem

Intestinal epithelial cells produce a large panel of antimicrobial peptides (AMPs), which are among the most potent effectors of innate immunity (Fig. 2). These molecules rapidly kill or inactivate microorganisms and provide protection against pathogens (Brandl et al. 2007). The host makes use of these innate immune defense mechanisms to control the gut microbial community by regulating the composition of the gut microbiota at the mucosal surface and the number and location of the microbes concerned (Gallo and Hooper 2012). The specialized secretory Paneth cells are the main source of AMPs in the small intestine (Bevins and Salzman 2011). They produce large amounts of diverse AMPs, such as lysozyme, α-defensins, secretory phospholipase A2, C-type lectins, such as regenerating islet-derived protein (REG) 3α (the equivalent of the mouse REG3γ) (Gallo and Hooper 2012). However, other IECs in the small intestine produce AMPs, including the enterocytes, which produce REG3 lectins (Vaishnava et al. 20011; Vaishnava et al. 2008). In the colon, in which the epithelium contains no lysozyme-secreting Paneth cells, the absorbing colonocytes produce β-defensins (O'Neil et al. 1999) and cathelicidins (Hase et al. 2002). In many cases, the genetic defects leading to Paneth cell dysfunction and a decrease in AMPs production are associated with susceptibility to intestinal inflammation (Cadwell et al. 2008; Kaser et al. 2008; Koslowski et al. 2012). Consistent with these findings, low levels of antimicrobial compounds, including α-defensin in particular, are observed in patients with IBD (Wehkamp et al. 2007).

AMPs to Keep Bacteria at a Safe Distance

It has been shown that commensal bacteria induce host epithelial AMPs production and that these molecules contain the bacteria within the intestine

(Fig. 2). Thus, in mice, the production of intestinal REG3γ is controlled by the gut microbiota (Cash et al. 2006) and this mechanism plays an essential role in keeping the bacteria at a safe distance from the epithelium. In REG3γ-deficient mice, larger numbers of bacteria reach the epithelium of the small intestine, where they trigger a deviant immune response with higher than normal levels of immunoglobulin A and T helper 1 cells (Vaishnava et al. 2011). Similarly, mice harboring a transgene encoding a toxin that eliminates Paneth cells also display higher levels of bacterial translocation to the epithelium of the small intestine (Vaishnava et al. 2008). Thus, the loss of REG3γ disrupts the ability of the host to keep commensal bacteria at a safe distance and to prevent bacterial translocation. IECs upregulate REG3γ production by sensing bacteria through Toll-like receptors (TLRs), a family of pattern recognition receptors. Indeed, an absence of myeloid differentiation primary response protein 88 (MyD88), an adaptor molecule common to several TLRs, prevents the induction of REG3γ in mice (Brandl et al. 2007; Vaishnava et al. 2008). It has recently been shown that antibiotic intake decreases REG3γ production in mice, which suggests that these drugs may also alter epithelial homeostasis (Reikvam et al. 2011).

Do Intestinal AMPs Control the Composition of the Intestinal Microbiota?

Commensal bacteria are able to expand within this antimicrobial compound-rich environment to such as extent that they outnumber eukaryotic cells by a factor of 10. Sensitivity to AMPs differs between bacterial strains and AMPs have recently been shown to shape the gut microbial ecosystem. Mice expressing a human α-defensin or lacking an enzyme required to process murine α-defensins display a significant shift in the dominant bacterial phyla with respect to their wild-type counterparts (Salzman et al. 2010). These two models display reciprocal modifications to the proportions of Firmicutes, the major Gram-positive phylum, and Bacteroidetes, the major Gram-negative phylum (Salzman et al. 2010). These changes to the composition of the microbial community of the small intestine also lead to changes in the mucosal immune response, including T-cell responses (Salzman et al. 2010). In Drosophila, the overproduction of AMPs also alters the commensal population within the intestine, favoring the outgrowth of a particular 'pathogenic' commensal community, leading to the death of the host (Ryu et al. 2008). Thus, AMPs contribute to the dynamics of the intestinal microbial community, with potential implications for host health.

Bacterial Incursions Close to the Intestinal Epithelium

Some members of the commensal microbiota escape host control mechanisms and reach areas close to the intestinal epithelium. It has recently been shown that resident microbes are not randomly distributed in the digestive environment. Instead, the gut microbial community displays a clear spatial organization (Nava et al. 2011). *Candidatus arthromitis*, known as segmented filamentous bacteria (SFB), is one of only a very small number of commensal bacteria known to come into direct contact with IECs. SFB inhabit the terminal part of the small intestine of various mammals, but they have not yet been demonstrated to be present in the human microbiota. Scanning electron microscopy of the intestine colonized with SFB and specific staining of SFB have shown that the SFB are anchored to ileal epithelial cells (Fig. 3; kindly provided by S. Da Silva, INRA, France) (Gaboriau-Routhiau et al. 2009). A full set of genes for flagellar assembly, fibronectin binding protein and other proteins predicted to act as adhesins has been identified in the genomes of SFB (Kuwahara et al. 2011; Prakash et al. 2011). Any or all of these genes may be involved in the tight attachment of the bacteria to the host epithelium. Several groups have shown that SFB play a crucial role in the maturation of gut immune function, particularly in the induction of a specific subset of proinflammatory Th17 cells (Gaboriau-Routhiau et al. 2009; Ivanov et al. 2009). The induction of proinflammatory T-cell responses has been shown to protect mice against *Citrobacter rodentium*

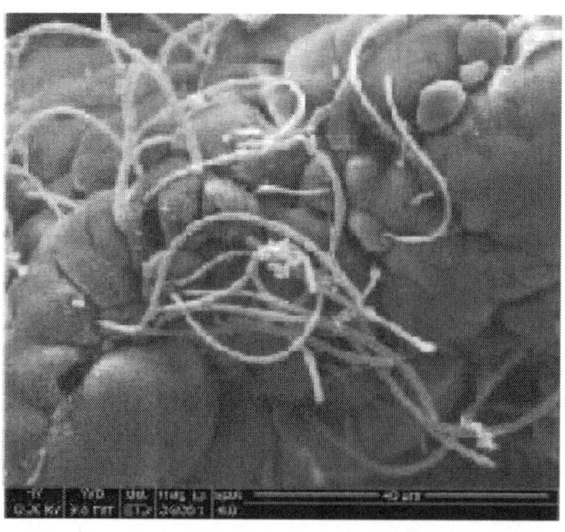

Figure 3. The Exception to Luminal Segregation: the example of SFB, which are found in close contact with the intestinal epithelium. Scanning electron microscopy of the surface of the rat ileum, showing the direct adhesion of segmented filamentous bacteria (SFB) to intestinal epithelial cells. Image courtesy of Stéphanie Da Silva "Laboratoire d'Ingénierie des Systèmes Biologiques et des Procédés, INRA, Toulouse, France". Photograph taken with a MEB FEG Quanta 250 FEI microscope at the CMEAB, Faculté de Médecine, Toulouse, France.

(Ivanov et al. 2009), but may also increase susceptibility to autoimmune diseases (Wu et al. 2010) or experimentally induced colitis (Stepankova et al. 2007). These data suggest that bacterial species that are able to get close to the host cells may have particularly strong effects on host health and that special attention should be paid to these microbes in cases of dysbiosis. Similarly, a recent study showed that the cecal and colonic crypts are not sterile, instead containing a microbiota dominated by aerobic bacteria. In particular, *Acinetobacter* has a strong tropism for crypts. The possible role of this crypt-specific core microbiota in crypt protection and homeostasis remain to be determined (Pedron et al. 2012).

Commensal and Probiotic Bacteria Strengthen Epithelial Barrier Function

On the apical side, epithelial cell-cell cohesion is ensured by tight junctions that bind the cells closely together and seal off the paracellular pathway (Fig. 2) (Suzuki 2012). The molecular composition of the tight junctions is complex, with multiple transmembrane proteins (claudin, occludin, members of junctional adhesion molecule family) linked to intracellular zonula occludens (ZO), which act as bridges to cytoskeletal actin and myosin filaments (Suzuki 2012). A breach in the intestinal epithelial barrier and underlying immune cells triggers the next line of antimicrobial defense. In mouse models, molecular or functional defects of tight junctions cause immune activation and contribute to the development of experimental colitis (Gilbert et al. 2012; Laukoetter et al. 2007; Su et al. 2009). The global barrier defects seen in active IBD are associated with a loss of tight junctions (Schulzke et al. 2009) and increases in intestinal permeability may precede the development of intestinal inflammation (Olson et al. 2006).

There is growing evidence, from *in vivo* and *in vitro* models, that commensal bacteria and probiotics may help to preserve an optimal epithelial barrier by targeting the production and/or localization of tight junction proteins. The treatment of epithelial cells with the probiotics *Escherichia coli* Nissle 1917 (Ukena et al. 2007; Zyrek et al. 2007), *Bifidobacterium infantis* (Ewaschuk et al. 2008), *Lactobacillus rhamnosus* and *Bifidobacterium lactis* (Mathias et al. 2010) strengthens tight junction structure through changes to the levels of the ZO-1, ZO-2 and occludin proteins. Monoxenic mice colonized with *E. coli* Nissle 1917 display an upregulation of ZO-1 in intestinal epithelial cells when compared to microbe-free mice (Ukena et al. 2007). Occludin and ZO-1 levels are also modified in the intestines of healthy human volunteers receiving *Lactobacillus plantarum* (Karczewski et al. 2010). A study of gene expression in a cultured cell line suggested that the effect of *L. plantarum* extends to many other tight junction-related genes

(Anderson et al. 2010). *Streptococcus thermophilus*, *Lactobacillus acidophilus* and the commensal bacterium *Bacteroides thetaiotaomicron* prevent the intestinal barrier dysfunction caused by the proinflammatory cytokines TNFα and IFNγ (Resta-Lenert and Barrett 2006).

Various mechanisms for the action of probiotics and commensal bacteria have been described. The ability of bacteria to modify tight junction proteins selectively may involve bacterial metabolites, as shown for *B. infantis* (Ewaschuk et al. 2008). The short-chain fatty acids generated by bacterial fermentation accelerate the assembly of tight junctions in the intestinal epithelium (Peng et al. 2009). A *Lactobacillus*-derived molecule, polyphosphate, has recently been shown to decrease oxidant-induced intestinal permeability in mouse small intestine (Segawa et al. 2011). Furthermore, DNA from the commensal bacteria *L. rhamnosus* and *Bifidobacterium longum* plays a role in the restructuring of the epithelial barrier after treatment with TNFα and IFNγ (Ghadimi et al. 2010). *E. coli* Nissle 1917-conditioned media and isolated lipopolysaccharide (LPS) reproduce the effect of the strain itself, increasing epithelial resistance (Stetinova et al. 2010).

Microbial Enhancement of the Epithelial Barrier and Turnover of Intestinal Epithelial Cells

Signals derived from the gut microbiota may modulate IEC proliferation, differentiation and apoptosis. The gut microbiota has been reported to increase intestinal cell proliferation and kinetic variables in the developing zebrafish (Cheesman et al. 2011), Drosophila (Buchon et al. 2009) and rodents (Alam et al. 1994). In pigs, the intestinal expression of genes involved in apoptotic and proliferative activities is modified in the presence of a complex microbiota (Danielsen et al. 2007; Willing and Van Kessel 2007) and the introduction of a nonpathogenic strain of *E. coli* into gnotobiotic piglets stimulates overall intestinal cell turnover (Danielsen et al. 2007; Willing and Van Kessel 2007). By contrast, the treatment of microbiota-depleted mice with antibiotics alters the expression of cell-cycle-related genes and decreases epithelial cell proliferation (Reikvam et al. 2011). Furthermore, the microbiota induces cell cycle-related proteins in a well-coordinated manner, with an initial increase in the amounts of proteins involved in cell proliferation followed by a compensatory induction of cell-cycle arrest-related proteins (Cherbuy et al. 2010). This structural maturation of the epithelium is associated with a profile of primocolonization in the context of the experimental colonization of gnotobiotic rodents (Tomas et al. 2012). The stimulation of IEC proliferation by commensal bacteria can be beneficial for the host. Indeed, it has been shown that microbially induced proliferation protects against epithelial injury after dextran sodium sulfate treatment (Pull et al. 2005; Rakoff-Nahoum et al. 2004).

Probiotics also drive cell proliferation in the intestinal epithelium. Indeed, in neonatal mice, human-derived *L. reuteri* strains stimulate IEC proliferation and the expression of intestinal genes involved in cell turnover and migration (Preidis et al. 2012). *S. thermophilus* also modulates cell-cycle arrest-related proteins, providing further evidence for a role of probiotics in epithelial maturation (Rul et al. 2011; Thomas et al. 2011).

The case of LGG

L. rhamnosus GG (LGG) has been widely studied as a probiotic bacterium. It improves crypt survival and decreases epithelial cell death levels after cytokine-induced damage (Yan and Polk 2002) or whole-body irradiation in mice (Ciorba et al. 2012). LGG maintains epithelial barrier function in the presence of hydrogen peroxide and increases the membrane translocation of ZO-1 and occludin (Seth et al. 2008). Two LGG-derived proteins (p40 and p75) have been implicated in the prevention of cytokine-induced epithelial cell apoptosis and the protection of epithelial barrier function against peroxide-induced damage (Seth et al. 2008; Yan et al. 2007). P40 protein has been administered in several different models of experimental colitis, with special hydrogel beads used to protect the protein against degradation, and this treatment has been shown to be effective for reducing inflammation (Yan et al. 2011). The identification of factors mediating the beneficial effects of probiotics illustrates the way in which an understanding of the molecular basis of microbe–host interactions may lead to the development of new effective therapeutic strategies.

How do Intestinal Epithelial Cells Sense Intestinal Microbiota?

Pattern recognition receptors (PRRs) are involved in the recognition of microbe-associated molecular patterns and play a key role in inducing innate effectors and inflammation. The recognition of ligands induces a signaling cascade that leads to the activation of intracellular pathways, such as the NF-κB pathway. IECs lineages express various PRRs, which sense the presence of microbes, and bacteria actively communicate with cells through these innate immune receptors. Several studies in human and murine models have suggested that PRR signaling in the gut is adapted to prevent continually, excessive activation by the resident microbiota (Abreu 2010). However, studies of knockout mice have shown that PRR signaling in the epithelium has a beneficial role in maintaining homeostasis and the barrier functions of the epithelium (Carvalho et al. 2012).

PRRs: systems for perceiving microbes

The members of the Toll-like receptor (TLR) family are the most widely studied PRRs, and more than 10 members of this family have been described in mice and humans. TLRs have a conserved transmembrane glycoprotein domain, with a ligand recognition domain in the extracellular region and a Toll/IL-1 receptor homology (TIR) domain in the cytoplasmic tail. The TIR domain is essential for interactions between homo- or heterodimeric TLR subunits and for the recruitment of cytoplasmic adapter proteins, such as MyD88, to initiate downstream signaling cascades (Abreu 2010). TLRs may be found at the surface of the cell or in endosomes. The various TLRs have similar structures, but each has its own set of ligands and they recognize different signals, such as LPS, lipoproteins, flagellin and bacterial nucleic acids (Wells et al. 2011). The other major family of PRRs studied is that of the nucleotide-binding oligomerization (NOD)-domain like receptors (NLRs). These proteins are located in the cytosol. NOD1 and NOD2 are the most strongly expressed members of this family in the intestine. Both recognize peptidoglycan fragments. NOD2 is essentially confined to the Paneth cells of the small intestine (Ogura et al. 2003).

TLR in intestinal epithelial cells

The TLR repertoires of the small intestine and colon tissues are highly diverse (Zarember and Godowski 2002). However, as IECs are constantly exposed to bacteria, TLR expression is regulated and spatially restricted and signaling is modulated by adaptors or negative regulators, to minimize the gut response to microbes and to prevent deleterious inflammatory responses in the intestine (Abreu 2010; Carvalho et al. 2012). Further investigations are required to complete our understanding of TLR expression and compartmentalization in the intestinal epithelial cells, but there is evidence to suggest that some of these receptors are expressed and localized exclusively in the crypt epithelial cells, far from the luminal commensal bacteria, as reported for TLR2 and TLR4 in the colonic epithelium (Fig. 4A) (Furrie et al. 2005). By contrast, TLR3 expression increases when the cells escape from the crypt, presumably to allow these cells to detect invasive pathogens (Fig. 4A) (Furrie et al. 2005). TLR4 is found in the Golgi apparatus and can induce an immune response only after the internalization of its ligand, the LPS (Hornef et al. 2002). Furthermore, some of accessory molecules required for TLR function, such as MD-2, are poorly expressed in IECs (Vamadevan et al. 2010). TLR9 is found on the surface of IECs and its response depends on its location on the membrane (apical or basolateral). Indeed, basolateral TLR9 activation elicits a classic TLR inflammatory response triggering NF-κB activation, whereas apical TLR9 activation induces a tolerogenic response

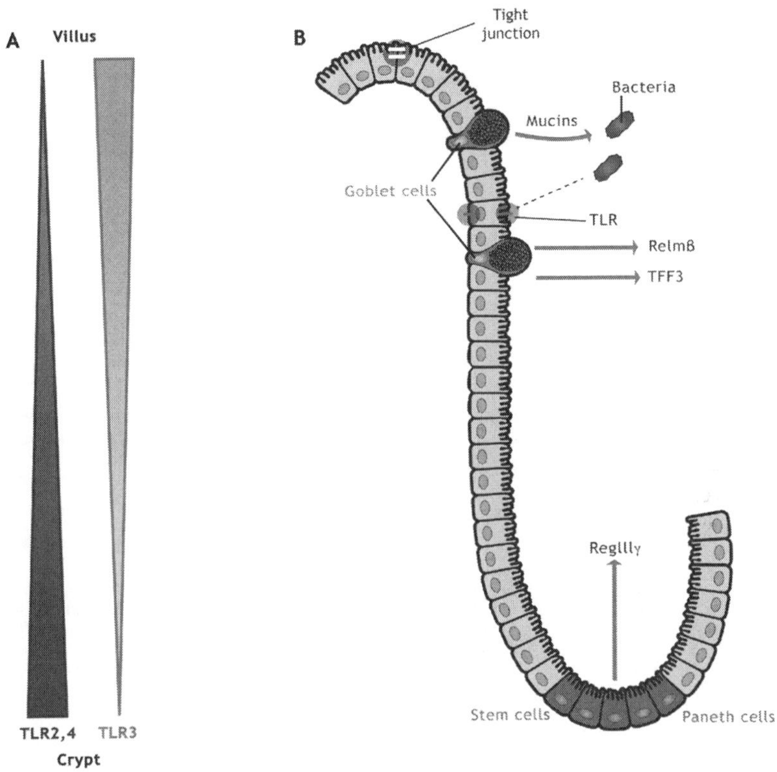

Figure 4. Epithelial Pattern Recognition Receptors for Perception of the Intestinal Microbiota. The sensing of microbe-associated molecular patterns is based on recognition by pattern recognition receptors (PRRs). Toll-like receptors (TLRs) and nucleotide oligomerization domain (NOD)-like receptors (NLRs) are the most important families of PRRs. TLR expression and distribution are controlled and signaling via these receptors is modulated so as to prevent deleterious inflammatory responses in the intestine. NOD2 is expressed predominantly in Paneth cells. PPR signaling plays a key role in epithelial homeostasis, as it stimulates the production of mucins, trefoil peptides (TFF3), resistin-like molecule β (RELMβ) and antimicrobial peptides, such as REG3γ, and strengthens tight junctions.

Color image of this figure appears in the color plate section at the end of the book.

(Lee et al. 2006). Alternatively, some TLRs may be located exclusively on the basolateral side of the cell for the detection of invading microbes, as shown for TLR5 (Gewirtz et al. 2001). TLR signaling in epithelial cells has also been shown to play a key role in the successful management of the initial wave of bacteria after birth. The recognition of ligands by a TLR may change between birth and adulthood. Indeed, in adulthood, TLR-3 recognizes viral dsRNA molecules, whereas, during the postnatal period, it recognizes Gram-positive bacteria, enabling the body to respond rapidly to primocolonization (Furrie et al. 2005). In neonatal mice, early colonization is accompanied by a transient extinction of the epithelial TLR signaling pathway that contributes to innate immune tolerance (Chassin et al. 2010). Furthermore, amniotic fluid inhibits TLR4 signaling in the developing intestinal epithelium, further attenuating the proinflammatory response in response to early microbial colonization (Good et al. 2012). Preterm infants with necrotizing enterocolitis have higher than normal levels of intestinal TLR4 expression (Gribar et al. 2009).

PRR signaling in intestinal homeostasis

The absence of MyD88, which acts as an adaptor for almost all the TLRs, in IECs results in the impaired expression of *muc2*, genes encoding AMPs and the gene encoding the polymeric immunoglobulin receptor required for the epithelial transport of IgA. This results in a defective epithelial barrier, higher levels of bacterial translocation to the mesenteric lymph nodes and greater susceptibility to colitis (Frantz et al. 2012). These data, together with those of previous studies, highlight the requirement of intestinal epithelial TLR signaling for gut homeostasis (Fig. 4B). TLR2 deficiency results in a loss of goblet cell-derived TFF3, contributing to the development of experimentally induced inflammation (Podolsky et al. 2009). TLR signaling enhances epithelial barrier function by modulating the tight junctions (Cario et al. 2007) and mediates the effect of LGG on restoration of the intestinal barrier after radiation injury (Ciorba et al. 2012). MyD88 signaling orchestrates the production of several antimicrobial factors, including RegIIIγ, RegIIIβ and CRP-ductin, in Paneth cells and of RELM β (Vaishnava et al. 2008) (Fig. 4B). This programme is activated by the microbiota, as the Paneth cells of microbe-free mice do not produce antimicrobial compounds, such as RegIIIγ (Cash et al. 2006). MyD88 signaling is required for intestinal cell proliferation in response to microbial signals in zebrafish (Cheesman et al. 2011) and for intestinal epithelial cell restoration after injury in mice (Rakoff-Nahoum et al. 2004). Similarly, NOD2 deficiency results in lower levels of AMP production by Paneth cells (Kobayashi et al. 2005), and intestinal crypts isolated from the terminal ilea of Nod2-deficient mice have a weaker antimicrobial effect than those from wild-type mice. The expression of *nod2* is induced by the gut microbiota (Petnicki-Ocwieja et al. 2009). An impairment of NLR and TLR function in epithelial cells is associated with chronic inflammatory diseases. The level of expression of TLRs, such as TLR4, in the colonic mucosa may be modified by disease (Cario and Podolsky 2000).

Intestinal Brush-border Alkaline Phosphatase, to Decrease the Proinflammatory Potential of Resident Gut Microbes

Intestinal alkaline phosphatase (IAP) is a brush border enzyme of IECs that regulates fat absorption. In addition to this role, IAP contributes to the detoxification of LPS, thereby minimizing the proinflammatory potential of the products of bacteria resident in the gut. Studies in zebrafish have shown that IAP expression is induced during the establishment of the microbiota (Bates et al. 2007). IAP deficiency is associated with hypersensitivity to LPS toxicity and inflammatory symptoms, neither of which are observed if zebrafish are reared in the absence of microbes (Bates et al. 2007). An active AP is also required to maintain the symbiotic association between the squid *Euprymna scolopes* and the luminous marine bacterium *Vibrio fischeri* (Rader et al. 2012).

In mice, IAP deficiency results in more severe colitis in a model of chemically induced disease, and this effect is counteracted by IAP supplementation (Ramasamy et al. 2011). Patients with IBD have lower levels of intestinal IAP than unaffected subjects (Tuin et al. 2009). IAP helps to restore the gut microbial community after antibiotic treatment, suggesting that it may increase the fitness of commensal bacteria. IAP loss in mice alters the composition of the gut microbial ecosystem (Malo et al. 2010).

The Gut Microbiota Shapes the Intestinal Epithelial Gene Expression Profile

The selection of mutualistic and symbiotic interactions during coevolution has resulted in the gut microbiota modifying the intestinal epithelium, a contribution that is very clear from studies of gnotobiotic animals.

A comparison of global host transcriptional responses between gnotobiotic and conventional animals has shown that the microbiota affects the expression of hundreds of genes in the host epithelium. The genes targeted by the gut microbiota encode proteins involved in diverse aspects of intestinal epithelial biology: nutrient processing and absorption, enzymes involved in cellular energy production, epithelial barrier function, responses to hormones, composition of the extracellular matrix or signal transduction pathways, mucus biosynthesis and secretion, cell proliferation and differentiation, and cell migration (Chowdhury et al. 2007; Comelli et al. 2008; Hooper et al. 2001). Conversely, the intake of a broad-spectrum antibiotic decreases the expression of gut microbiota induced-genes (Reikvam et al. 2011). Global microRNA (miRNA) expression profiling data have suggested that miRNA, non-coding RNA molecules involved in the regulation of gene expression, may be involved in the microbiota-mediated regulation of host gene expression (Dalmasso et al. 2011, Singh et al. 2012) as suggested for miRNA involved in the regulation of genes encoding proteins with functions in glycosylation, cell-cell junctions and mucus layer formation (Singh et al. 2012).

As discussed above, some commensal bacteria have evolved a capacity to make use of the carbohydrate chains and glycosylated molecules produced by host IECs, and they effectively use epithelial terminal fucose as a nutrient (Koropatkin et al. 2012). *B. thetaiotaomicron*, one of the predominant bacterial species, induces fucosyltransferase expression in epithelial cells, leading to the production of fucose (Bry et al. 1996). In turn, fucose availability regulates the expression of bacterial genes (Hooper et al. 1999).

The Gut Microbiota Modifies Intestinal Epithelial Metabolism

Short-chain fatty acids (SCFA) are produced principally in the lumen of the colon, by the fermentation of undigested polysaccharides. One such molecule, butyrate, plays a major role in the physiology of the colonic epithelium, as it is metabolized by the epithelial cells, for which it constitutes a major energy source (Clausen and Mortensen 1994; Roediger 1982). In gnotobiotic conditions, the absence of butyrate production results in metabolic starvation and colonic mucosal atrophy (Cherbuy et al. 1995) and the stimulation of autophagy (Donohoe et al. 2011). The intestinal microbiota contributes to the metabolic specialization of the colonic epithelium in the use of butyrate, and *in vivo* data have shown that butyrate can modulate the expression of an enzyme involved in its own metabolism (Cherbuy et al. 2004). Furthermore, *in vivo* studies have repeatedly shown that butyrate has many biological effects, most of which are beneficial to the host. For example, butyrate has a trophic effect and has been reported to improve tissue development and repair (Guilloteau et al. 2010).

From the Health-promoting Function of Commensal Microbes to the Health Promising Function of Probiotics

There are 10 times as many bacteria as human cells in the digestive tract, maintained in conditions that should ensure homeostasis. We are only just beginning to understand how this dense bacterial load is managed without triggering overtly damaging responses. Our awareness of the richness and essential nature of these complex and dynamic consortia of microorganisms is gradually increasing, but social and dietary changes and medical advances are challenging the balance between the host and the microbiota. Modern hygiene, lifestyle and medical practices in Western countries are damaging to our microbiota. It is increasingly being argued that the loss of ancestral microbial lineages that have coevolved with their hosts is linked to the rapid emergence of metabolic and autoimmune diseases in humans (Blaser and Falkow 2009). This has led to growing interest in "bugs as drugs" approaches to manipulating the gut microbiota to restore a healthy community (Reid et al. 2011).

Probiotics are being used to harness positive interactions between the host and the microbes, to favor return to homeostasis (Bron et al. 2011). Studies providing insight in this area will be challenging and basic research on mutualistic microbe–host interactions will improve our understanding of the benefits of probiotics. As a dominant force behind ecosystem stability, the crosstalk between the gut microbiota and the epithelium constitutes a major resource for the design and development of new probiotics or therapeutic drugs. The exploitation of these interactions is in its infancy and is likely to lead to concrete benefits for human health. The successful use of a factor released from the probiotic strain LGG in inflammatory models illustrates the promising nature of this approach (Yan et al. 2011).

References Cited

Abreu, M.T. 2010. Toll-like receptor signalling in the intestinal epithelium: how bacterial recognition shapes intestinal function. Nat. Rev. Immunol. 10: 131–144.

Alam, M., T. Midtvedt and A. Uribe. 1994. Differential cell kinetics in the ileum and colon of germfree rats. Scand. J. Gastroenterol. 29: 445–451.

An, G., B. Wei, B. Xia, J.M. McDaniel, T. Ju, R.D. Cummings, J. Braun and L. Xia. 2007. Increased susceptibility to colitis and colorectal tumors in mice lacking core 3-derived O-glycans. J. Exp. Med. 204: 1417–1429.

Anderson, R.C., A.L. Cookson, W.C. McNabb, Z. Park, M.J. McCann, W.J. Kelly and N.C. Roy. 2010. Lactobacillus plantarum MB452 enhances the function of the intestinal barrier by increasing the expression levels of genes involved in tight junction formation. BMC Microbiol. 10: 316.

Atuma, C., V. Strugala, A. Allen and L. Holm. 2001. The adherent gastrointestinal mucus gel layer: thickness and physical state *in vivo*. Am. J. Physiol. Gastrointest. Liver Physiol. 280: G922–929.

Barker, N., J.H. van Es, J. Kuipers, P. Kujala, M. van den Born, M. Cozijnsen, A. Haegebarth, J. Korving, H. Begthel, P.J. Peters and H. Clevers. 2007. Identification of stem cells in small intestine and colon by marker gene Lgr5. Nature. 449: 1003–1007.

Bates, J.M., J. Akerlund, E. Mittge and K. Guillemin. 2007. Intestinal alkaline phosphatase detoxifies lipopolysaccharide and prevents inflammation in zebrafish in response to the gut microbiota. Cell Host Microbe. 2: 371–382.

Bergstrom, K.S., V. Kissoon-Singh, D.L. Gibson, C. Ma, M. Montero, H.P. Sham, N. Ryz, T. Huang, A. Velcich, B.B. Finlay, K. Chadee and B.A. Vallance. 2010. Muc2 protects against lethal infectious colitis by disassociating pathogenic and commensal bacteria from the colonic mucosa. PLoS Pathog. 6: e1000902.

Bevins, C.L. and N.H. Salzman. 2011. Paneth cells, antimicrobial peptides and maintenance of intestinal homeostasis. Nat. Rev. Microbiol. 9: 356–368.

Blaser, M.J. and S. Falkow. 2009. What are the consequences of the disappearing human microbiota? Nat. Rev. Microbiol. 7: 887–894.

Brandl, K., G. Plitas, B. Schnabl, R.P. DeMatteo and E.G. Pamer. 2007. MyD88-mediated signals induce the bactericidal lectin RegIII gamma and protect mice against intestinal Listeria monocytogenes infection. J. Exp. Med. 204: 1891–1900.

Bron, P.A., P. van Baarlen and M. Kleerebezem. 2011. Emerging molecular insights into the interaction between probiotics and the host intestinal mucosa. Nat. Rev. Microbiol. 10: 66–78.

Bry, L., P.G. Falk, T. Midtvedt and J.I. Gordon. 1996. A model of host-microbial interactions in an open mammalian ecosystem. Science. 273: 1380–1383.

Buchon, N., N.A. Broderick, S. Chakrabarti and B. Lemaitre. 2009. Invasive and indigenous microbiota impact intestinal stem cell activity through multiple pathways in Drosophila. Genes Dev. 23: 2333–2344.

Cadwell, K., J.Y. Liu, S.L. Brown, H. Miyoshi, J. Loh, J.K. Lennerz, C. Kishi, W. Kc, J.A. Carrero, S. Hunt, C.D. Stone, E.M. Brunt, R.J. Xavier, B.P. Sleckman, E. Li, N. Mizushima, T.S. Stappenbeck and H.W.t. Virgin. 2008. A key role for autophagy and the autophagy gene Atg16l1 in mouse and human intestinal Paneth cells. Nature. 456: 259–263.

Cario, E., G. Gerken and D.K. Podolsky. 2007. Toll-like receptor 2 controls mucosal inflammation by regulating epithelial barrier function. Gastroenterology. 132: 1359–1374.

Cario, E. and D.K. Podolsky. 2000. Differential alteration in intestinal epithelial cell expression of toll-like receptor 3 (TLR3) and TLR4 in inflammatory bowel disease. Infect. Immun. 68: 7010–7017.

Carvalho, F.A., J.D. Aitken, M. Vijay-Kumar and A.T. Gewirtz. 2012. Toll-like receptor-gut microbiota interactions: perturb at your own risk! Annu. Rev. Physiol. 74: 177–198.

Cash, H.L., C.V. Whitham, C.L. Behrendt and L.V. Hooper. 2006. Symbiotic bacteria direct expression of an intestinal bactericidal lectin. Science. 313: 1126–1130.

Chassin, C., M. Kocur, J. Pott, C.U. Duerr, D. Gutle, M. Lotz and M.W. Hornef. 2010. miR-146a mediates protective innate immune tolerance in the neonate intestine. Cell Host Microbe. 8: 358–368.

Cheesman, S.E., J.T. Neal, E. Mittge, B.M. Seredick and K. Guillemin. 2011. Epithelial cell proliferation in the developing zebrafish intestine is regulated by the Wnt pathway and microbial signaling via Myd88. Proc. Natl. Acad. Sci. USA 108 Suppl. 1: 4570–4577.

Cherbuy, C., C. Andrieux, E. Honvo-Houeto, M. Thomas, C. Ide, N. Druesne, C. Chaumontet, B. Darcy-Vrillon and P.H. Duee. 2004. Expression of mitochondrial HMGCoA synthase and glutaminase in the colonic mucosa is modulated by bacterial species. Eur. J. Biochem. 271: 87–95.

Cherbuy, C., B. Darcy-Vrillon, M.T. Morel, J.P. Pegorier and P.H. Duee. 1995. Effect of germfree state on the capacities of isolated rat colonocytes to metabolize n-butyrate, glucose, and glutamine. Gastroenterology. 109: 1890–1899.

Cherbuy, C., E. Honvo-Houeto, A. Bruneau, C. Bridonneau, C. Mayeur, P.H. Duee, P. Langella and M. Thomas. 2010. Microbiota matures colonic epithelium through a coordinated induction of cell cycle-related proteins in gnotobiotic rat. Am. J. Physiol. Gastrointest. Liver Physiol. 299: G348–357.

Cho, J.H. and S.R. Brant. 2011. Recent insights into the genetics of inflammatory bowel disease. Gastroenterology. 140: 1704–1712.

Chowdhury, S.R., D.E. King, B.P. Willing, M.R. Band, J.E. Beever, A.B. Lane, J.J. Loor, J.C. Marini, L.A. Rund, L.B. Schook, A.G. Van Kessel and H.R. Gaskins. 2007. Transcriptome profiling of the small intestinal epithelium in germfree versus conventional piglets. BMC Genomics. 8: 215.

Ciorba, M.A., T.E. Riehl, M.S. Rao, C. Moon, X. Ee, G.M. Nava, M.R. Walker, J.M. Marinshaw, T.S. Stappenbeck and W.F. Stenson. 2012. Lactobacillus probiotic protects intestinal epithelium from radiation injury in a TLR-2/cyclo-oxygenase-2-dependent manner. Gut. 61: 829–838.

Clausen, M.R. and P.B. Mortensen. 1994. Kinetic studies on the metabolism of short-chain fatty acids and glucose by isolated rat colonocytes. Gastroenterology. 106: 423–432.

Comelli, E.M., R. Simmering, M. Faure, D. Donnicola, R. Mansourian, F. Rochat, I. Corthesy-Theulaz and C. Cherbut. 2008. Multifaceted transcriptional regulation of the murine intestinal mucus layer by endogenous microbiota. Genomics. 91: 70–77.

Dalmasso, G., H.T. Nguyen, Y. Yan, H. Laroui, M.A. Charania, S. Ayyadurai, S.V. Sitaraman and D. Merlin. 2011. Microbiota modulate host gene expression via microRNAs. PLoS One. 6: e19293.

Danielsen, M., H. Hornshoj, R.H. Siggers, B.B. Jensen, A.G. van Kessel and E. Bendixen. 2007. Effects of bacterial colonization on the porcine intestinal proteome. J. Proteome. Res. 6: 2596–2604.

de Lau, W., P. Kujala, K. Schneeberger, S. Middendorp, V.S. Li, N. Barker, A. Martens, F. Hofhuis, R.P. Dekoter, P.J. Peters, E. Nieuwenhuis and H. Clevers. 2012. Peyer's Patch M Cells Derived from Lgr5+ Stem Cells Require SpiB and Are Induced by RankL in Cultured "Miniguts". Mol. Cell Biol. 32: 3639–3647.

Donohoe, D.R., N. Garge, X. Zhang, W. Sun, T.M. O'Connell, M.K. Bunger and S.J. Bultman. 2011. The microbiome and butyrate regulate energy metabolism and autophagy in the mammalian colon. Cell Metab. 13: 517–526.

Ewaschuk, J.B., H. Diaz, L. Meddings, B. Diederichs, A. Dmytrash, J. Backer, M. Looijer-van Langen and K.L. Madsen. 2008. Secreted bioactive factors from Bifidobacterium infantis enhance epithelial cell barrier function. Am. J. Physiol. Gastrointest. Liver Physiol. 295: G1025–1034.

Frantz, A.L., E.W. Rogier, C.R. Weber, L. Shen, D.A. Cohen, L.A. Fenton, M.E. Bruno and C.S. Kaetzel. 2012. Targeted deletion of MyD88 in intestinal epithelial cells results in compromised antibacterial immunity associated with downregulation of polymeric immunoglobulin receptor, mucin-2, and antibacterial peptides. Mucosal Immunol. 5: 501–512.

Fu, J., B. Wei, T. Wen, M.E. Johansson, X. Liu, E. Bradford, K.A. Thomsson, S. McGee, L. Mansour, M. Tong, J.M. McDaniel, T.J. Sferra, J.R. Turner, H. Chen, G.C. Hansson, J. Braun and L. Xia. 2011. Loss of intestinal core 1-derived O-glycans causes spontaneous colitis in mice. J. Clin. Invest. 121: 1657–1666.

Furrie, E., S. Macfarlane, G. Thomson and G.T. Macfarlane. 2005. Toll-like receptors-2, -3 and -4 expression patterns on human colon and their regulation by mucosal-associated bacteria. Immunology. 115: 565–574.

Gaboriau-Routhiau, V., S. Rakotobe, E. Lecuyer, I. Mulder, A. Lan, C. Bridonneau, V. Rochet, A. Pisi, M. De Paepe, G. Brandi, G. Eberl, J. Snel, D. Kelly and N. Cerf-Bensussan. 2009. The key role of segmented filamentous bacteria in the coordinated maturation of gut helper T cell responses. Immunity. 31: 677–689.

Gallo, R.L. and L.V. Hooper. 2012. Epithelial antimicrobial defence of the skin and intestine. Nat. Rev. Immunol. 12: 503–516.

Garrett, W.S., J.I. Gordon and L.H. Glimcher. 2010. Homeostasis and inflammation in the intestine. Cell. 140: 859–870.

Gerbe, F., C. Legraverend and P. Jay. 2012. The intestinal epithelium tuft cells: specification and function. Cell Mol. Life Sci. 69: 2907–2917.

Gerbe, F., J.H. van Es, L. Makrini, B. Brulin, G. Mellitzer, S. Robine, B. Romagnolo, N.F. Shroyer, J.F. Bourgaux, C. Pignodel, H. Clevers and P. Jay. 2011. Distinct ATOH1 and Neurog3 requirements define tuft cells as a new secretory cell type in the intestinal epithelium. J. Cell Biol. 192: 767–780.

Gewirtz, A.T., T.A. Navas, S. Lyons, P.J. Godowski and J.L. Madara. 2001. Cutting edge: bacterial flagellin activates basolaterally expressed TLR5 to induce epithelial proinflammatory gene expression. J. Immunol. 167: 1882–1885.

Ghadimi, D., M. Vrese, K.J. Heller and J. Schrezenmeir. 2010. Effect of natural commensal-origin DNA on toll-like receptor 9 (TLR9) signaling cascade, chemokine IL-8 expression, and barrier integritiy of polarized intestinal epithelial cells. Inflamm. Bowel Dis. 16: 410–427.

Gilbert, S., R. Zhang, L. Denson, R. Moriggl, K. Steinbrecher, N. Shroyer, J. Lin and X. Han. 2012. Enterocyte STAT5 promotes mucosal wound healing via suppression of myosin light chain kinase-mediated loss of barrier function and inflammation. EMBO Mol. Med. 4: 109–124.

Good, M., R.H. Siggers, C.P. Sodhi, A. Afrazi, F. Alkhudari, C.E. Egan, M.D. Neal, I. Yazji, H. Jia, J. Lin, M.F. Branca, C. Ma, T. Prindle, Z. Grant, S. Shah, D. Slagle, 2nd, J. Paredes, J. Ozolek, G.K. Gittes and D.J. Hackam. 2012. Amniotic fluid inhibits Toll-like receptor 4 signaling in the fetal and neonatal intestinal epithelium. Proc. Natl. Acad. Sci. USA 109: 11330–11335.

Goto, Y. and H. Kiyono. 2012. Epithelial barrier: an interface for the cross-communication between gut flora and immune system. Immunol. Rev. 245: 147–163.

Gribar, S.C., C.P. Sodhi, W.M. Richardson, R.J. Anand, G.K. Gittes, M.F. Branca, A. Jakub, X.H. Shi, S. Shah, J.A. Ozolek and D.J. Hackam. 2009. Reciprocal expression and signaling of TLR4 and TLR9 in the pathogenesis and treatment of necrotizing enterocolitis. J. Immunol. 182: 636–646.

Guilloteau, P., L. Martin, V. Eeckhaut, R. Ducatelle, R. Zabielski and F. Van Immerseel. 2010. From the gut to the peripheral tissues: the multiple effects of butyrate. Nutr. Res. Rev. 23: 366–384.

Gum, J.R., J.C. Byrd, J.W. Hicks, N.W. Toribara, D.T. Lamport and Y.S. Kim. 1989. Molecular cloning of human intestinal mucin cDNAs. Sequence analysis and evidence for genetic polymorphism. J. Biol. Chem. 264: 6480–6487.

Hase, K., L. Eckmann, J.D. Leopard, N. Varki and M.F. Kagnoff. 2002. Cell differentiation is a key determinant of cathelicidin LL-37/human cationic antimicrobial protein 18 expression by human colon epithelium. Infect. Immun. 70: 953–963.

Heazlewood, C.K., M.C. Cook, R. Eri, G.R. Price, S.B. Tauro, D. Taupin, D.J. Thornton, C.W. Png, T.L. Crockford, R.J. Cornall, R. Adams, M. Kato, K.A. Nelms, N.A. Hong, T.H. Florin, C.C. Goodnow and M.A. McGuckin. 2008. Aberrant mucin assembly in mice causes endoplasmic reticulum stress and spontaneous inflammation resembling ulcerative colitis. PLoS Med. 5: e54.

Hooper, L.V. and A.J. Macpherson. 2010. Immune adaptations that maintain homeostasis with the intestinal microbiota. Nat. Rev. Immunol. 10: 159–169.

Hooper, L.V., M.H. Wong, A. Thelin, L. Hansson, P.G. Falk and J.I. Gordon. 2001. Molecular analysis of commensal host-microbial relationships in the intestine. Science. 291: 881–884.

Hooper, L.V., J. Xu, P.G. Falk, T. Midtvedt and J.I. Gordon. 1999. A molecular sensor that allows a gut commensal to control its nutrient foundation in a competitive ecosystem. Proc. Natl. Acad. Sci. USA. 96: 9833–9838.

Hornef, M.W., T. Frisan, A. Vandewalle, S. Normark and A. Richter-Dahlfors. 2002. Toll-like receptor 4 resides in the Golgi apparatus and colocalizes with internalized lipopolysaccharide in intestinal epithelial cells. J. Exp. Med. 195: 559–570.

Hoskins, L.C. and E.T. Boulding. 1981. Mucin degradation in human colon ecosystems. Evidence for the existence and role of bacterial subpopulations producing glycosidases as extracellular enzymes. J. Clin. Invest. 67: 163–172.

Ivanov, II, K. Atarashi, N. Manel, E.L. Brodie, T. Shima, U. Karaoz, D. Wei, K.C. Goldfarb, C.A. Santee, S.V. Lynch, T. Tanoue, A. Imaoka, K. Itoh, K. Takeda, Y. Umesaki, K. Honda and D.R. Littman. 2009. Induction of intestinal Th17 cells by segmented filamentous bacteria. Cell. 139: 485–498.

Johansson, M.E., J.M. Larsson and G.C. Hansson. 2011. The two mucus layers of colon are organized by the MUC2 mucin, whereas the outer layer is a legislator of host-microbial interactions. Proc. Natl. Acad. Sci. USA. 108 Suppl. 1: 4659–4665.

Johansson, M.E., M. Phillipson, J. Petersson, A. Velcich, L. Holm and G.C. Hansson. 2008. The inner of the two Muc2 mucin-dependent mucus layers in colon is devoid of bacteria. Proc. Natl. Acad. Sci. USA. 105: 15064–15069.

Karam, S.M. 1999. Lineage commitment and maturation of epithelial cells in the gut. Front Biosci. 4: D286–298.

Karczewski, J., F.J. Troost, I. Konings, J. Dekker, M. Kleerebezem, R.J. Brummer and J.M. Wells. 2010. Regulation of human epithelial tight junction proteins by Lactobacillus plantarum *in vivo* and protective effects on the epithelial barrier. Am. J. Physiol. Gastrointest. Liver Physiol. 298: G851–859.

Kaser, A., A.H. Lee, A. Franke, J.N. Glickman, S. Zeissig, H. Tilg, E.E. Nieuwenhuis, D.E. Higgins, S. Schreiber, L.H. Glimcher and R.S. Blumberg. 2008. XBP1 links ER stress to intestinal inflammation and confers genetic risk for human inflammatory bowel disease. Cell. 134: 743–756.

Kim, Y.S. and S.B. Ho. 2010. Intestinal goblet cells and mucins in health and disease: recent insights and progress. Curr. Gastroenterol. Rep. 12: 319–330.

Kobayashi, K.S., M. Chamaillard, Y. Ogura, O. Henegariu, N. Inohara, G. Nunez and R.A. Flavell. 2005. Nod2-dependent regulation of innate and adaptive immunity in the intestinal tract. Science. 307: 731–734.

Korinek, V., N. Barker, P. Moerer, E. van Donselaar, G. Huls, P.J. Peters and H. Clevers. 1998. Depletion of epithelial stem-cell compartments in the small intestine of mice lacking Tcf-4. Nat. Genet. 19: 379–383.

Koropatkin, N.M., E.A. Cameron and E.C. Martens. 2012. How glycan metabolism shapes the human gut microbiota. Nat. Rev. Microbiol. 10: 323–335.

Koslowski, M.J., Z. Teltschik, J. Beisner, E. Schaeffeler, G. Wang, I. Kubler, M. Gersemann, R. Cooney, D. Jewell, W. Reinisch, S. Vermeire, P. Rutgeerts, M. Schwab, E.F. Stange and J. Wehkamp. 2012. Association of a functional variant in the Wnt co-receptor LRP6 with early onset ileal Crohn's disease. PLoS Genet. 8: e1002523.

Kuwahara, T., Y. Ogura, K. Oshima, K. Kurokawa, T. Ooka, H. Hirakawa, T. Itoh, H. Nakayama-Imaohji, M. Ichimura, K. Itoh, C. Ishifune, Y. Maekawa, K. Yasutomo, M. Hattori and T. Hayashi. 2011. The lifestyle of the segmented filamentous bacterium: a non-culturable gut-associated immunostimulating microbe inferred by whole-genome sequencing. DNA Res. 18: 291–303.

Laukoetter, M.G., P. Nava, W.Y. Lee, E.A. Severson, C.T. Capaldo, B.A. Babbin, I.R. Williams, M. Koval, E. Peatman, J.A. Campbell, T.S. Dermody, A. Nusrat and C.A. Parkos. 2007. JAM-A regulates permeability and inflammation in the intestine *in vivo*. J. Exp. Med. 204: 3067–3076.

Lee, J., J.H. Mo, K. Katakura, I. Alkalay, A.N. Rucker, Y.T. Liu, H.K. Lee, C. Shen, G. Cojocaru, S. Shenouda, M. Kagnoff, L. Eckmann, Y. Ben-Neriah and E. Raz. 2006. Maintenance of colonic homeostasis by distinctive apical TLR9 signalling in intestinal epithelial cells. Nat. Cell Biol. 8: 1327–1336.

Mahowald, M.A., F.E. Rey, H. Seedorf, P.J. Turnbaugh, R.S. Fulton, A. Wollam, N. Shah, C. Wang, V. Magrini, R.K. Wilson, B.L. Cantarel, P.M. Coutinho, B. Henrissat, L.W. Crock, A. Russell, N.C. Verberkmoes, R.L. Hettich and J.I. Gordon. 2009. Characterizing a model human gut microbiota composed of members of its two dominant bacterial phyla. Proc. Natl. Acad. Sci. USA. 106: 5859–5864.

Malo, M.S., S.N. Alam, G. Mostafa, S.J. Zeller, P.V. Johnson, N. Mohammad, K.T. Chen, A.K. Moss, S. Ramasamy, A. Faruqui, S. Hodin, P.S. Malo, F. Ebrahimi, B. Biswas, S. Narisawa, J.L. Millan, H.S. Warren, J.B. Kaplan, C.L. Kitts, E.L. Hohmann and R.A. Hodin. 2010. Intestinal alkaline phosphatase preserves the normal homeostasis of gut microbiota. Gut. 59: 1476–1484.

Marcobal, A., M. Barboza, E.D. Sonnenburg, N. Pudlo, E.C. Martens, P. Desai, C.B. Lebrilla, B.C. Weimer, D.A. Mills, J.B. German and J.L. Sonnenburg. 2011. Bacteroides in the infant gut consume milk oligosaccharides via mucus-utilization pathways. Cell Host Microbe. 10: 507–514.

Mathias, A., M. Duc, L. Favre, J. Benyacoub, S. Blum and B. Corthesy. 2010. Potentiation of polarized intestinal Caco-2 cell responsiveness to probiotics complexed with secretory IgA. J. Biol. Chem. 285: 33906–33913.

McDole, J.R., L.W. Wheeler, K.G. McDonald, B. Wang, V. Konjufca, K.A. Knoop, R.D. Newberry and M.J. Miller. 2012. Goblet cells deliver luminal antigen to CD103+ dendritic cells in the small intestine. Nature. 483: 345–349.

Meyer-Hoffert, U., M.W. Hornef, B. Henriques-Normark, L.G. Axelsson, T. Midtvedt, K. Putsep and M. Andersson. 2008. Secreted enteric antimicrobial activity localises to the mucus surface layer. Gut. 57: 764–771.

Nava, G.M., H.J. Friedrichsen and T.S. Stappenbeck. 2011. Spatial organization of intestinal microbiota in the mouse ascending colon. Isme J. 5: 627–638.

O'Neil, D.A., E.M. Porter, D. Elewaut, G.M. Anderson, L. Eckmann, T. Ganz and M.F. Kagnoff. 1999. Expression and regulation of the human beta-defensins hBD-1 and hBD-2 in intestinal epithelium. J. Immunol. 163: 6718–6724.

Ogura, Y., S. Lala, W. Xin, E. Smith, T.A. Dowds, F.F. Chen, E. Zimmermann, M. Tretiakova, J.H. Cho, J. Hart, J.K. Greenson, S. Keshav and G. Nunez. 2003. Expression of NOD2 in Paneth cells: a possible link to Crohn's ileitis. Gut. 52: 1591–1597.

Olson, T.S., B.K. Reuter, K.G. Scott, M.A. Morris, X.M. Wang, L.N. Hancock, T.L. Burcin, S.M. Cohn, P.B. Ernst, F. Cominelli, J.B. Meddings, K. Ley and T.T. Pizarro. 2006. The primary defect in experimental ileitis originates from a nonhematopoietic source. J. Exp. Med. 203: 541–552.

Ouwehand, A.C., M. Derrien, W. de Vos, K. Tiihonen and N. Rautonen. 2005. Prebiotics and other microbial substrates for gut functionality. Curr. Opin. Biotechnol. 16: 212–217.

Pedron, T., C. Mulet, C. Dauga, L. Frangeul, C. Chervaux, G. Grompone and P.J. Sansonetti. 2012. A crypt-specific core microbiota resides in the mouse colon. MBio. 3.

Peng, L., Z.R. Li, R.S. Green, I.R. Holzman and J. Lin. 2009. Butyrate enhances the intestinal barrier by facilitating tight junction assembly via activation of AMP-activated protein kinase in Caco-2 cell monolayers. J. Nutr. 139: 1619–1625.

Petnicki-Ocwieja, T., T. Hrncir, Y.J. Liu, A. Biswas, T. Hudcovic, H. Tlaskalova-Hogenova and K.S. Kobayashi. 2009. Nod2 is required for the regulation of commensal microbiota in the intestine. Proc. Natl. Acad. Sci. USA. 106: 15813–15818.

Podolsky, D.K., G. Gerken, A. Eyking and E. Cario. 2009. Colitis-associated variant of TLR2 causes impaired mucosal repair because of TFF3 deficiency. Gastroenterology. 137: 209–220.

Prakash, T., K. Oshima, H. Morita, S. Fukuda, A. Imaoka, N. Kumar, V.K. Sharma, S.W. Kim, M. Takahashi, N. Saitou, T.D. Taylor, H. Ohno, Y. Umesaki and M. Hattori. 2011. Complete genome sequences of rat and mouse segmented filamentous bacteria, a potent inducer of th17 cell differentiation. Cell Host Microbe. 10: 273–284.

Preidis, G.A., D.M. Saulnier, S.E. Blutt, T.A. Mistretta, K.P. Riehle, A.M. Major, S.F. Venable, M.J. Finegold, J.F. Petrosino, M.E. Conner and J. Versalovic. 2012. Probiotics stimulate enterocyte migration and microbial diversity in the neonatal mouse intestine. Faseb J. 26: 1960–1969.

Pull, S.L., J.M. Doherty, J.C. Mills, J.I. Gordon and T.S. Stappenbeck. 2005. Activated macrophages are an adaptive element of the colonic epithelial progenitor niche necessary for regenerative responses to injury. Proc. Natl. Acad. Sci. USA 102: 99–104.

Rader, B.A., N. Kremer, M.A. Apicella, W.E. Goldman and M.J. McFall-Ngai. 2012. Modulation of symbiont lipid A signaling by host alkaline phosphatases in the squid-vibrio symbiosis. MBio. 3.

Rakoff-Nahoum, S., J. Paglino, F. Eslami-Varzaneh, S. Edberg and R. Medzhitov. 2004. Recognition of commensal microflora by toll-like receptors is required for intestinal homeostasis. Cell. 118: 229–241.

Ramasamy, S., D.D. Nguyen, M.A. Eston, S.N. Alam, A.K. Moss, F. Ebrahimi, B. Biswas, G. Mostafa, K.T. Chen, K. Kaliannan, H. Yammine, S. Narisawa, J.L. Millan, H.S. Warren, E.L. Hohmann, E. Mizoguchi, H.C. Reinecker, A.K. Bhan, S.B. Snapper, M.S. Malo and R.A. Hodin. 2011. Intestinal alkaline phosphatase has beneficial effects in mouse models of chronic colitis. Inflamm. Bowel Dis. 17: 532–542.

Reid, G., J.A. Younes, H.C. Van der Mei, G.B. Gloor, R. Knight and H.J. Busscher. 2011. Microbiota restoration: natural and supplemented recovery of human microbial communities. Nat. Rev. Microbiol. 9: 27–38.

Reikvam, D.H., A. Erofeev, A. Sandvik, V. Grcic, F.L. Jahnsen, P. Gaustad, K.D. McCoy, A.J. Macpherson, L.A. Meza-Zepeda and F.E. Johansen. 2011. Depletion of murine intestinal microbiota: effects on gut mucosa and epithelial gene expression. PLoS One. 6: e17996.

Resta-Lenert, S. and K.E. Barrett. 2006. Probiotics and commensals reverse TNF-alpha- and IFN-gamma-induced dysfunction in human intestinal epithelial cells. Gastroenterology. 130: 731–746.

Roediger, W.E. 1982. Utilization of nutrients by isolated epithelial cells of the rat colon. Gastroenterology. 83: 424–429.

Rul, F., L. Ben-Yahia, F. Chegdani, L. Wrzosek, S. Thomas, M.L. Noordine, C. Gitton, C. Cherbuy, P. Langella and M. Thomas. 2011. Impact of the metabolic activity of Streptococcus thermophilus on the colon epithelium of gnotobiotic rats. J. Biol. Chem. 286: 10288–10296.

Ryu, J.H., S.H. Kim, H.Y. Lee, J.Y. Bai, Y.D. Nam, J.W. Bae, D.G. Lee, S.C. Shin, E.M. Ha and W.J. Lee. 2008. Innate immune homeostasis by the homeobox gene caudal and commensal-gut mutualism in Drosophila. Science. 319: 777–782.

Salzman, N.H., K. Hung, D. Haribhai, H. Chu, J. Karlsson-Sjoberg, E. Amir, P. Teggatz, M. Barman, M. Hayward, D. Eastwood, M. Stoel, Y. Zhou, E. Sodergren, G.M. Weinstock, C.L. Bevins, C.B. Williams and N.A. Bos. 2010. Enteric defensins are essential regulators of intestinal microbial ecology. Nat. Immunol. 11: 76–83.

Sato, T., J.H. van Es, H.J. Snippert, D.E. Stange, R.G. Vries, M. van den Born, N. Barker, N.F. Shroyer, M. van de Wetering and H. Clevers. 2011. Paneth cells constitute the niche for Lgr5 stem cells in intestinal crypts. Nature. 469: 415–418.

Schuijers, J. and H. Clevers. 2012. Adult mammalian stem cells: the role of Wnt, Lgr5 and R-spondins. Embo J. 31: 2685–2696.

Schulzke, J.D., S. Ploeger, M. Amasheh, A. Fromm, S. Zeissig, H. Troeger, J. Richter, C. Bojarski, M. Schumann and M. Fromm. 2009. Epithelial tight junctions in intestinal inflammation. Ann. N Y Acad. Sci. 1165: 294–300.

Segawa, S., M. Fujiya, H. Konishi, N. Ueno, N. Kobayashi, T. Shigyo and Y. Kohgo. 2011. Probiotic-derived polyphosphate enhances the epithelial barrier function and maintains intestinal homeostasis through integrin-p38 MAPK pathway. PLoS One. 6: e23278.

Seth, A., F. Yan, D.B. Polk and R.K. Rao. 2008. Probiotics ameliorate the hydrogen peroxide-induced epithelial barrier disruption by a PKC- and MAP kinase-dependent mechanism. Am. J. Physiol. Gastrointest. Liver Physiol. 294: G1060–1069.

Shulzhenko, N., A. Morgun, W. Hsiao, M. Battle, M. Yao, O. Gavrilova, M. Orandle, L. Mayer, A.J. Macpherson, K.D. McCoy, C. Fraser-Liggett and P. Matzinger. 2011. Crosstalk between B lymphocytes, microbiota and the intestinal epithelium governs immunity versus metabolism in the gut. Nat. Med. 17: 1585–1593.

Singh, N., E.A. Shirdel, L. Waldron, R.H. Zhang, I. Jurisica and E.M. Comelli. 2012. The murine caecal microRNA signature depends on the presence of the endogenous microbiota. Int. J. Biol. Sci. 8: 171–186.

Sonnenburg, J.L., J. Xu, D.D. Leip, C.H. Chen, B.P. Westover, J. Weatherford, J.D. Buhler and J.I. Gordon. 2005. Glycan foraging *in vivo* by an intestine-adapted bacterial symbiont. Science. 307: 1955–1959.

Stepankova, R., F. Powrie, O. Kofronova, H. Kozakova, T. Hudcovic, T. Hrncir, H. Uhlig, S. Read, Z. Rehakova, O. Benada, P. Heczko, M. Strus, P. Bland and H. Tlaskalova-Hogenova. 2007. Segmented filamentous bacteria in a defined bacterial cocktail induce intestinal inflammation in SCID mice reconstituted with CD45RBhigh CD4+ T cells. Inflamm. Bowel Dis. 13: 1202–1211.

Stetinova, V., L. Smetanova, J. Kvetina, Z. Svoboda, Z. Zidek and H. Tlaskalova-Hogenova. 2010. Caco-2 cell monolayer integrity and effect of probiotic Escherichia coli Nissle 1917 components. Neuro. Endocrinol. Lett. 31 Suppl. 2: 51–56.

Strugala, V., P.W. Dettmar and J.P. Pearson. 2008. Thickness and continuity of the adherent colonic mucus barrier in active and quiescent ulcerative colitis and Crohn's disease. Int. J. Clin. Pract. 62: 762–769.

Su, L., L. Shen, D.R. Clayburgh, S.C. Nalle, E.A. Sullivan, J.B. Meddings, C. Abraham and J.R. Turner. 2009. Targeted epithelial tight junction dysfunction causes immune activation and contributes to development of experimental colitis. Gastroenterology. 136: 551–563.

Suzuki, T. 2012. Regulation of intestinal epithelial permeability by tight junctions. Cell Mol. Life Sci.

Swidsinski, A., J. Weber, V. Loening-Baucke, L.P. Hale and H. Lochs. 2005. Spatial organization and composition of the mucosal flora in patients with inflammatory bowel disease. J. Clin. Microbiol. 43: 3380–3389.

Thomas, M., L. Wrzosek, L. Ben-Yahia, M.L. Noordine, C. Gitton, D. Chevret, P. Langella, C. Mayeur, C. Cherbuy and F. Rul. 2011. Carbohydrate metabolism is essential for the colonization of Streptococcus thermophilus in the digestive tract of gnotobiotic rats. PLoS One. 6: e28789.

Tobisawa, Y., Y. Imai, M. Fukuda and H. Kawashima. 2010. Sulfation of colonic mucins by N-acetylglucosamine 6-O-sulfotransferase-2 and its protective function in experimental colitis in mice. J. Biol. Chem. 285: 6750–6760.

Tomas, J., L. Wrzosek, N. Bouznad, S. Bouet, C. Mayeur, M.L. Noordine, E. Honvo-Houeto, P. Langella, M. Thomas and C. Cherbuy. 2012. Primocolonization is associated with colonic epithelial maturation during conventionalization. Faseb J. (In Press).

Tuin, A., K. Poelstra, A. de Jager-Krikken, L. Bok, W. Raaben, M.P. Velders and G. Dijkstra. 2009. Role of alkaline phosphatase in colitis in man and rats. Gut. 58: 379–387.

Ukena, S.N., A. Singh, U. Dringenberg, R. Engelhardt, U. Seidler, W. Hansen, A. Bleich, D. Bruder, A. Franzke, G. Rogler, S. Suerbaum, J. Buer, F. Gunzer and A.M. Westendorf. 2007. Probiotic Escherichia coli Nissle 1917 inhibits leaky gut by enhancing mucosal integrity. PLoS One. 2: e1308.

Vaishnava, S., C.L. Behrendt, A.S. Ismail, L. Eckmann and L.V. Hooper. 2008. Paneth cells directly sense gut commensals and maintain homeostasis at the intestinal host-microbial interface. Proc. Natl. Acad. Sci. USA 105: 20858–20863.

Vaishnava, S., M. Yamamoto, K.M. Severson, K.A. Ruhn, X. Yu, O. Koren, R. Ley, E.K. Wakeland and L.V. Hooper. 2011. The antibacterial lectin RegIIIgamma promotes the spatial segregation of microbiota and host in the intestine. Science. 334: 255–258.

Vamadevan, A.S., M. Fukata, E.T. Arnold, L.S. Thomas, D. Hsu and M.T. Abreu. 2010. Regulation of Toll-like receptor 4-associated MD-2 in intestinal epithelial cells: a comprehensive analysis. Innate. Immun. 16: 93–103.

van der Flier, L.G. and H. Clevers. 2009. Stem cells, self-renewal, and differentiation in the intestinal epithelium. Annu. Rev. Physiol. 71: 241–260.

Van der Sluis, M., B.A. De Koning, A.C. De Bruijn, A. Velcich, J.P. Meijerink, J.B. Van Goudoever, H.A. Buller, J. Dekker, I. Van Seuningen, I.B. Renes and A.W. Einerhand. 2006. Muc2-deficient mice spontaneously develop colitis, indicating that MUC2 is critical for colonic protection. Gastroenterology. 131: 117–129.

Wehkamp, J., G. Wang, I. Kubler, S. Nuding, A. Gregorieff, A. Schnabel, R.J. Kays, K. Fellermann, O. Burk, M. Schwab, H. Clevers, C.L. Bevins and E.F. Stange. 2007. The Paneth cell alpha-defensin deficiency of ileal Crohn's disease is linked to Wnt/Tcf-4. J. Immunol. 179: 3109–3118.

Wei, X., Z. Yang, F.E. Rey, V.K. Ridaura, N.O. Davidson, J.I. Gordon and C.F. Semenkovich. 2012. Fatty acid synthase modulates intestinal barrier function through palmitoylation of mucin 2. Cell Host Microbe. 11: 140–152.

Wells, J.M., O. Rossi, M. Meijerink and P. van Baarlen. 2011. Epithelial crosstalk at the microbiota-mucosal interface. Proc. Natl. Acad. Sci. USA 108 Suppl 1: 4607–4614.

Willing, B.P. and A.G. Van Kessel. 2007. Enterocyte proliferation and apoptosis in the caudal small intestine is influenced by the composition of colonizing commensal bacteria in the neonatal gnotobiotic pig. J. Anim. Sci. 85: 3256–3266.

Wu, H.J., Ivanov, II, J. Darce, K. Hattori, T. Shima, Y. Umesaki, D.R. Littman, C. Benoist and D. Mathis. 2010. Gut-residing segmented filamentous bacteria drive autoimmune arthritis via T helper 17 cells. Immunity. 32: 815–827.

Yan, F., H. Cao, T.L. Cover, M.K. Washington, Y. Shi, L. Liu, R. Chaturvedi, R.M. Peek, Jr., K.T. Wilson and D.B. Polk. 2011. Colon-specific delivery of a probiotic-derived soluble protein ameliorates intestinal inflammation in mice through an EGFR-dependent mechanism. J. Clin. Invest. 121: 2242–2253.

Yan, F., H. Cao, T.L. Cover, R. Whitehead, M.K. Washington and D.B. Polk. 2007. Soluble proteins produced by probiotic bacteria regulate intestinal epithelial cell survival and growth. Gastroenterology. 132: 562–575.

Yan, F. and D.B. Polk. 2002. Probiotic bacterium prevents cytokine-induced apoptosis in intestinal epithelial cells. J. Biol. Chem. 277: 50959–50965.

Zarember, K.A. and P.J. Godowski. 2002. Tissue expression of human Toll-like receptors and differential regulation of Toll-like receptor mRNAs in leukocytes in response to microbes, their products, and cytokines. J. Immunol. 168: 554–561.

Zyrek, A.A., C. Cichon, S. Helms, C. Enders, U. Sonnenborn and M.A. Schmidt. 2007. Molecular mechanisms underlying the probiotic effects of Escherichia coli Nissle 1917 involve ZO-2 and PKCzeta redistribution resulting in tight junction and epithelial barrier repair. Cell Microbiol. 9: 804–816.

4

Pathogen-Host Cell Interactions at the Intestinal Level

Lessons from Cultured Human Fully-Differentiated Colon Cancer Caco-2 and T84 Epithelial Cell Lines

Vanessa Liévin-Le Moal and *Alain L. Servin**

Introduction

The mucosal surface of the intestinal tract is the largest body surface in contact with the external environment (200–300 m^2), and is lined by a simple columnar epithelium that is folded to form a number of invaginations, or crypts. Epithelial intestinal cells are constantly renewed from a source of multipotent stem cells located in the crypts of Lieberkühnn. Five highly specialized cell phenotypes are present in the intestinal epithelium: enterocytes, also known as fluid-transporting cells, neuroendocrine cells, mucus-secreting cells, also known as goblet cells, Paneth cells, and M cells, respectively (Louvard et al. 1992). M cells are located in the epithelium overlying mucosa-associated lymphoid tissues, such as Peyer's patches, where they act as the antigen-sampling cells of the mucosal immune

CNRS, UMR 8076 « BioCIS». « Chimiothérapie antiparasitaire ». Team Faculté de Pharmacie, Châtenay-Malabry, 92296. France; and Université Paris-Sud, Faculté de Pharmacie, Châtenay-Malabry, 92296. France.
* Corresponding author: Alain.servin@u-psud.fr

system, and play a pivotal role in the pathogenesis of several enterovirulent microorganisms. The integrity of the layer of cultured intestinal fully-differentiated epithelial cells is maintained by intercellular junctional complexes composed of tight junctions (TJs), adherent junctions (AJs), and desmosomes.

The intestinal epithelium acts as a physical and chemical barrier that protects the host from attack by potentially harmful enteric microorganisms. Enteric microbial pathogens have developed membrane-associated functional structures and virulence factors that hijack the host's cellular molecules and signaling pathways. In the first step in the infectious process, some enteric bacterial pathogens adhere to the brush border of intestinal cells. Host defense systems against the unwelcome intrusion of pathogenic enteric microorganisms also include adaptive and innate immunity. The intestinal epithelium provides a surface where the host can sense the microbial environment in order to trigger strong cellular defense responses when required, by releasing host cell signaling molecules such as cytokines and chemokines. These in turn trigger the recruitment of leukocytes, and initiate the attraction of immune cells. The intestinal mucosa also has a surface coating of mucus that is secreted by the specialized goblet cells, also known as mucin-secreting cells. Moreover, the intestinal epithelium also provides antimicrobial peptides and antimicrobial molecules which, together with resident microbiota, provide the front line of defense against pathogens (Lievin-Le Moal and Servin 2006).

Microbial pathogenesis has primarily been investigated using undifferentiated, non-intestinal, cultured epithelial cells. Examination of the mechanisms of pathogenicity of enterovirulent bacteria has long been complicated, because human enterocytes and goblet cells are difficult to isolate and to maintain in culture for prolonged periods, and tend to give rather variable results depending on the particular donor. The establishment in 1983 of human colon carcinoma cell lines that are able to express the functional and structural characteristics of mature enterocytes or goblet-cells in culture (Zweibaum et al. 1991), has considerably facilitated the *in vitro* study of microbial pathogenesis. These cell lines were initially used to study basic questions related to the organization and functions of human polarized intestinal cells. Indeed, when fully-differentiated in culture, these cells display a polarized organization and contain highly-regulated tight and junctional domains resulting in their forming a cell monolayer that mimics an intestinal epitheliumin culture. Moreover, specifically in each domain, these cells expressed the major differentiation-associated proteins supporting specific intestinal functions. Human, parental, fully-differentiated intestinal cells, and cell subpopulations and clones have been used since 1987 as models for investigating the cellular and molecular mechanisms, by which enteric microbial pathogens hijack membrane-associated proteins, signaling pathways and intracellular traffic functions, which in turn promote structural and functional lesions, and host cellular responses in enterocytes or mucin-secreting cells. The aim of this review is to analyze the insights that have been obtained using cultured parental Caco-2 cell line and clones, and T84 cells into the mechanisms of pathogenesis of the human enterovirulent bacteria that interacted with host intestinal cells at the brush border.

Human Colon Cancer Caco-2 and T84 Cell Lines

Parental Caco-2 cell lines and clones

In 1972, J. Fogh (Fogh and Trempe 1975) isolated and cultured cells from human colonic adenocarcinomas. Parental colon cancer Caco-2 cells (Fogh et al. 1977) spontaneously differentiate in culture (Standard culture conditions: DMEM, 25 mM glucose, supplemented with 20% inactivated fetal calf serum, and 1% non-essential amino acids at 37°C in a 10% CO_2/90% air atmosphere, 15 days in culture, culture medium changed daily) and are characterized by the appearance of the structural and functional properties of the enterocytes of the small intestine (Pinto et al. 1983). Several different clones of Caco-2 cells have been established: clones C2BBe 1 and 2 (Peterson and Mooseker 1992), clone 40 (Woodcook et al. 1991), clone cl1 (Costa de Beauregard et al. 1995), clone NGI3 (Tian and Quaroni 1999), clones #1, #20 and #21 (Horie et al. 2003). Moreover, twenty-six Caco-2 clones, including the most currently used Caco-2/TC7 clone, have been isolated from early and late passages of parental Caco-2 cells (Chantret et al. 1994). It is important to note that the fully differentiated Caco-2 cells and Caco-2 clones, even though they are of colonic origin, express apical and baso-lateral proteins that display the specific functions characteristic of the mature enterocytes of the small intestine (Zweibaum et al. 1991). The differentiation process in the culture of Caco-2 cells closely mimics the differentiation of enterocytes that occurs along the crypt-villus axis of the small intestine. The differentiation process of parental Caco-2 cells and clones is growth related; exponentially-dividing cells are undifferentiated, and differentiation begins once confluence has been reached, i.e., when the cells stop dividing. Fully-differentiated parental Caco-2 cells and clones are highly polarized, expressing a dense and well-ordered apical brush border which carpets the entire apical surface. Parental Caco-2 cells provide the best model for investigating the cell contact-induced assembly of the intestinal brush border (Peterson and Mooseker 1993), and have been used to identify the mechanisms of the sorting and surface delivery of apical and basolateral proteins in human enterocytes (Le Bivic et al. 1990; Matter et al. 1990). The brush border membrane of these cells is endowed with human functional proteins found

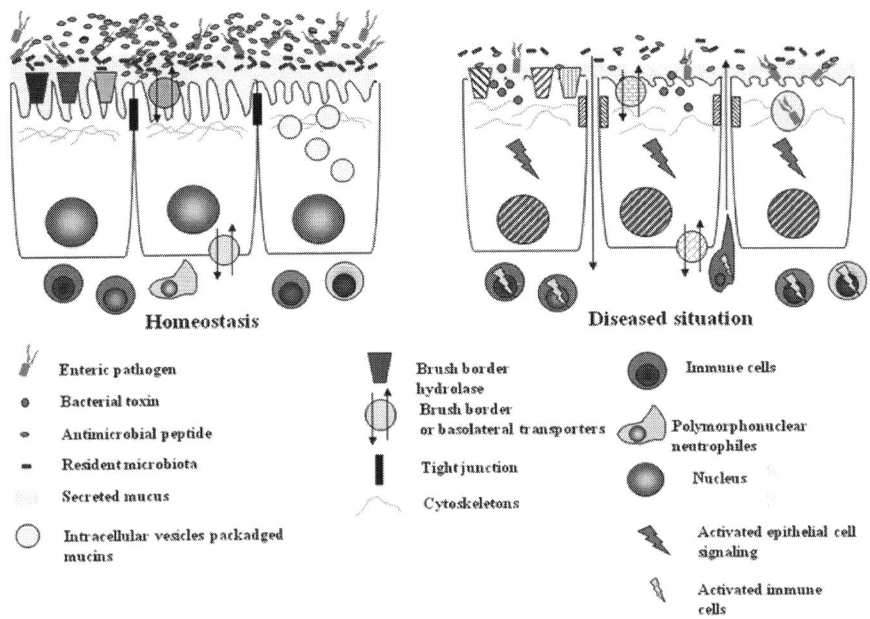

Figure 1. The Intestinal Epithelial Barrier. The intestinal epithelium consists of a single layer of highly polarized epithelial cells. The tight junction, a component of the apical junctional complex, seals the paracellular space between epithelial cells. Specific structural proteins composed the cytoskeleton and microtubule networks playing pivotal roles in the polarized organization of intestinal cells, the structural organization of the brush border, and the intracellular traffic of functional proteins to be addressed into the apical, lateral and basal domains of the polarized cells. The brush border at the apical domain and basolateral cell domain contain proteins and transporters exerting specific intestinal functions. The intestinal epithelial barrier plays an essential role in maintaining immune homeostasis. The lamina propria, located beneath the basement membrane, contains immune cells and neutrophils. Intestinal microbiota resides in the lumen, outside the mucus layer. Intestinal functions are modulated by quorum sensing-dependent cross-talk between the microbiota and host cells. Secreted mucins, combined with membrane-bound mucins, act as a physicochemical barrier and protect the epithelial cell surface against undesirable and harmful pathogens. Secreted antimicrobial peptides form the first chemical defense system against unwanted enteric pathogens.

Enteric pathogens use their adhesive factors to interact with the brush border membrane and hijack membrane-bound molecules as receptors for triggering signaling events. Invasive pathogens cross the epithelial cell membrane, penetrate into the host cells and pursue sophisticated intracellular lifestyles either in vacuoles or directly within the cell cytoplasm. Enteric toxins are secreted by both adhering and non-adhering pathogens, and either bind to membrane-bound receptors, thus activating signaling pathways, or are internalized. Enteric pathogens, acting either directly or via their toxins, disrupt the epithelial barrier, increasing paracellular permeability and inducing pro-inflammatory responses. Adhering and invading enteric pathogens trigger structural and functional lesions in epithelial cells, and even cell death.

Color image of this figure appears in the color plate section at the end of the book.

in the enterocytes of the human small intestine, including sucrase-isomaltase (SI), lactase, APN, and dipeptidylpeptidase IV (DPP IV), but not maltase-glucoamylase, angiotensin I-converting enzyme, and p-aminobenzoic acid peptide hydrolase (Zweibaum et al. 1991; Pinto et al. 1983). The mosaic pattern of brush border-associated hydrolases in these cells results from the differing patterns of microvilli, and the fact that the level of brush-border associated functional-protein expression can vary from one cell to another. In parental Caco-2 cells, the level of SI increases 7-fold from early (passage 29) to late (passage 198) passages in culture (Chantret et al. 1994). Other intestinal functional proteins are expressed in fully-differentiated parental Caco-2 cell line and clones, including hexose transporters SGLT1, GLUT1, GLUT2, GLUT3 and GLUT5, Na^+/H^+ exchanger, human intestinal $H^+/$ peptide cotransport system, hPEPT1, H^+-coupled dipeptide transporter, cholesteryl ester transfer protein (LTP-I), PAR2, and receptors including the epidermal growth factor receptor (EGFR), insulin-like growth factor-binding proteins (IGF-BP-2, IGF-BP-3, and IGF-BP-4), vasoactive intestinal peptide receptor, nuclear hormone receptor peroxisome proliferator-activated receptor (PPAR) gamma. In and out efflux systems controlling the intestinal transport of drugs have been studied using fully-differentiated Caco-2 cell line and clones. The functional intestinal proteins are expressed by these cells in their appropriate membrane domains as a function of the number of days in culture, and parallel cell polarization and brush border assembly. The fully-differentiated parental Caco-2 cell line and the clone formed in a culture cell monolayer that mimics the intestinal epithelial barrier, with a well-organized and regulated cell-to-cell junctional domain. In particular, the TJ is a highly developed structure, which acts as a "fence" separating the apical and basolateral domains of polarized cells, thereby segregating cell surface proteins and lipids into separate domains. Moreover, TJs also act as a "gates" providing a permeable barrier between the mucosal and serous environments and that enables paracellular vectorial transport to occur across the layer of cells. TJs specifically containing zonula occludens (ZO) proteins (ZO-1, -2 and -3), occludin and claudins (claudin-1, -2, -3, -4, -12, but not claudin-5), allow positive transepithelial resistance (TER). Moreover, the fully-differentiated Caco-2 cell monolayer forms fluid-filled, blister-like domes that are highly dynamic structures formed as the result of both the paracellular and transcellular pathways of ion and water transport that determine the net apical to basolateral vectorial-transportation of water (Grasset et al. 1984). The parental, fully-differentiated Caco-2 cell line and its clones have been extensively used to investigate the cellular and molecular mechanisms by which human enterovirulent bacteria and viruses create structural and functional cell lesions and trigger cellular responses in the intestinal barrier. Fully-differentiated Caco-2 cells and their clones have

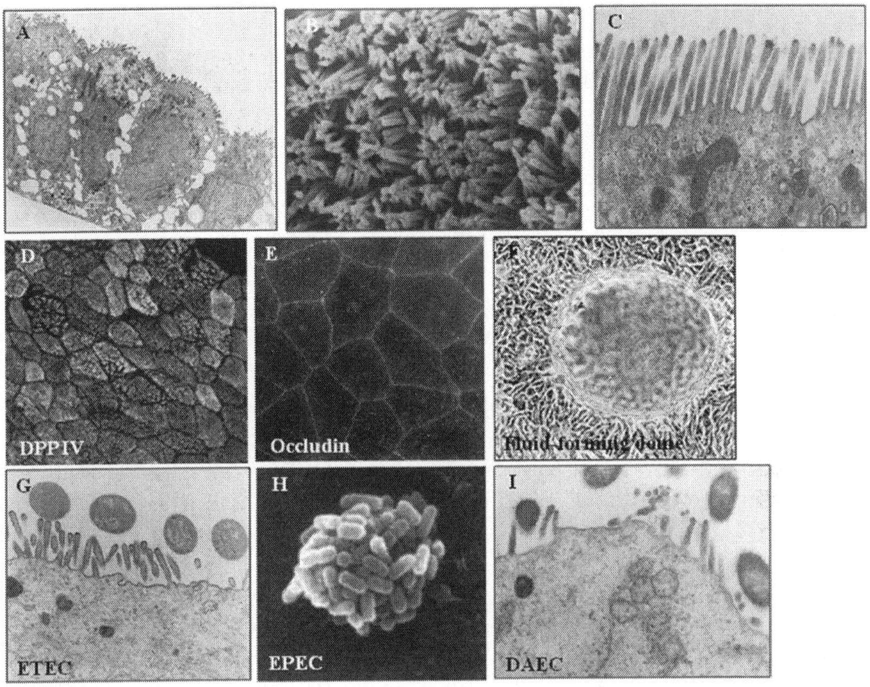

Figure 2. Human colon cancer Caco-2/TC7 Cells Spontaneously Differentiating in Culture into Enterocyte-like Cells of the Small Intestine. In A, transmission electron micrograph showing the polarized organisation of fully-differentiated cells forming a cell monolayer (15 days in culture). In B, scanning electron micrograph showing the well-organized and dense brush border. In C, transmission electron micrograph showing the well-ordered brush border microvilli. In D, immunofluorescent labeling showing the mosaic pattern of expression of functional brush border-associated DPP IV. In E, immunofluorescent labeling showing the tight junction-associated occludin localizing at the cell-to-cell contact of the cell monolayer. In F, microscopy phase-contrast micrograph showing a fluid-forming dome in the cell monolayer. In G to I, transmission (G and I) and scanning (H) electron micrographs showing interaction of ETEC with microvilli (G), typical, well-formed and compact EPEC microcolonies at the brush border (H), and brush border disappearance in Afa/Dr DAEC-infected cells (I).

Color image of this figure appears in the color plate section at the end of the book.

also been used to investigate how bacterial species of the human intestinal microbiota control the pathogenesis of human enteropathogens.

In the mammalian gut mucosa, Peyer's patches, which are also known as follicle-associated epithelium, are areas of aggregated and isolated lymphoid follicles separated from the intestinal lumen by a single layer of columnar epithelial cells. In these randomly distributed and particular intestinal cell areas, a relatively small number of highly specialized epithelial cells known as M (for microfold or membranous) cells are present (Kraehenbuhl and

Neutra 2000; Nicoletti 2000). The phenotype of the human M cell has long remained elusive as a result of the difficulty of isolating these sparsely occurring intestinal cell areas. M cells play a central role in triggering mucosal immune responses by transporting intact foreign antigens and commensal bacteria down to the underlying lymphoid tissue. This allows the mucosal immune system to confront the vast range of antigens that enter the body through the gut mucosa, and it reacts by mounting specific mucosal and systemic immune responses. M cells are also major sites of adhesion and invasion for several invasive enteric pathogens (Jepson and Clark 1998). Moreover, after having crossed the follicle-associated epithelium the invading bacteria face phagocytic cells, including the macrophages that are present in the follicle dome. For several enteroinvasive pathogens, the outcome of infection depends on their ability to survive in the presence of macrophages or even within macrophages after phagocytosis, and how these invasive species manage to survive (Sansonetti and Phalipon 1999). A model of M-like cells has been obtained by co-culturing freshly isolated BALB/c mouse Peyer's patch lymphocytes with parental, fully-differentiated Caco-2 cells triggering the phenotypic conversion of Caco-2 cells into cells expressing the morphological and functional properties of M-cells (Kerneis et al. 1997). More stable models of M-like cells have been obtained by co-culturing fully-differentiated parental Caco-2 cells (Gullberg et al. 2000) or Caco-2 cl1 clone cells (El Bahi et al. 2002) with the human Burkitt's lymphoma Raji B cell line. Another model has been devised which consists of fully-differentiated Caco-2 cl1 clone cells cultured in the presence of freshly-isolated human blood lymphocytes (Pielage et al. 2007). These M cell-like models have been used to investigate M cell particle transport, and the interaction and entry into M cells of enteroinvasive pathogens such as *Listeria monocytogenes*, *Yersinia enterocolitica*, and *Salmonella* spp., and the non-invasive enteropathogenic pathogen *Escherichia coli*. A complex *in vitro* co-culture model of the inflamed intestinal mucosa has been constructed comprising macrophages and dendritic cells derived from peripheral blood monocytes embedded in a collagen layer on a Transwell filter insert with fully-differentiated Caco-2 cells seeded on top (Leonard et al. 2010).

T84 cell line

The T84 cell line is composed of colonic epithelial cells derived from a human colonic carcinoma (Dharmsathaphorn et al. 1984). The cells grew to confluence as a monolayer attached to the surface of a culture dish or permeable supports (1:1 (v/v) mixture of Dulbecco-Vogt modified Eagle medium and Ham's F-12 medium supplemented with 6% fetal calf serum, pH 7.5 at 37°C in a 10% CO_2/90% air atmosphere). After 14 days in culture during which the culture medium was daily changed, it was observed that these cells are highly polarized, displaying short microvilli at the apical membrane facing the medium, a basolateral membrane; TJs and desmosomes have been identified between adjacent cells. The presence of a high level of positive TERs resulting from tight-formed and regulated functional TJs expressing ZO-1, occludin and claudins, have made to measure unidirectional fluxes of inulin, mannitol, and sodium from apical domain to basolateral domain, and activities of membrane-associated transporters allowing paracellular passages (Madara and Dharmsathaphorn 1985). Mounted in a Ussing chamber, this cell line provides an excellent model system for studying electrolyte transport processes and the functions of voltage-dependent channels by means of electrical circuit analysis (Madara et al. 1987). For example, T84 cells have been used to investigate the regulation of chloride secretion, and transport systems Na^+/H^+ exchanger 1, 2 and 4, HCO_3^- secretion, Na^+-independent equilibrating nucleoside transporters, ENT1 and ENT2, the bumetanide-sensitive $Na^+:K^+:2Cl^-$ cotransporter and chloride channel, and the cystic fibrosis transmembrane conductance regulator (CFTR). As a result of these properties, T84 cells have been extensively used to investigate the impact of enterovirulent bacterial pathogens and enteric viruses, as well as of bacterial toxins, on the structural organization and functioning of human colonic cells. This cell line offers the best model for investigating how enterovirulent pathogens induce the transmigration of polymorphonuclear leukocytes (PMNLs) across the intestinal barrier (Nash et al. 1987).

Campylobacter jejuni

Campylobacter is a microaerophilic, curved, Gram-negative rod bacterium exhibiting motility, and is carried in the intestine of many wild and domestic animals, particularly avian species including poultry (Silva et al. 2011). It is recognized as being the leading cause of bacterial, food borne, diarrheal disease worldwide. These can range from mild to serious infections of children and the elderly, and can include permanent neurological symptoms. *Campylobacter* spp. expresses a set of virulence factors including flagella (for swimming within mucus post infection and host–bacterium interactions), chemotactic proteins, adhesive factors including CadF-recognizing fibronectin, Peb1, Cj1496c, JlpA also activating NF-κB and p38 mitogen-activated protein (MAP) kinase, and lipoprotein CapA, implicated as a possible adhesin, the secreted protein CiaB that is required for the invasion of epithelial cells, and cytolethal distending toxin (Young et al. 2007). *Campylobacter* spp. lacks T3SS and invades intestinal epithelial cells, but the mechanisms that control cell-entry are not fully understood. It is clear that cell-entry involves a microtubule-dependent mechanism since pseudopods entrapping cell-associated bacteria contain microtubules.

In addition, several *Campylobacter* strains also require microfilament polymerization for cell-entry. Internalized *C. jejuni* localized within vacuoles move along microtubules to the perinuclear region of the cell. The role of these internalized bacteria in pathogenesis, possibly by evasion of the immune system or by establishing a protected reservoir, remains to be been determined.

In fully-differentiated Caco-2 cell monolayers grown on microporous membrane filters, Konkel et al. (Konkel et al. 1992) first observed that *C. jejuni* translocated across the cell monolayers by passing both through and between cells. When infecting T84 cell monolayer at the basal domain, *C. jejuni* translocation appeared to occur via a paracellular route rather than a transcellular route (Monteville and Konkel 2002). *C. jejuni* isolates expressing ganglioside-like lipooligosaccharides (LOSs) are highly adherent onto fully-differentiated Caco-2 and T84 cells, in contrast to *C. jejuni* isolates, which lack these structures (Louwen et al. 2012). Moreover, the ganglioside-like LOS facilitates the entry of *C. jejuni* into these cells. *Campylobacter* isolates, which possess *flaA*, *cadF*, *pldA*, *ciaB*, *cdtA*, *cdtB*, and *cdtC* genes, but not the *virB11* gene, are able to penetrate fully-differentiated T84 cells (Zheng et al. 2006). Cell association and invasion into Caco-2 cells are inhibited by various sugars, including D-glucose, D-mannose, and D-fucose, and intracellular *C. jejuni* localized within membrane-bound vacuoles (Russell and Blake 1994). Mannose and/or N-acetylneuraminic acid may be involved in the initial interaction step allowing *C. jejuni* to bind to fully-differentiated Caco-2 cells (Day et al. 2009). An autotransporter protein *capA* insertion mutant displays significantly reduced capacity for associating with and invasion of fully-differentiated Caco-2 cells (Ashgar et al. 2007). During *C. jejuni* translocation across fully-differentiated Caco-2 cell monolayers, the absence of any change in TER and inulin paracellular passage indicates that the bacteria translocate through the cytoplasm of the invaded cells rather than via intercellular spaces (Bras and Ketley 1999). Examination of the role of flagella and swimming motility in the interactions of *C. jejuni* with fully-differentiated Caco-2 cells, reveals that unlike the wild-type strain, both the flagellated, nonmotile (flaA flaB+ Mot-) and nonflagellated and nonmotile (flaA flaB Mot-) mutants were unable to translocate across the cell monolayers (Russell and Blake 1994; Grant et al. 1993). Moreover, when the viscosity of the intestinal mucus is mimicked, *C. jejuni* displays impaired straight-swimming motility punctuated by tumbling (Szymanski et al. 1995).

The serine protease HtrA of *C. jejuni* plays a role in the basolateral invasion of fully-differentiated Caco-2 by cleaving E-cadherin (Hoy et al. 2012). The genome sequence of *C. jejuni* NCTC 11168 contains a gene encoding an orthologue of LuxS, which is required for autoinducer-2 (AI-2). A luxS mutant shows unchanged ability to invade fully-differentiated

Caco-2 cell monolayers even though it exhibits decreased motility in semisolid media, suggesting that quorum sensing may play a role in the regulation of motility (Elvers and Park 2002). When *C. jejuni* is grown in iron-limited media in the presence of norepinephrine (NE), the motility of bacteria and their ability to enter fully-differentiated Caco-2 cells both increase (Cogan et al. 2007). In addition, in the presence of NE *C. jejuni* causes greater disruption of cultured epithelial cell monolayers than in the absence of NE.The roles of environmental stress factors including temperature shifts, nutrient starvation and decreased atmospheric oxygen concentration on *C. jejuni* pathogenicity has been evaluated in fully-differentiated Caco-2 cell monolayers (Mills et al. 2012; Klancnik et al. 2009; Mihaljevic et al. 2007). Nutrient insufficiency and temperature elevation that transiently impair bacteria growth have been shown to affect the adhesion and invasion properties of *C. jejuni*. Oxidative stress affects neither the binding of cells nor their invasion, whereas oxygen exposure or microaerobic conditions increase the invasion capability and intracellular survival of *C. jejuni*.

C. jejuni causes no change in the short-circuit current after infection of fully-differentiated Caco-2 cell monolayers mounted in Ussing chambers, but does reduce dome formation, a marker of fluid-transport across the monolayer, and this is accompanied by a decrease in TER correlated with an altered distribution of the TJ protein occludin (MacCallum et al. 2005). Apical infection of fully-differentiated T84 monolayers causes a decrease in TER, accompanied by redistribution of the TJ protein occludin from an intercellular to an intracellular location (Chen et al. 2006). When it infects the basolateral domain *C. jejuni* causes a more rapid decrease in TER, but a comparable redistribution of TJ proteins (Chen et al. 2006).

Salmonella spp.

Salmonellae are Gram-negative bacteria that cause gastroenteritis and enteric fever (Grassl and Finlay 2008). *Salmonella* serotypes associated with gastroenteritis orchestrate a strong intestinal inflammatory response and the deleterious structural and functional cell injuries that result in severe secretory disease. *Salmonella* species share the ability to invade the host by inducing their own uptake (Valdez et al. 2009), and they survive and multiply within the epithelial cells lining the intestinal epithelium (Ramsden et al. 2007). To become virulent *Salmonella* requires the coordinated expression of complex arrays of virulence factors (Wagner et al 2010; Chevance and Hughes 2008; Galan 2001; Ohl and Miller 2001). Central to the pathogenesis of *Salmonella* are the type III secretion apparatus (T3SS) systems encoded by the *Salmonella* pathogenicity islands SPI-1 (T3SS-1) and SPI-2 (T3SS-2). Both T3SS-1 and T3SS-2 are composed of approximately 20–30 proteins with a major subset of these proteins having a structural role in forming the supramolecular injection apparatus, known as the needle complex, composed of three distinct substructures: a multi-ring base, an inner rod, and a needle (Galan 2001). Another set of proteins form the translocon forming a pore into the host cell membrane permitting the delivery of bacterial effectors into the host cell cytoplasm. T3SS-1 and T3SS-2 are responsible for the secretion and translocation of a set of bacterial proteins known as effectors into host cells altering host cell organization and permit bacterial entry and survival. These secretion systems are thought to play distinct yet overlapping functions. For example, whereasT3SS-1 plays a key function in bacterial invasion T3SS-2 is thought to be required for bacterial survival in host cells. However, recent observations suggest that T3SS-1 and T3SS-2 depend on each other to function efficiently. T3SS-1 and T3SS-2 are responsible for delivering a series of bacterial effectors into host cells (Agbor and McCormick 2011; Coburn et al. 2007; Galan and Wolf-Watz 2006). The *Salmonella* translocated effectors are unique proteins with very specialized functions. However, recent evidence indicates that several modular proteins express functionally-distinct domains that may act at different stages in the infection process. Several secreted effectors, including SipA, SipB, SopA, SopB, SopD, SopE, and SopE2, play critical roles in manipulating the host cell for bacteria invasion. Other effectors, including SopA, SopB, SseF, SseG, SopD2, SifA and PipB, control the assembly, maturation and positioning of intracellular vacuoles created for bacterial survival and replication. Moreover, other effectors including AvrA, SopE, SopE2, SopB and SptB, and SpvC, are involved in the promotion or control of inflammatory responses. Emerging evidence indicates that these effectors are modular proteins consisting of distinct functional domains/ motifs that are utilized by the bacteria to activate intracellular signaling pathways modifying host cell function. Despite the fact that the target cells of *Salmonella* are the polarized epithelial cell phenotypes lining the intestinal epithelium, the virulence factors of *Salmonella* and their roles in cell infection and cell responses have been identified using non-polarized and non-intestinal Hep-2, HeLa cell lines, non-polarized, intestinal INT407 cells lines, polarized, non-intestinal Madin Darby Canine Kidney (MDCK) cells, or parental, non-polarized intestinal HT-29 cell line.

The adhesion of *Salmonella* at the brush border of fully-differentiated Caco-2 cells seems to result from various different adhesive factors including the SPI4-dependent, very large, non-fimbrial adhesin SiiE (Gerlach et al. 2007) Std fimbria (Chessa et al. 2008). It has been suggest that a Galβ (1-3) GalNAc epitope located in the glycocalyx is involved in the early recognition events between *S. typhimurium* and Caco-2 BBe clone cells (Giannasca et al. 1996). Adherence of Std fimbriated *S. typhimurium* to fully-differentiated Caco-2 cells was blocked by the H type 2 oligosaccharide (Fucα1-2Galβ1-4GlcNAc) (Chessa et al. 2009). Sialic acid plays a role in the adhesion of

S. typhi onto fully-differentiated Caco-2 cells (Sakarya et al. 2010). SadA, a purported trimeric autotransporter adhesin of *S. typhimurium*, appears to be involved in cell aggregation, and biofilm formation, and to increase adhesion to fully-differentiated Caco-2 (Raghunathan et al. 2011). In addition, it has been observed that flagella of *S. enteritidis* were involved in the association with and invasion of fully-differentiated cells (van Asten et al. 2004; Van Asten et al. 2000; Dibb-Fuller et al. 1999).

The first reports describing the interaction and cellular consequences of *Salmonella enterica* serovar Typhimurium (*S. typhimurium*) with fully-differentiated Caco-2 cells were made by Finlay's group (Mills and Finlay 1994; Betts and Finlay 1992; Leung and Finlay 1991; Finlay and Falkow 1990). Finlay and Falkow (Finlay and Falkow 1990) report that during the infection of a fully-differentiated parental Caco-2 cell line, after *Salmonella* has interacted with the well-organized apical microvilli, it induces a dramatic localized reorganization of the F-actin cytoskeleton at the site of bacterial attachment followed by the disruption of the brush border including elongation of the microvilli, and penetrates through the monolayer and appears in the basolateral medium. Using three of the four classes of mutants that remain virulent in mice, Betts and Finlay (Betts and Finlay 1992) observed that *S. typhimurium* invasiveness requires intact motility, and at least six distinct genetic loci. F-actin filament rearrangement and morphological changes at the apical domain of Caco-2 cells are essential for the entry of *S. typhimurium* (Francis et al. 1992; Finlay and Falkow 1990). Moreover, *Salmonella* translocation has been observed in the co-culture system of parental Caco-2 cells and the human lymphoblast-like Raji B cell line mimicking M cells (Martinez-Argudo et al. 2007) independently of factors encoded by T3SS-1 and T3SS-2 (Martinez-Argudo and Jepson 2008) and involving the host cell membrane-associated caveolin-1 (Lim et al. 2009). These observations in fully-differentiated Caco-2 cells have been validated in a human, intestinal, *in vitro*, organ culture system in which *S. typhimurium* was shown to interact with border-expressing cells and induce membrane ruffles (Haque et al. 2004).

The dramatic reorganization of the F-actin cytoskeleton known as a ruffle allows *Salmonella* to enter epithelial host cells results from the action of a group of bacterial proteins delivered to the host cells by the *Salmonella* T3SS. The T3SS-1 effector SipA that binds directly to F-actin to modulate the dynamics of actin in facilitating bacterial entry is found to be preferentially associated with peripheral cortical F-actin filaments, but not stress fibers, in infected fully-differentiated Caco-2 cells (Higashide et al. 2002). T3SS-1 effectors SopE2 and SopB are required for invasion, while the SipA protein accelerates entry into epithelial cells. The T3SS-1 effectors SopA or SopD have been observed to contribute to the penetration of *S. typhimurium* into nonpolarized T84 cells, whereas in contrast, SopA, SopB, SopD, and SopE2 increase the invasiveness of the SipA-positive strain towards polarized T84 cells (Raffatellu et al. 2005). CorA, the primary or "housekeeping" Mg^{2+} channel, appears to be involved in the expression of several *S. typhimurium* virulence factors, since a *corA* mutant strain loses its swimming motility, and expresses a lower level of InvH and SipC accompanied by reduced ability to invade fully-differentiated Caco-2 cells (Papp-Wallace et al. 2008).

Internalized *S. typhimurium* are enclosed within the cytoplasm in membrane-bound vacuoles (Mills and Finlay 1994; Finlay and Falkow 1990). It has been noted that *S. typhi* and *S. typhimurium* use similar mechanisms to invade and carry out intracellular trafficking in Caco-2 cells (Kops et al. 1996; Mills and Finlay 1994). The biogenesis of the *Salmonella*-containing vacuole within mammalian cells has been intensively studied over recent years in various different polarized epithelial cells. The ability of *S. choleraesuis* to survive within fully-differentiated Caco-2 cells was found to be dependent on *sodC* genes, *sodC2* being more important than *sodC1* (Ammendola et al. 2005). Investigating the role of T3SS-2 in cell-entry, survival and proliferation within the cells has revealed that entry is more efficient into fully-differentiated Caco-2 and T84 cells than into the non-polarized, epithelial, cell line HeLa (Holzer and Hensel 2012). Moreover, while T3SS-dependent intracellular proliferation has been observed in HeLa cells, the intracellular replication in the fully-differentiated cells was severely restricted, and not affected by the deletion of SPI2 (Holzer and Hensel 2012). Analysis of the transcriptome of *S. typhimurium* within fully-differentiated Caco-2 cells has been carried out and compared with *Salmonella* gene expression inside macrophages (Hautefort et al. 2008). Upregulation of the *mgtBC*, *pstACS*, and *iro* genes for magnesium, phosphate and iron uptake, and SPI2 pathogenicity island has been observed. Moreover, the invasion-associated SPI1 pathogenicity island and the genes involved in flagellar biosynthesis were expressed inside epithelial cells at later stages of the infection, while they were always downregulated in macrophage-like cells. Further, the ability of *Salmonella* to sense and to adapt to the intracellular environment of different types of host cells has also received attention.

S. typhimurium replicating in the *Salmonella*-containing vacuole within undifferentiated Caco-2 cells utilized glucose, but not glucose 6-phosphate, and other phosphorylated carbohydrates, gluconate, or fatty acids as major carbon substrates (Gotz et al. 2010). Carbohydrate phosphotransferase systems were encoded by the genes *ptsG/crr*, *manXYZ*, *fruBA*, *malX/crr*, *scrA/crr* and *bglF* and glucose 6-phosphate was taken up by both pathogens via the UhpT transporter, which is under the control of a complex two-component system (*uhpABC*). *S. typhimurium* mutants that display defective uptake of glucose and mannose (*DptsG*, *manXYZ*) and glucose 6-phosphate (*DuhpT*) have a reduced capacity for intracellular replication in undifferentiated

Caco-2 cells, and the triple mutants *DptsG*, *manXYZ*, and *uhpT* were still able to replicate within the vacuoles (Gotz and Goebel 2010).

Fully-differentiated Caco-2 cells grown on permeable filters and infected apically with either *S. choleraesuis* or *S. typhimurium* showed a loss of TER (Finlay and Falkow 1990). *S. typhimurium* delocalized the TJ-associated ZO-1 and occludin proteins in fully-differentiated Caco-2/TC7 clone cells (Boyle et al. 2006). SopB, SopE, SopE2, and SipA are the T3SS-1-secreted effectors that are responsible for the *S. typhimurium*-induced disruption of TJ structure and function through stimulation of host cell, Rho-family GTPases (Boyle et al. 2006).

McCormick et al. (McCormick et al. 1993) were the first to report that after apical *S. typhimurium* infection of fully-differentiated T84 cell monolayer, PMNLs subsequently placed on the basolateral domain of the monolayers started a transepithelial migration without any impairment of monolayer integrity as assessed by TER and measurements of ion transport. Examination of the potential of various *Salmonella* serotypes, including *S. typhimurium*, *S. enteritidis*, *S. pullorum*, *S. arizonae*, *S. typhi*, and *S. paratyphi* to induce the transepithelial migration of PMNLs has revealed that the strains or serotypes that elicit diffuse enteritis in humans induce transmigration, whereas those that do not elicit diffuse enteritis in humans do not (McCormick et al. 1995a). In contrast, the ability to enter the fully-differentiated T84 cells did not differentiate between the strains or serotypes that induce diffuse enteritis and those which do not (McCormick et al. 1995a). The *Salmonella*-induced PMNLs transepithelial migration was not attributable to the classical pathway by which an enteropathogenic bacterium induces the migration of PMNLs, because despite the fact that infected cells release the potent chemotactic interleukin (IL)-8, this cytokine is not solely responsible for the *Salmonella*-induced PMNL transepithelial transmigration elicited, suggesting that one or more chemotactic factor may be involved (McCormick et al. 1998, 1993).

After analyzing the phenomenon, McCormick et al. (McCormick et al. 1995b) have provided evidence that the primary role of basolateral secretion of IL-8 by the fully-differentiated T84 cells is the recruitment of PMNLs through the matrix to the subepithelial space, rather than directing the final movement of PMNLs across the epithelium. In fully-differentiated T84 cell monolayers, *S. typhimurium* recruits PMNLs and induces their transepithelial migration by the co-ordinated production of two potent chemoattractant PMNLs, the basolateral IL-8 and the apical hepoxilin A3 (HXA3), which is a member of the eicosanoid class of lipids (Mrsny et al. 2004; Gewirtz et al. 1999; McCormick et al. 1998). The *S. typhimurium* effector protein, SipA has been found to be both necessary and sufficient for the epithelial transmigration of PMNLs to occur in fully-differentiated T84 monolayers (Lee et al. 2000). In particular, SipA facilitates the apical release of HXA3 via an increase in the protein expression of the ATP-binding cassette (ABC) transporter, and multidrug resistance-associated protein 2 (MRP2) (Mrsny et al. 2004). Mechanistically, SipA activates ezrin, a member of the ERM protein family that regulates the localization and functionality of intestinal apical transporters including the Na^+/H^+ exchanger-3, and the CFTR and MRP2 in polarized epithelial cells, which in turn increases the apical expression of MRP2 controlling the release of HXA3, and hence the induction of PMNL transepithelial migration (Agbor et al. 2011). In addition, by initiating an ADP-ribosylation factor-6- and the phospholipase D-dependent lipid-signaling cascade SipA directs the specific activation of protein kinase C-α (PKC-α) and subsequent PMNLs migration (Silva et al. 2004).

Enterovirulent *Escherichia coli*

Six classes of human enterovirulent *E. coli* have been defined, including enterotoxigenic *E. coli* (ETEC), enteropathogenic *E. coli* (EPEC), enterohemorrhagic, *E. coli* (EHEC), enteroaggregative *E. coli* (EAEC), enteroinvasive *E. coli* (EIEC), and diffusely adherent *E. coli* (DAEC) (Viswanathan et al. 2009; Kaper et al. 2004; Nataro and Kaper 1998). All these pathogenic *E. coli* interact with the brush border of mature enterocytes and remain extracellular, but only EIEC is capable of entering and replicating within enterocytes. Largely on the basis of *in vitro* experiments using appropriate cultured human intestinal cell lines representing models of human enterocytes or colonic cells, we can briefly summarize the situation as being that the diarrhea promoted by enterovirulent *E. coli* is the result of the fact that ETEC adheres to small bowel enterocytes and induce watery diarrhea by the secretion of heat-labile and/or heat-stable enterotoxins; EPEC also adheres to small bowel enterocytes and induces the characteristic attaching and effacing lesion resulting from the destruction of the microvillus, which in turn loses its associated intestinal functions; EHEC also induces the attaching and effacing lesion, but in the colon, and also produces Shiga toxin; EAEC adheres to the epithelia of the small and large bowel, forming a thick biofilm, and produce secretory enterotoxins and cytotoxins; EIEC invades the colonic epithelial cells, resides within intracellular vacuoles, lyses the phagosome, moves into the cell cytoplasm, and directs the cell-to-cell spread by penetrating into neighboring cells; DAEC through the interaction of adhesins with its specific brush border-associated receptors, elicits cascades of signal transductions that lead to the structural destruction of microvilli without forming an attaching and effacing lesion, which in turn results in the loss of intestinal function, and promotes inflammatory responses.

Enterotoxigenic *Escherichia coli* (ETEC)

ETEC strains are a principal cause of diarrhea in young children in developing countries, and also in travelers to these areas (Qadri et al. 2005). ETEC produces diarrheal disease by adherence to the brush border of mature enterocytes and the expression of a heat-labile (LT) and/or heat-stable (ST) enterotoxin (Isidean et al. 2011; Fleckenstein et al. 2010; Johnson and Nolan 2009).

ETEC strains attach to the microvilli via filamentous bacterial surface structures known as colonization factors (CFs); more than 20 different CFs have been described. A family of ETEC adhesive factors includes colonization factor antigens (CFAs), *coli* surface antigens (CSAs), and PCFO71. On the basis of the complete DNA sequences of the gene clusters encoding CFA/I, CS1, and CS2, on the fact that the four-gene bioassembly operons of CFA/I, CS1, and CS2 are similarly organized, encoding in order of a periplasmic chaperone, a major fimbrial subunit, an outer membrane protein (OMP), and a minor fimbrial subunit, and based on the primary sequence of the major fimbrial subunit, CFA/I and related fimbriae, the adhesive factors of ETEC have been grouped together as class 5 fimbriae (Anantha et al. 2004). ETEC strains expressing colonization factor antigen I (CFA/I), CFA/II, CFA/III; and antigen 2230 adhered to isolated human enterocytes, but not to the non-polarized or polarized, non-intestinal epithelial HeLa, HEp-2, human colorectal adenocarcinoma cell line (HRT 18), human duodenal adenocarcinoma (Hutu 80), Madin Darby Bovine Kidney (MDBK), MDCK, or Vero cells (Darfeuille-Michaud et al. 1990). In contrast, human ETEC strains harbour colonization factor antigen I (CFA/I) (Darfeuille-Michaud et al. 1990), CFA/II (Darfeuille-Michaud et al. 1990), CFA/III (Taniguchi et al. 2001; Darfeuille-Michaud et al. 1990), and antigen 2230 (Darfeuille-Michaud et al. 1990). Other ETEC adhesive factors conferring adhesion to fully-differentiated Caco-2 cells have been identified, including adhesive factor PCFO20 (Viboud et al. 1993); antigens 8786 (Aubel et al. 1991), CS4 (Anantha et al. 2004), CS6 (Tobias et al. 2008), CS14 (Anantha et al. 2004), CS17 (Anantha et al. 2004), CS19 (Grewal et al. 1997), CS20 (Valvatne et al. 1996), CS22 (Pichel et al. 2001; Pichel et al. 2000), CS23 (Del Canto et al. 2012), and CS31A (Jallat et al. 1994b); PCF071 (Anantha et al. 2004); the type IV long pilus encoded by *ingA* (Mazariego-Espinosa et al. 2010); high-molecular-weight glycosylated protein, EtpA (Fleckenstein et al. 2006), and the putative EtpB transporter (Fleckenstein et al. 2006). Adhesion of ETEC strains develops at the brush border without modifying the structure of the microvilli (Darfeuille-Michaud et al. 1990). The expression of ETEC CFAs receptors at the brush border of fully-differentiated HT-29 Glc$^{-/+}$ and Caco-2 cells develops as a function of the cell differentiation, and appears to be controlled by the presence of glucose during cell culture (Bernet-Camard et al. 1997; Kerneis et al. 1992).

To cause diarrhea, ETEC produces and delivers LT and/or ST enterotoxins, which elicit watery, cholera-like diarrhea. Fully-differentiated Caco-2 and T84 cells have been used to demonstrate that ETECs use their adhesive factors to deliver toxins efficiently to the brush border expressing the receptors for ETEC enterotoxins (Dorsey et al. 2006; Wimer-Mackin et al. 2001). LT enterotoxin, like *Vibrio cholera* enterotoxin, stimulates cAMP adenylatecyclase in fully-differentiated Caco-2 cells (Bowman and Clements 2001; Guidry et al. 1997; Cieplak et al. 1995; Grant et al. 1994; Orlandi et al. 1994). In contrast to cholera toxin, LT enterotoxin does not fractionate with the caveolae-like, detergent-insoluble membranes of fully-differentiated T84 cells (Wimer-Mackin et al. 2001; Wolf et al. 1998). The apical entry, but not the basolateral entry of LT enterotoxin into fully-differentiated T84 cells leads to the production and activation of a toxin A through the proteolytic action of a serine protease (Lencer et al. 1997). In fully-differentiated Caco-2 cells, both cholera toxin and LT enterotoxin inhibit the activity of the H^+/peptide cotransporter (Muller et al. 1996). Cholera toxin and LT enterotoxin transcriptionally downregulate two human antimicrobial peptides, cathelicidin hCAP-18/LL-37 and HBD-1, by activating several intracellular signaling pathways involving protein kinase A, Erk1/2 and Cox-2 downstream of cAMP accumulation and inducible cAMP early repressor (ICER) in fully-differentiated HT-29, Caco-2 and T84 cell lines (Chakraborty et al. 2008).

ST enterotoxin stimulates cGMP adenylate cyclase in fully-differentiated T84 cells (Forte et al. 1992; Lin et al. 1992; Huott et al. 1988). When ST enterotoxin binds to fully-differentiated Caco-2 cells, this results in activation of the receptor and is followed by the production of high intracellular levels of cGMP (Ghanekar et al. 2003; Vaandrager et al. 1992). The characteristics of the ST enterotoxin B (STb) binding to the cell membrane (Chao and Dreyfus 1997; Garrett and Visweswariah 1996; de Sauvage et al. 1992; Guarino et al. 1987), the characterization and partial purification of the receptor for the ST enterotoxin (Visweswariah et al. 1994), the regulation of the receptor (Bakre et al. 2000; Bakre and Visweswariah 1997; Crane and Shanks 1996; Crane et al. 1992; Visweswariah et al. 1992; Levine et al. 1991; Crane et al. 1990; Weikel et al. 1990; Crane et al. 1989) and the internalization of the receptor (Urbanski et al. 1995) have all been extensively investigated in fully-differentiated T84 cells. ST enterotoxin induces cGMP-chloride secretion in fully-differentiated Caco-2 cells (Buccigrossi et al. 2011; Albano et al. 2005; Gali et al. 2001; Parkinson et al. 1997). ST enterotoxin increases net secretory transepithelial volume flux, short-circuit current values, and net secretory chloride flux in fully-differentiated T84 cells (Toriano et al. 2001). Na^+-K^+-$2Cl^-$ cotransport and Cl^- secretion induced by ST enterotoxin has been found to depend on the organization of the F-actin cytoskeleton in fully-differentiated T84 cells (Matthews et al. 1993). Both the LT and STa enterotoxins activate the exocytosis of syntaxin 3, an intestine-specific soluble N-ethylmaleimide-sensitive factor attachment receptor, and the functional CFTR, which colocalize onto the apical domain of fully-differentiated Caco-2 BBe clone cells (Collaco et al. 2010).

Enteroinvasive *Escherichia coli* (EIEC)

EIEC strains are capable of causing diarrhea (Nataro and Kaper 1998). They are biochemically, genetically, and pathogenically closely related to *Shigella* spp. Like *Shigella,* EIEC infects the colonic mucosa. The pathogenesis of EIEC has not yet been fully elucidated. EIEC can be distinguished from *Shigella* by a few minor biochemical tests, but these pathotypes share several identical virulence factors (Johnson and Nolan 2009; Le Gall et al. 2005; Parsot 2005; Kaper et al. 2004; Nataro and Kaper 1998). The genes required for EIEC cell-entry are carried on a 140-MDa plasmid designated *pInv*. Prominent among these genes are the *mxi* and *spa* loci, which encode a T3SS that permits the secretion of multiple proteins, including the effectors of the invasion phenotype Ipa proteins such as IpaA, IpaB, IpaC, and IpaD. Escape of internalized AIEC from the endocytic vacuole into the cytoplasm is followed by an actin-dependent motility movement, which leads the bacteria to pass into neighboring cells.

EIEC growing in the cytosol of undifferentiated, confluent Caco-2 cells utilize glucose, but not glucose 6-phosphate, other phosphorylated carbohydrates, gluconate or fatty acids as major carbon substrates (Gotz et al. 2010; Gotz and Goebel 2010). Moreno et al. (Moreno et al. 2009) have pertinently compared the intracellular lifestyle of EIEC at the cellular and molecular levels within fully-differentiated Caco-2 cells to those of *S. flexneri*. In a plaque assay of confluent, undifferentiated Caco-2 cells, the levels of EIEC and *S. flexneri* were similar for time points 1, 2 and 3 h post-infection, but after 4 and 5 h post-infection fewer intracellular EIEC were found, suggesting that intracellular proliferation was lower than that of *S. flexneri*. EIEC displays significantly less efficient cell-to-cell passage in the plaque assay than *S. flexneri,* because the plaques formed by EIEC were four times smaller than those formed by *S. flexneri*. When examined at the molecular level, the expression of regulatory gene *virB*, and the invasion of *ipaA, ipaB, ipaC, ipaD* virulence genes by EIEC during the invasion of Caco-2 cells were lower than for *S. flexneri* infection. Similarly, the expression of the regulatory genes *virB* and *virF*, of the invasion *ipaA, ipaB, ipaC, ipaD* virulence genes, and of *iscA*, which is necessary for actin tail assembly, and *iscB* which is involved in escaping autophagy in *S. flexneri*, in EIEC during cell-to-cell spreading was lower than in *S. flexneri* infection. In contrast, the regulatory gene *virF* was the only gene to be more expressed by EIEC than by *S. flexneri*. In addition, it is interesting to note that in a model of kerato conjunctivitis developed in the guinea-pig, *S. flexneri* infection results in a pronounced and prolonged proinflammatory response, whereas EIEC induces a milder form of the disease (Moreno et al. 2009).

The existence of a particular group of EIEC strains has recently emerged from studies that have investigated mucosa-associated bacteria in patients with inflammatory bowel disease (IBD), including ulcerative colitis (UC) and Crohn's disease (CD). Darfeuille-Michaud et al. (1998) examining *E. coli* strains isolated from ileal specimens of CD patients with chronic lesions or early recurrent lesion were the first to report that several CD-associated strains displayed mannose-resistant adherence properties onto fully-differentiated Caco-2 cells and human undifferentiated Intestinal-407 (INT407) cells, and harbored any of the virulence factor-encoding genes of enterovirulent *E. coli* involved in acute enteric diseases, such as ETEC, EIEC, EHEC, EaggEC, and EPEC (Darfeuille-Michaud et al. 2004; Boudeau et al. 2001; Masseret et al. 2001; Darfeuille-Michaud et al. 1998). On the basis of the particular characteristics of CD-associated *E. coli* strains, it has been proposed that the CD-associated *E. coli* strains may constitute a new, potentially pathogenic group of *E. coli* designated AIEC, for adherent-invasive *E. coli* (Boudeau et al. 1999), which has: (1) the ability to adhere to fully-differentiated Caco-2 and/or undifferentiated INT407 cells with an adhesion index equal or superior to 1 bacterium per cell (Darfeuille-Michaud et al. 2004; Boudeau et al. 2001; Masseret et al. 2001; Darfeuille-Michaud et al. 1998), (2) the ability to invade human intestinal undifferentiated INT407 with an invasion index equal or superior to 0.1% of the original inoculums (Boudeau et al. 1999), (3) the involvement of host cell actin polymerization and microtubule recruitment in bacterial uptake (Boudeau et al. 1999), (4) the ability to survive and to replicate within macrophages (Bringer et al. 2005; Glasser et al. 2001), and (5) no known cell invasive determinants (Darfeuille-Michaud et al. 2004; Boudeau et al. 2001; Masseret et al. 2001; Darfeuille-Michaud et al. 1998). Following the isolation of the prototype AIEC LF-82 strain (Darfeuille-Michaud et al. 2004; Darfeuille-Michaud et al. 1998), other CD- and UC-associated AIEC strains have been isolated and characterized (Negroni et al. 2012; Schippa et al. 2012; Raso et al. 2011; Sepehri et al. 2011; Martinez-Medina et al. 2009a; Petersen et al. 2009; Sepehri et al. 2009; Eaves-Pyles et al. 2008). Analysis of predicted coding sequences (CDSs) within the LF82 flexible genome indicates the presence of four putative pathogenic islands carrying virulence-related genes, and a close relationship with the avian pathogenic strain APEC_01, meningitis-associated strain S88, and urinary-isolated strain UTI89 (Miquel et al. 2010). It has been hypothesized that consistent with the fact that most of the *E. coli* strains colonizing the intestinal mucosa in patients with inflammatory bowel disease belong to the B2 and D phylogroups, the LF82 genome has apparently evolved from those of extra intestinal pathogenic *E. coli* (ExPEC) B2 strains, by acquiring isolated or clustered genes from *Salmonella* and *Yersinia* or CDSs located on the pLF82 plasmid and at various loci on the chromosome. Multi-locus sequence typing and fimH sequence analysis suggest that AIEC isolated from IBD patients have not evolved from a single ancestral background, but instead constitute a group of bacteria that have been able to take advantage of an "IBD microenvironment", and probably shares common genes with ExPEC (Sepehri et al. 2009). ExPEC and AIEC strains share similar virulence gene sets and certain strains are phylogenetically related to the B2 phylogroup, but the majority of ExPEC strains do not behave like AIEC strains, confirming that AIEC possesses virulence-specific features that are principally detectable phenotypically (Martinez-Medina et al. 2009a; Martinez-Medina et al. 2009b). Analysis of the complete genome of AIEC strains NRG857c (O83:H1), UM146, and HM605, confirmed a phylogenetic linkage between AIEC and ExPEC which causes urinary tract infections and neonatal meningitis (Clarke et al. 2011; Krause et al. 2011; Nash et al. 2010). Interestingly, comparing the NRG857c AIEC genome with other pathogenic and commensal *E. coli* has made it possible to identify some distinctive genetic features of the AIEC pathotype, including 41 genomic islands, and unique genes that are found only in *E. coli* strains that exhibit the adherent and invasive phenotype (Nash et al. 2010). It has been noted that the HM605 genome is more divergent than the genomes of LF82 and NRG857c (Clarke et al. 2011). Analysis of the genomes of *E. coli* isolates obtained from patients with UC and CD (Vejborg et al. 2011), and of the genomes of AIEC strains LF82 (Miquel et al. 2010), UM146 (Krause et al. 2011), NRG857c (O83:H1) (Nash et al. 2010), and HM605 (Clarke et al. 2011), reveals that the genome of AIEC is characterized by the absence of known virulence factors composing the T3SS of enteroinvasive *Salmonella*, *Yersinia* and *Shigella*, and by the presence of a number of ExPEC-related virulence determinants such as the *pap*, *sfa*, *cdt*, *sat* and *hly* genes, and genes of ExPEC-associated genomic islands, but importantly, with little or no evidence of group-specific determinants. However, it impossible to exclude the possibility that AIEC expresses unidentified AIEC-specific genes that could in fact be IBD-specific. Indeed, it has been observed that the pathogenesis of many enterovirulent pathogens is often multifactorial, and involves the coordinated engagement of an assortment of diverse virulence factors.

AIEC strains isolated from ileal specimens of CD patients with chronic lesions, including strain LF82 and early recurrent lesions have the ability to adhere to fully-differentiated Caco-2 cells with an adhesion index of 3 to 14 bacteria per cell (Darfeuille-Michaud et al. 2004). LF82 adheres to fully-differentiated Caco-2 and T84 cells, and in Caco-2 cells a clear colocalization has been observed at the brush border between CEACAM6 and adherent bacteria (Barnich et al. 2007). The role of OmpCin the adhesion and cell-entry of LF82 has been demonstrated using fully-differentiated T84 cells (Rolhion et al. 2007). AIEC strains LF82 and O83:H1 displayed a similar ability to adhere and to invade fully-differentiated Caco-2BB2 and T84 cells (Eaves-Pyles et al. 2008). The flagellum plays a pivotal role in the adhesion and invasion of fully-differentiated Caco-2BB2 and T84 cells by AIEC O83:H1, since a non-flagellated *E. coli*, O83:H1, losses both its adhesiveness and its invasive properties (Eaves-Pyles et al. 2008). In fully-differentiated Caco-2 cells, LF82 infection promotes the apical over expression of endoplasmic reticulum (ER) -localized stress response chaperone Gp96, which acts as a receptor for LF82 OMVs and in turn plays a role in the internalization of LF82 bacteria (Rolhion et al. 2010). The transcriptional activation of the protein produced by the nucleotide-binding oligomerization domain containing 1 gene has been found in fully-differentiated Caco-2 cells infected with strain LF82 (Huebner et al. 2009). Using the *in vitro* model composed of fully-differentiated Caco-2-cl1 cells co-cultured with the Raji B cell line, Chassaing et al. (Chassaing et al. 2011) have observed that LF82 bacteria interact with M-like cells, and that a large number of LF82 bacteria translocate, a phenomenon that was enhanced when the bacteria are grown in the presence of sodium choleate (Chassaing et al. 2012).

Apical infection of fully-differentiated T84 cell monolayers with AIEC strain LF82 led to a reduction in TER, and an increased paracellular flux accompanied by a redistribution of the TJs-associated ZO-1 protein. Interestingly, the basolateral AIEC infection resulted in more severe disruption of the epithelial barrier (Wine et al. 2009). Moreover, a flagella-dependent AIEC O83:H1-induced basolateral secretion of IL-8 and CCL20 had been observed in fully-differentiated Caco-2BBe and T84 (Eaves-Pyles et al. 2008). In addition, both AIEC strains O83:H1 and LF82 stimulate the transepithelial migration of DC and PMNLs in fully-differentiated Caco-2BBe and T84 cell monolayers (Eaves-Pyles et al. 2008). In the human myeloid cell line, PLB-985, differentiated in fully mature neutrophils, strain LF82 induces the extension of the neutrophil extracellular traps (NETs) (Marin-Esteban et al. 2012).

Enteropathogenic *Escherichia coli* (EPEC)

EPEC is the first type of *E. coli* to have been shown to be associated with human disease (Chen and Frankel 2005; Kaper et al. 2004; Nataro and Kaper 1998). It is a frequent cause of infantile diarrhea in the developing world. The three different patterns of adhesion of diarrheagenic *E. coli* to epithelial cells are diffuse adherence (DA), in which the bacteria cover the whole surface of the cell, localized adherence (LA), in which attachment is limited to one or a few sites of the cell surface, and aggregative adherence (AA), in which adhering bacteria aggregate in an unorganized fashion (Nataro et al. 1987). tEPEC strains produce localized adherence, forming typical, well-formed and compact microcolonies (Scaletsky et al. 1984). Recently, two sub-classes of EPEC have been defined and designated typical EPEC (tEPEC) and atypical EPEC (aEPEC) (do Valle et al. 1997; Goncalves et al. 1997).

tEPEC

Knutton et al. (Knutton et al. 1989) were the first to describe the presence of well-formed and dense tEPEC microcolonies in fully-differentiated Caco-2 cells. LA of tEPEC is mediated by the bundle forming pilus (BFP), a type IV pilus encoded by pEAF, which interconnects bacteria within the dense microcolonies, in turn promoting their particular microcolony organization into the brush border of fully-differentiated Caco-2 cells (Tobe and Sasakawa 2002; Tobe and Sasakawa 2001). tEPEC microcolonies have been also observed on fully-differentiated T84 cells, and the formation of tEPEC microcolonies also increases, paralleling the establishment of the brush border during the cell differentiation of Caco-2 cells (Gabastou et al. 1995).

The main intestinal cell lesion produced by tEPEC and aEPEC strains is attaching and effacing (A/E) histopathology, characterized by the effacement of the brush border microvilli in enterocytes after intimate adherence of the bacterium to the epithelial cell membrane and signaling events (Garmendia et al. 2005). tEPEC and aEPEC both use a T3SS to translocate an assortment of effectors proteins into the colonized enterocytes. These proteins are encoded both within and outside the locus of enterocyte effacement (LEE) leading to the formation of A/E lesions. The LEE is organized into 5 operons (*LEE1* to *LEE5*), where *LEE1* to *LEE3* encode T3SS proteins and the LEE-encoded regulator (Ler), *LEE4* encodes the secreted proteins that form the external part of the T3SS used to translocate effector proteins to the host cell, and *LEE5* encodes the adhesin intimin and its translocated receptor Tir. The membrane-inserted Tir is phosphorylated at the C-terminal domain, and recruits the cellular protein Nck, where it activates the neural Wiskott-Aldrich syndrome protein (N/WASP), leading to Arp2/3 complex-mediated actin polymerization. Knutton et al. (Knutton et al. 1989) observed that in fully-differentiated Caco-2 cells, the presence of concentrated cellular F-actin at the sites of attachment accompanying A/E lesion of microvilli and after effacement of the microvillus EPEC bacteria are seen localized on a pedestal-like structure. tEPEC BFP, and the translocated T3SS effector and adhesin intimin have been identified as necessary for intimate adhesion to fully-differentiated Caco-2 cells to occur (Cleary et al. 2004; Phillips et al. 2000). The tEPEC T3SS effectors EspA, EspB, and EspD are required for the formation of the A/E lesions in fully-differentiated Caco-2 cells (Knappstein et al. 2004). EspB binds to myosin and blocks its interaction with actin in fully-differentiated Caco-2 cells (Iizumi et al. 2007). tEPEC recruits and subsequently sequesters cell-surface nucleolin around bacteria present in tEPEC microcolonies in fully-differentiated Caco-2 cells, a phenomenon unrelated to tEPEC-induced pedestal formation or microvillus effacement, but apparently related to the tEPEC-induced disruption of epithelial barrier function (Dean and Kenny 2011). tEPEC use the clathrin-coated pits components Eps15 and epsin1, but not adaptor protein 2 (AP-2), for clathrin-dependent pedestal formation in fully-differentiated Caco-2 cells (Lin et al. 2011).

As a consequence of the tEPEC-induced A/E lesions described above leading to the disappearance of the brush border microvilli, but also independently of the A/E lesion, to functional alterations develop in tEPEC-infected, fully-differentiated intestinal cells. tEPEC stimulates electrolyte transport in fully-differentiated Caco-2 cell monolayers through the action of T3SS translocated effectors EspA, EspB, and EspD (Collington et al. 1998b; Collington et al. 1998a). tEPEC affects the Na^+/H^+ exchanger (NHE) isoforms differently in fully-differentiated Caco-2 and T84 cells, increasing the activity of apical isoform NHE2, decreasing the activity of apical NHE3, and increasing the activity of the basolateral isoform NHE1 through a T3SS-dependent mechanism (Hecht et al. 2004). There is no change in the basolateral K^+ channel or $Na^+/K^+/2Cl^-$cotransport activity in tEPEC-infected, fully-differentiated T84 cells (Hecht and Koutsouris 1999). Independently of the A/E lesions, but requiring tEPEC adhesion, a signal transduction cascade involving Ca^{2+}, as well as PKC-α and PKC-€, is responsible for the increase in NHE2 activity in fully-differentiated Caco-2 cells (Hodges et al. 2006). tEPEC rapidly inactivates the Na^+/glucose cotransporter SGLT-1 in fully-differentiated Caco-2 cells by multiple mechanisms that are both dependent and independent of the A/E lesion (Dean et al. 2006). The T3SS effector Map induces brush border elongation in fully-differentiated Caco-2 cells, and proteolysis of Na^+/H^+ exchanger regulatory factor 1 (NHERF1) (Simpson et al. 2006). A T3SS-dependent inhibition of the short-chain fatty acid absorption mediated by apical monocarboxylate transporter 1 (MCT1) develops in tEPEC-infected, fully-differentiated Caco-2 cells (Borthakur et al. 2006). The tEPEC T3SS effectors EspG and EspG2 participate in inhibiting the apical Cl^-/OH^- exchange activity in fully-differentiated Caco-2 and T84 cells (Gill et al. 2007). A tEPEC T3SS-dependent inhibition of the uptake of the water-soluble vitamin B1 (thiamin) has been observed in fully-differentiated Caco-2 cells (Ashokkumar et al. 2009). The apical activity of the serotonin transporter playing a role in regulating the availability of serotonin in the gut is decreased following tEPEC infection of fully-differentiated Caco-2 cells without affecting its expression at the brush border (Esmaili et al. 2009).

Deletion of the *escN*, *espA*, *espB*, and *espD* genes encoding for elements of T3SS, and the gene encoding for BFP reveals that both early and intimate tEPEC attachment are needed for EPEC to have their inhibitory effect on the apical, sodium-dependent bile acid transporter in fully-differentiated Caco-2 cells (Annaba et al. 2012).

tEPEC causes a decrease in TER in the monolayers of fully-differentiated Caco-2 and T84 cells (Philpott et al. 1996; Spitz et al. 1995; Canil et al. 1993), without affecting the desmosomes (Guttman et al. 2007) or inducing host cell death (Flynn and Buret 2008; Viswanathan et al. 2008). tEPEC infection of fully-differentiated Caco-2 cells results in a decrease in the cell resting membrane potential (Stein et al. 1996). Activation of myosin light chain kinase (MLC20) leads to TER alteration in fully-differentiated T84 cells infected with tEPEC (Yuhan et al. 1997). Intimate attachment of tEPEC involving the binding of intimin to Tir is necessary for the alteration at the TJs to occur in fully-differentiated Caco-2 and T84 cells (Miyake et al. 2005; Czerucka et al. 2001). Intimate attachment of tEPEC onto fully-differentiated Caco-2 cells perturbs the fence function of TJs by delocalizing basolateral proteins, including β1-integrin and Na^+/K^+ ATPase, to the apical membrane (Muza-Moons et al. 2004). In fully-differentiated T84 cell monolayers, the tEPEC-induce structural alterations in TJs, rearrangements in TJ-associated ZO-1, in the occludin and claudin-1 proteins, and the loss of TER occur as a result of the action of the T3SS translocated EspF effector (Muza-Moons et al. 2004). Intimate EPEC attachment, and the T3SS translocated effectors EspF and Map (mitochondria-associated protein) are to be required for EPEC to disrupt the TJ integrity of fully-differentiated Caco-2 cells, and increase the paracellular movement of molecules (Dean and Kenny 2004; Viswanathan et al. 2004). The T3SS translocated effector protein, NleA is involved in TJ disruption during tEPEC infection of fully-differentiated Caco-2 cells through the inhibition of COPII-dependent host cell protein trafficking (Thanabalasuriar et al. 2010a; Thanabalasuriar et al. 2010b). In fully-differentiated Caco-2 cells, infection by tEPEC induces a decrease in TER, rearrangement in the distribution of TJ-associated ZO-1 and occludin, and the delocalization of occludin and flottilin-1, but not of claudin-1 and -3, from lipid rafts, thus allowing the cell-entry of bacteria into the infected cells (Li et al. 2008). tEPEC OMPs in fully-differentiated Caco-2 cell monolayers induces changes in adherent junctions, leading to the dissociation of the cadherin/β-catenin complex and the cytoplasmic redistribution of β-catenins through the activation of the PKC-α signaling (Malladi et al. 2004). The tEPEC T3SS EspG effector, as well as its homologue Orf3, plays a role in microtubule disruption in fully-differentiated Caco-2 cells, but EspG alone has no effect on TER, whereas EspG together with Orf3 decreases TER (Tomson et al. 2005). EPEC expressed two different types of EspG (consisting of EspG α and EspG β) and EspG2 that are responsible for microtubule disruption in fully-differentiated Caco-2 cells (Smollett et al. 2006). EspG1 and its homologue EspG2 both contribute to a loss of barrier function via an undefined mechanism that may be linked to the disruption of microtubule network (Glotfelty and Hecht 2012). Without affecting TJ organization, the transepithelial migration of PMNLs across fully-differentiated T84 cells (Michail et al. 2003) is promoted when the tEPEC T3SS effector NleE activates Erk1/2 MAPK and NF-κB (Zurawski et al. 2008). tEPEC EspF-induced caspase 3, 8 and 9 cleavage develops in fully-differentiated Caco-2 cells (Flynn and Buret 2008; Nougayrede et al. 2007).

EspC is a toxin belonging to the same protein family known as the Serine Protease Autotransporters of the *Enterobacteriaceae* (SPATE) toxins (Henderson and Nataro 2001) that are secreted via the type-V secretion pathway (Henderson et al. 2004). EspC toxin acts within the cytosol of epithelial cells disrupting the architecture of the F-actin cytoskeleton (Navarro-Garcia and Elias 2011; Navarro-Garcia 2010; Navarro-Garcia et al. 2010). A T3SS-dependent cytopathic effect of tEPEC in fully-differentiated Caco-2 cells is characterized by the recruitment of focal adhesion plates and the formation of stress fibers, irreversible cell cycle arrest at the G (2) /M transition, and sustained inhibitory phosphorylation of the mitosis inducer, CDK1, has been found to be triggered by a cycle inhibiting factor (Cif) belonging to a group of bacterial toxins and effector proteins known as the cyclomodulins (Taieb et al. 2006).

Transport of tEPEC occurs at similarly low levels across both native fully-differentiated Caco-2 and M-like Caco-2/Raji-cocultured monolayers (Martinez-Argudo et al. 2007). Since translocation rates are markedly higher for tEPEC that lack either functional T3SS or the effector protein EspF, it is conceivable that T3SS effectors may down-regulate tEPEC translocation.

aEPEC

It has been defined that the basic difference between tEPEC and aEPEC is the presence of the EPEC adherence factor plasmid (pEAF) in tEPEC, and its absence plus the lack of Shiga-toxin genes in aEPEC (Hernandes et al. 2009; Trabulsi et al. 2002). Localized adherence-like (LAL), DA, and AA patterns have been observed for aEPEC strains on fully-differentiated Caco-2 cells (Sampaio et al. 2011; Vieira et al. 2001). Despite promoting LAL, AA, or DA,aEPEC strains are able to induce A/E lesions of microvilli in fully-differentiated Caco-2 cells without F-actin condensation (Rocha et al. 2011).

Enterohemorrhagic *Escherichia coli* (EHEC)

EHEC is a non-invasive, gastrointestinal pathogen that causes hemorrhagic colitis, non-bloody diarrhea, acute intestinal inflammation, and hemolytic uremic syndrome (Kaper et al. 2004; Nataro and Kaper 1998). EHEC strains have evolved from EPEC strains by acquiring bacteriophages encoding Shiga-like toxins (Stxs), but there are clear differences between EPEC and EHEC pathogenesis (Spears et al. 2006; Wick et al. 2005). Strains of EHEC belonging to serogoup O157 are most commonly associated with severe human diseases. Like EPEC, EHEC strains promote the A/E histopathology that results in the LEE pathogenicity island that encodes a T3SS, and effector proteins that are homologous to those produced by EPEC (Garmendia et al. 2005). Effacement of the brush border microvilli of intestinal cells follows the intimate adherence of the bacterium to the epithelial cell membrane, the accumulation of polymerized actin beneath the adhering bacteria, and the effacement of microvilli so that the bacterium localizes onto a pedestal-like structure. A significant divergence exists in the structure of EHEC Tir effector proteins and those of EPEC, and between the resultant pathogenic mechanisms utilized for the EHEC-induced host cell actin polymerization (Spears et al. 2006).

EHEC forms dense and localized EHEC microcolonies at the brush border of fully-differentiated T84 and Caco-2 cells (Winsor et al. 1992; Knutton et al. 1989). OmpA and the long polar fimbriae of EHEC are two of the factors that allow EHEC to adhere to the brush border of fully-differentiated Caco-2 cells (Torres et al. 2004; Torres and Kaper 2003). EHEC encodes adherence-associated loci involved in the initial diffuse adherence, and the intimin-Tir interaction is required for the subsequent development of EHEC microcolonies at the brush border of fully-differentiated Caco-2 cells (Tatsuno et al. 2000). The role of LEE gene transcription for the adhesion of EHEC has been investigated using fully-differentiated Caco-2 cells (Nakanishi et al. 2006; Tobe et al. 2005; Abe et al. 2002; Tatsuno et al. 2000). The genes *yhiE, yhiF,* and *ToxB* act as negative or positive regulators of the expression of the genes required for T3SS-dependent adhesion of EHEC within fully-differentiated Caco-2 cells (Tatsuno et al. 2003; Tatsuno et al. 2001).

In fully-differentiated T84 cell monolayers, EHEC strains decrease TER and increase the transepithelial passage of fluids by inducing lower expression of ZO-1, occludin, and claudin-2, but not of claudin-1 or claudin-4 (Howe et al. 2005; Philpott et al. 1998), and without affecting the desmosomes (Guttman et al. 2007). In contrast to EPEC, for which EspF decreases TER, EHEC EspF fails to develop this activity in fully-differentiated Caco-2 cell monolayers (Viswanathan et al. 2004).

EHEC produce Stxs (verocytotoxin) (Melton-Celsa et al. 2012). Stxs are classical A_1B_5 toxins composed of two subunits, a pentameric array of binding subunits, and a single active subunit. There are two main families of Stx: Stx1 and Stx2. The Stx1 family consists of Stx1, Stx1c, and Stx1d, while the Stx2 group consists of Stx2, Stx2c, Stx2c2, Stx2d, Stx2e, and Stx2f. Stx causes inhibition of protein synthesis and apoptosis in fully-differentiated Caco-2, but not in T84 cells (Schuller et al. 2004). In fully-differentiated Caco-2 cells infected with EHEC, the previously observed EHEC-induced down-regulation of IFN-γ-induced iNOS mRNA expression and NO production (Vareille et al. 2008; Jandu et al. 2006), results of the action of Stx in inhibiting the IFN-γ-mediated proinflammatory pathway by decreasing Stat-1 tyrosine phosphorylation (Ho et al. 2012). EHEC Stx acts as a modulator of the innate immune response of human enterocytes, since in fully-differentiated T84 cells, Stx inhibits the NF-κB-dependent CCL20 and IL-8 gene transcription and chemokine production via a PI3K/Akt-dependent signaling pathway (Gobert et al. 2007). In fully-differentiated T84 cells, EHEC Stx1 causes galectin-3 depletion by increasing the apical expression and secretion of galectin-3,which in turn impairs trafficking and leads to apical protein mistargeting of several brush border structural proteins and transporters, including villin, DPP IV and sodium-proton exchanger 2 (Laiko et al. 2010; Hecht et al. 1999).

The LEE-negative EHEC strain O113:H21 is internalized within fully-differentiated Caco-2 cells and the intracellular bacteria are located within a membrane-bound vacuole, whereas in contrast, the EHEC strain O157:H7 remains extracellular and intimately attached to the epithelial cell surface (Luck et al. 2005). EHEC translocating at a low level into fully-differentiated Caco-2-cl1 monolayers shows a high level of translocation in the *in vitro* model composed of fully-differentiated Caco-2-cl1 cells co-cultured with the Raji B cell line (Etienne-Mesmin et al. 2011).

Enteroaggregative *Escherichia coli* (EAEC)

EAEC strains are emerging as significant diarrheal pathogens in various population groups, and are most commonly associated with pediatric diarrhea in developing countries (Estrada-Garcia and Navarro-Garcia 2012; Harrington et al. 2006; Kaper et al. 2004; Nataro and Kaper 1998). EAEC is also linked to diarrhea in adults, including AIDS patients and travelers, and has been a cause of food-borne outbreaks in the industrialized countries. EAEC pathogenesis starts with adherence to the epithelial cell lining in the terminal ileum and colon in an aggregative pattern in which bacteria adhere to each other in a 'stacked-brick' configuration (Nataro et al. 1987) involving several different hydrophobic aggregative adherence fimbriae, including aggregative adherence fimbriae I and II (AAF/I and AAF/II), which are related to the Afa/Dr family of adhesions (Harrington et al. 2006). Some strains of EAEC may then produce cytotoxins, including the enteroaggregative *E. coli* ST (EAST1), a partial homologue of the ETEC STa toxin and *Shigella* enterotoxin 1 (ShET1) (Harrington et al. 2006). Moreover, several EAEC strains produce toxins known as Pet (plasmid-encoded toxin) and Pic (protein involved in intestinal colonization), which belong to the subfamily of serine protease autotransporters of *Enterobacteriaceae* (SPATEs) toxins (Dutta et al. 2002; Henderson and Nataro 2001) that are secreted through the type-V secretion pathway (Henderson et al. 2004).

The cytotoxic and proinflammatory effects of EAEC strains have mainly been investigated using polarized, fully-differentiated, colon cancer T84 cells. EAEC strain 042, which adheres to the brush border of human, cultured, fully-differentiated T84 monolayers but not to fully-differentiated Caco-2 cells, promotes apical membrane of damaged cells that are characterized by the loss of microvilli and subnuclear vacuolization (Abe et al. 2001; Henderson et al. 1999; Nataro et al. 1996). Severely affected cells detached throughout the monolayer and display signs of cell death (Nataro et al. 1996). EAEC has been observed inside fully-differentiated T84 cells (Abe et al. 2001). EAEC infecting fully-differentiated T84 cell monolayers induce an AAF/II-dependent decrease in TER and delocalization of the TJ proteins claudin-1 and, to a lesser degree, occludin (Strauman et al. 2010). Exposing fully-differentiated T84 cell monolayers mounted in Ussing flux chambers to the purified EAST1 toxin results in an increase in the short circuit current (Veilleux et al. 2008). In contrast with the cytotoxic effects observed in fully-differentiated T84 cells infected with EAEC strains, no cytotoxic effect occurs in fully-differentiated Caco-2 cells, despite the fact that EAEC adhered to the brush border (Abe et al. 2001).

The Pet toxin secreted by EAEC belongs to the SPATE family of proteins (Dutta et al. 2002; Henderson and Nataro 2001). Pet disrupts the architecture of the F-actin cytoskeleton of epithelial cells as a result of toxin cleavage of the actin-binding protein fodrin (Navarro-Garcia and Elias 2011; Navarro-Garcia 2010; Navarro-Garcia et al. 2010). Pic (protein involved in colonization) toxin is a second SPATE protein produced by EAEC that does not damage epithelial cells, cleave fodrin, or degrade host defense proteins embedded in the mucus layer, but which does exert a mucinolytic activity, possibly to enable EAEC to penetrate the intestinal mucus layer during EAEC colonization (Navarro-Garcia and Elias 2011). The cytotoxic effect of EAEC, investigated in fully-differentiated T84 cells, is clearly independent of Pet (Henderson et al. 1999).

Diffusely-adhering *Escherichia coli* (DAEC)

Diffusely-adhering *E. coli* (DAEC) have been identified from their DA pattern on cultured epithelial cells (Nataro et al. 1987; Scaletsky et al. 1984). Two classes of DAEC strains have recently been defined: the typical

DAEC (tDAEC) and the atypical DAEC (aDAEC), each being subdivided into two sub-classes of strains (Servin 2005). The tDAEC include *E. coli* strains harboring AfaE-I, AfaE-II, AfaE-III, AfaE-V, Dr, Dr-II, F1845 or NFA-I adhesins and which: (1) have identical genetic organization, (2) have the ability to bind to human DAF and (3) the ability to promote DAF clustering. tDAEC includes two sub-classes of strains. The first sub-class includes the AfaE-III, Dr and F1845 adhesins that bind to both human DAF and CEACAMs (Afa/$Dr_{DAF/CEACAMs}$ tDAEC), and the second sub-class includes AfaE-I and Dr-II adhesins that do not bind to human CEACAMs (Afa/Dr_{DAF} tDAEC). The aDAEC class also includes two sub-classes of strains. The first sub-class includes *E. coli* strains harboring Afa/Dr adhesins or other adhesins which: (1) have an identical genetic organization and (2) do not bind onto human DAF. Currently, sub-class-1 of aDAEC includes strains that express AfaE-VII, AfaE-VIII, AAF-I, AAF-II and AAF-III adhesins. The second sub-class includes *E. coli* strains that harbor adhesins promoting diffuse adhesion with or without tEPEC pathogenicity islands. Currently, sub-class-2 of aDAEC includes AIDA-I-expressing *E. coli*, CF16K-expressing *E. coli*, CS31A-expressing *E. coli* (negative for genetic determinants of tEPEC), and ET5, 3431, B6, and 0181 *E. coli* strains (positive for genetic determinants of tEPEC).

Afa/Dr tDAEC

The Afa/$Dr_{DAF/CEACAMs}$ tDAEC and Afa/Dr_{DAF} tDAEC strains are generally related to the B2 phylogenetic group (Le Bouguenec and Servin 2006; Servin 2005). These *E. coli* strains have been found to be associated with urinary tract infections (UTIs) (pyelonephritis, cystitis, and asymptomatic bacteriuria) and with diarrhea in infants but not in adults. The structural assembly genes coding for Afa/$Dr_{DAF/CEACAMs}$ and Afa/Dr_{DAF} tDAEC adhesins are organized in a similar way, including operons of at least five genes. Genes *A* to *D*, which encode accessory proteins, are highly conserved in the different family members, whereas gene *E*, which encodes the major adhesin subunits AfaE-I, AfaE-II, AfaE-III, AfaE-V, DraE, and DaaE, is more divergent. These adhesins are assembled via the chaperone-usher pathway (Zav'yalov et al. 2010; Zavialov et al. 2007). In human cultured urinary epithelial cells and fully-differentiated intestinal Caco-2 and T84 cells, the Afa/$Dr_{DAF/CEACAMs}$ and Afa/Dr_{DAF} adhesins act as the major virulence factors of tDAEC by cell signaling through the recognition of membrane-bound receptors such as accelerating factor (DAF) (Nowicki et al. 1993; Nowicki et al. 1990; Nowicki et al. 1988) and CEACAMs (CEACAM1, CEA, and CEACAM6) (Berger et al. 2004; Guignot et al. 2000b). The major adhesin sub-units of tDAEC are AfaE-I, AfaE-V, DraE, and DaaE (Van Loy et al. 2002), and DraE expresses two separate adhesion domains the first recognizing DAF and the second recognizing the N-domain of CEACAMs (Korotkova et al. 2006; Anderson et al. 2004; Pettigrew et al. 2004).

tDAEC bearing the F1845 adhesin bind diffusely at the brush border of fully-differentiated Caco-2 and T84 cells recognizing a membrane-bound receptor that take place at the apical cell membrane during the enterocytic differentiation of the cells and paralleling the appearance of brush border hydrolase, SI (Bernet-Camard et al. 1996a; Kerneis et al. 1991). *E. coli* strains bearing Dr or Afa-I or C1845 adhesins adhere to both undifferentiated and fully-differentiated Caco-2 cells (Bernet-Camard et al. 1996a; Kerneis et al. 1994). tDAEC harboring Afa/$Dr_{DAF/CEACAMs}$ Dr or F1845 adhesins promote DAF and CEA receptor clustering around bacteria adhering to the brush border of fully-differentiated Caco-2 (Guignot et al. 2000b).

Bernet-Camard et al. (Bernet-Camard et al. 1996b) were the first to report that in fully-differentiated Caco-2 cells infected with a wild-type F1845-positive tDAEC, the recognition of DAF by an Afa/$Dr_{DAF/CEACAMs}$ adhesin is followed by the disassembly of the apical F-actin cytoskeleton, which in turn affects the structural organization of the brush border microvilli characterized by elongation and nucleation of the microvilli that detach from the cells to remain attached to the bacteria. At late times post-infection, the brush border disappeared from the infected cells. The disappearance of the brush border observed in tDAEC-infected, fully-differentiated Caco-2 cells resembles that observed in tEPEC- or aEPEC-infected, fully-differentiated Caco-2 cells but results from a different mechanism, since tDAEC does not express LEE. tDAEC expressing the Afa/$Dr_{DAF/CEACAMs}$ Dr-II adhesin promotes apical cytoskeleton rearrangements, and as a result of the presence of a functional hemolysin promoting cell apoptosis and cell lysis in fully-differentiated Caco-2 cells (Guignot et al. 2000a). Accompanying the Dr or F1845 adhesin-induced apical F-actin cytoskeleton disassembly in fully-differentiated Caco-2 cells, the distribution of brush border-associated functional proteins SI, DPPIV, glucose transporter SGLT1, and fructose transporter GLUT5 was dramatically altered (Peiffer et al. 2000b), and the brush border sucrase and DPPIV enzyme activities are decreased (Peiffer et al. 2001). In parallel, the infected, fully-differentiated Caco-2 cells infected with tDAEC expressing Dr or F1845 adhesin display a blockade of the biosynthesis of SI and DPP IV without affecting mRNAs levels or enzyme stability (Peiffer et al. 2001). Interestingly, when the cells were infected with recombinant *E. coli* strains expressing the Dr or F1845 adhesins no decrease in sucrase and DPP IV enzyme activities or inhibition of enzyme biosynthesis was observed, indicating that other pathogenic factor(s) distinct from the Afa/$Dr_{DAF/CEACAMs}$ adhesins may play crucial role(s) in this tDAEC mechanism of pathogenicity.

Infection of fully-differentiated Caco-2/TC7 cell monolayers by tDAEC expressing F1845 adhesin is followed by an increase in the paracellular permeability without any decrease in TER (Peiffer et al. 2000a). The distribution of TJ-associated occludin and ZO-1 proteins is markedly modified, whereas the distribution of the zonula adherens-associated E-cadherin is not modified. Guignot et al. have identified the secreted autotransporter toxin, Sat, belonging to the subfamily of SPATEs, as being the virulence factor of tDAEC that promotes TJ injuries in fully-differentiated Caco-2/TC7 cell monolayers (Guignot et al. 2007). Interestingly, the authors indicated that the *sat* gene is very prevalent in UTI-associated tDAEC strains (88% positive), generally absent from tDAEC strains collected from the stools of children without diarrhea (16% positive), and present in about half of the tDAEC strains collected from the stools of children with diarrhea (46% positive).

After recognizing the brush border DAF, tDAEC expressing Dr or F1845 adhesins promote PMNL epithelial transmigration across fully-differentiated T84 monolayers (Betis et al. 2003a). The F1845-induced transepithelial migration of PMNLs in turn triggers the synthesis of tumor necrosis factor alpha and IL-1β, which promote up-expression of the brush border-associated DAF, thus increasing the brush border adhesion of tDAEC (Betis et al. 2003b).

PMNLs that have transmigrated across fully-differentiated T84 monolayers display elevated global caspase activity, indicating apoptosis and a decreased capacity to phagocyte tDAEC (Brest et al. 2004). The neutrophil response to the wild-type strain expressing F1845 adhesin has also been examined using freshly isolated PMNLs and the human myeloid cell line PLB-985, which differentiates into fully mature neutrophils. Infection promotes the release of NETs, composed of a nuclear DNA backbone associated with antimicrobial peptides, histones, and proteases, which entrap and kill F1845-positive bacteria (Marin-Esteban et al. 2012). In a co-culture model of fully-differentiated Caco-2/TC7 cells exposed to PLB-985 cells previously infected with F1845-positive bacteria, the apical F-actin cytoskeleton of Caco-2/TC7 cells is dramatically damaged suggesting that activated NETs may damage the intestinal epithelium. tDAEC bearing F1845 adhesin induce the production of bioactive vascular endothelial growth factor (VEGF) in fully-differentiated T84 cells through the recognition of DAF and subsequent Src protein kinase activation upstream of the activation of the Erk1/2 and Akt signaling pathways (Cane et al. 2007). tDAEC expressing F1845 adhesin induce hypoxia-induced factor 1-α (HIF-

1α), thus triggering the production of IL-8 and VEGF in fully-differentiated T84 cells. Concomitantly, infected, fully-differentiated T84 cells show a loss of E-cadherin and cytokeratin 18 and an increase in fibronectin, indicating that bacteria may induce an epithelium to mesenchyme transition-like phenotype (Cane et al. 2010) and suggesting that tDAEC could play a role in angiogenesis, inflammation and some aspects of intestinal cancer initiation (Hofman and Vouret-Craviari 2012). Following adhesion of tDAEC expressing F1845 adhesin at the brush border of fully-differentiated Caco-2 cells triggers an antibacterial cell response, since at a late time after adhesion, dramatic alterations of the adhering bacteria occur indicating bacterial lysis, a phenomenon that correlates with the apical presence of antimicrobial peptides (Bernet-Camard et al. 1996c).

Dr- and F1845-positive tDAEC apically enter at a low level fully-differentiated Caco-2 cells (Guignot et al. 2001). The apical entry of Dr-positive bacteria is greatly enhanced after the TJs open. When they infect the basal domain of fully-differentiated Caco-2/TC7 cells, a highly significantly greater level of internalized Dr-positive bacteria was found than during apical infection. Consistent with this, it was also observed that apical cell-entry decreases dramatically after confluence has triggered their differentiation process (Guignot et al. 2001). Lipid rafts play a role in the cell-entry of Dr- and F1845-positive tDAEC into Caco-2/TC7 cells since the ganglioside GM1 and VIP21/caveolin are recruited around adhering bacteria and extracting membrane cholesterol with methyl-beta-cyclodextrin inhibits the bacterial cell-entry (Kansau et al. 2004; Guignot et al. 2001). Cell-entry of tDAEC is triggered by the major tDAEC adhesin sub-units DraE and AfaE-III (Guignot et al. 2009; Korotkova et al. 2008).

aDAEC

The diarrhea-causing aDAEC strain 2787 (O126:H27) expresses the AIDA-I adhesin and displays diffuse cellular adhesion to non-intestinal and non-polarized epithelial Hep-2 and HeLa cells (Benz and Schmidt 1989). AIDA-I belongs to a subgroup of auto transporter proteins including the TibA adhesin/invasin associated with some ETEC strains, and the Ag43 auto aggregation factor found in the majority of *E. coli* strains (Klemm et al. 2006). Despite its EPEC serotype, strain 2787 is not positive for the probe of the *eae* gene coding for EPEC intimin, does not secrete Esp proteins, and is not able to induce A/E lesions (Beinke et al. 1998). AIDA-I adhesin recognizes a 119-kDa non-GPI membrane-bound glycoprotein, and induces ligand-receptor clustering in non-intestinal and non-polarized epithelial cells (Laarmann and Schmidt 2003). It has been noted that despite the intestinal origin of the diarrheic AIDA-I-positive strain 2787, no information is available about the interaction, and cell responses in fully-differentiated Caco-2 and T84 cells.

An adhesin known as CF16K that triggers DA in fully-differentiated Caco-2 cells has been found in 9.8% of DAEC strains isolated in France (Jallat et al. 1994a). The CS31A (*coli* surface-associated antigen) adhesin expressed by human diarrheic *E. coli* allows DA to occur in fully-differentiated Caco-2 cells (Jallat et al. 1994b). It is interesting to note that an association of CS31A and another adhesive factor of the Afa/Dr family has been found in 70% of the diarrheic *E. coli* isolates examined.

Acknowledgements

Our sincere apologies to colleagues whose work has not been cited in this manuscript because of space constraints and because of the focus of the review, which does not purport to be exhaustive.

Competing interest's statement

The authors declare no competing financial interests.

References Cited

Abe, C.M., S. Knutton, M.Z. Pedroso, E. Freymuller and T.A. Gomes. 2001. An enteroaggregative *Escherichia coli* strain of serotype O111:H12 damages and invades cultured T84 cells and human colonic mucosa. FEMS Microbiol. Lett. 203: 199–205.

Abe, H., I. Tatsuno, T. Tobe, A. Okutani and C. Sasakawa. 2002. Bicarbonate ion stimulates the expression of locus of enterocyte effacement-encoded genes in enterohemorrhagic *Escherichia coli* O157:H7. Infect. Immun. 70: 3500–3509.

Agbor, T.A. and B.A. McCormick. 2011. *Salmonella* effectors: important players modulating host cell function during infection. Cell Microbiol. 13: 1858–1869.

Agbor, T.A., Z.C. Demma, K.L. Mumy, J.D. Bien and B.A. McCormick. 2011. The ERM protein, ezrin, regulates neutrophil transmigration by modulating the apical localization of MRP2 in response to the SipA effector protein during *Salmonella Typhimurium* infection. Cell Microbiol. 13: 2007–2021.

Albano, F., G. de Marco, R.B. Canani, P. Cirillo, V. Buccigrossi, R.A. Giannella and A. Guarino. 2005. Guanylin and *E. coli* heat-stable enterotoxin induce chloride secretion through direct interaction with basolateral compartment of rat and human colonic cells. Pediatr. Res. 58: 159–163.

Ammendola, S., M. Ajello, P. Pasquali, J.S. Kroll, P.R. Langford, G. Rotilio, P. Valenti and A. Battistoni. 2005. Differential contribution of sodC1 and sodC2 to intracellular survival and pathogenicity of *Salmonella enterica* serovar Choleraesuis. Microbes Infect. 7: 698–707.

Anantha, R.P., A.L. McVeigh, L.H. Lee, M.K. Agnew, F.J. Cassels, D.A. Scott, T.S. Whittam and S.J. Savarino. 2004. Evolutionary and functional relationships of colonization factor antigen i and other class 5 adhesive fimbriae of enterotoxigenic *Escherichia coli*. Infect. Immun. 72: 7190–7201.

Anderson, K.L., J. Billington, D. Pettigrew, E. Cota, P. Simpson, P. Roversi, H.A. Chen, P. Urvil, L. du Merle, P.N. Barlow, M.E. Medof, R.A. Smith, B. Nowicki, C. Le Bouguenec, S.M.

Lea and S. Matthews. 2004. An atomic resolution model for assembly, architecture, and function of the Dr adhesins. Mol. Cell. 15: 647–657.

Annaba, F., Z. Sarwar, R.K. Gill, A. Ghosh, S. Saksena, A. Borthakur, G.A. Hecht, P.K. Dudeja and W.A. Alrefai. 2012. Enteropathogenic *Escherichia coli* inhibits ileal sodium-dependent bile acid transporter ASBT. Am. J. Physiol. Gastrointest. Liver Physiol. 302: G1216–1222.

Ashgar, S.S., N.J. Oldfield, K.G. Wooldridge, M.A. Jones, G.J. Irving, D.P. Turner and D.A. Ala'Aldeen. 2007. CapA, an autotransporter protein of *Campylobacter jejuni*, mediates association with human epithelial cells and colonization of the chicken gut. J. Bacteriol. 189: 1856–1865.

Ashokkumar, B., J.S. Kumar, G.A. Hecht and H.M. Said. 2009. Enteropathogenic *Escherichia coli* inhibits intestinal vitamin B1 (thiamin) uptake: studies with human-derived intestinal epithelial Caco-2 cells. Am. J. Physiol. Gastrointest. Liver Physiol. 297: G825–833.

Aubel, D., A. Darfeuille-Michaud and B. Joly. 1991. New adhesive factor (antigen 8786) on a human enterotoxigenic *Escherichia coli* O117:H4 strain isolated in Africa. Infect. Immun. 59: 1290–1299.

Bakre, M.M. and S.S. Visweswariah. 1997. Dual regulation of heat-stable enterotoxin-mediated cGMP accumulation in T84 cells by receptor desensitization and increased phosphodiesterase activity. FEBS Lett. 408: 345–349.

Bakre, M.M., S. Sopory and S.S. Visweswariah. 2000. Expression and regulation of the cGMP-binding, cGMP-specific phosphodiesterase (PDE5) in human colonic epithelial cells: role in the induction of cellular refractoriness to the heat-stable enterotoxin peptide. J. Cell Biochem. 77: 159–167.

Barnich, N., F.A. Carvalho, A.L. Glasser, C. Darcha, P. Jantscheff, M. Allez, H. Peeters, G. Bommelaer, P. Desreumaux, J.F. Colombel and A. Darfeuille-Michaud. 2007. CEACAM6 acts as a receptor for adherent-invasive *E. coli*, supporting ileal mucosa colonization in Crohn disease. J. Clin. Invest. 117: 1566–1574.

Beinke, C., S. Laarmann, C. Wachter, H. Karch, L. Greune and M.A. Schmidt. 1998. Diffusely adhering *Escherichia coli* strains induce attaching and effacing phenotypes and secrete homologs of Esp proteins. Infect. Immun. 66: 528–539.

Benz, I. and M.A. Schmidt. 1989. Cloning and expression of an adhesin (AIDA-I) involved in diffuse adherence of enteropathogenic *Escherichia coli*. Infect. Immun. 57: 1506–1511.

Berger, C.N., O. Billker, T.F. Meyer, A.L. Servin and I. Kansau. 2004. Differential recognition of members of the carcinoembryonic antigen family by Afa/Dr adhesins of diffusely adhering *Escherichia coli* (Afa/Dr DAEC). Mol. Microbiol. 52: 963–983.

Bernet-Camard, M.F., M.H. Coconnier, S. Hudault and A.L. Servin. 1996a. Differential expression of complement proteins and regulatory decay accelerating factor in relation to differentiation of cultured human colon adenocarcinoma cell lines. Gut. 38: 248–253.

Bernet-Camard, M.F., M.H. Coconnier, S. Hudault and A.L. Servin. 1996b. Pathogenicity of the diffusely adhering strain *Escherichia coli* C1845: F1845 adhesin-decay accelerating factor interaction, brush border microvillus injury, and actin disassembly in cultured human intestinal epithelial cells. Infect. Immun. 64: 1918–1928.

Bernet-Camard, M.F., M.H. Coconnier, S. Hudault and A.L. Servin. 1996c. Differentiation-associated antimicrobial functions in human colon adenocarcinoma cell lines. Exp. Cell Res. 226: 80–89.

Bernet-Camard, M.F., F. Duigou, S. Kerneis, M.H. Coconnier and A.L. Servin. 1997. Glucose up-regulates expression of the differentiation-associated brush border binding site for enterotoxigenic *Escherichia coli* colonization factor antigen I in cultured human enterocyte-like cells. Infect. Immun. 65: 1299–1306.

Betis, F., P. Brest, V. Hofman, J. Guignot, M.F. Bernet-Camard, B. Rossi, A. Servin and P. Hofman. 2003a. The Afa/Dr adhesins of diffusely adhering *Escherichia coli* stimulate interleukin-8 secretion, activate mitogen-activated protein kinases, and promote polymorphonuclear transepithelial migration in T84 polarized epithelial cells. Infect. Immun. 71: 1068–1074.

Betis, F., P. Brest, V. Hofman, J. Guignot, I. Kansau, B. Rossi, A. Servin and P. Hofman. 2003b. Afa/Dr diffusely adhering *Escherichia coli* infection in T84 cell monolayers induces increased neutrophil transepithelial migration, which in turn promotes cytokine-dependent upregulation of decay-accelerating factor (CD55), the receptor for Afa/Dr adhesins. Infect. Immun. 71: 1774–1783.

Betts, J. and B.B. Finlay. 1992. Identification of *Salmonella typhimurium* invasiveness loci. Can. J. Microbiol. 38: 852–857.

Borthakur, A., R.K. Gill, K. Hodges, K. Ramaswamy, G. Hecht and P.K. Dudeja. 2006. Enteropathogenic *Escherichia coli* inhibits butyrate uptake in Caco-2 cells by altering the apical membrane MCT1 level. Am. J. Physiol. Gastrointest. Liver Physiol. 290: G30–35.

Boudeau, J., N. Barnich and A. Darfeuille-Michaud. 2001. Type 1 pili-mediated adherence of *Escherichia coli* strain LF82 isolated from Crohn's disease is involved in bacterial invasion of intestinal epithelial cells. Mol. Microbiol. 39: 1272–1284.

Boudeau, J., A.L. Glasser, E. Masseret, B. Joly and A. Darfeuille-Michaud. 1999. Invasive ability of an *Escherichia coli* strain isolated from the ileal mucosa of a patient with Crohn's disease. Infect. Immun. 67: 4499–4509.

Bowman, C.C. and J.D. Clements. 2001. Differential biological and adjuvant activities of cholera toxin and *Escherichia coli* heat-labile enterotoxin hybrids. Infect. Immun. 69: 1528–1535.

Boyle, E.C., N.F. Brown and B.B. Finlay. 2006. *Salmonella enterica* serovar Typhimurium effectors SopB, SopE, SopE2 and SipA disrupt tight junction structure and function. Cell Microbiol. 8: 1946–1957.

Bras, A.M. and J.M. Ketley. 1999. Transcellular translocation of *Campylobacter jejuni* across human polarised epithelial monolayers. FEMS Microbiol. Lett. 179: 209–215.

Brest, P., F. Betis, N. Cuburu, E. Selva, M. Herrant, A. Servin, P. Auberger and P. Hofman. 2004. Increased rate of apoptosis and diminished phagocytic ability of human neutrophils infected with Afa/Dr diffusely adhering *Escherichia coli* strains. Infect. Immun. 72: 5741–5749.

Bringer, M.A., N. Barnich, A.L. Glasser, O. Bardot and A. Darfeuille-Michaud. 2005. HtrA stress protein is involved in intramacrophagic replication of adherent and invasive *Escherichia coli* strain LF82 isolated from a patient with Crohn's disease. Infect. Immun. 73: 712–721.

Buccigrossi, V., C. Armellino, E. Ruberto, M.V. Barone, G.D. Marco, C. Esposito and A. Guarino. 2011. Polar effects on ion transport and cell proliferation induced by GC-C ligands in intestinal epithelial cells. Pediatr. Res. 69: 17–22.

Cane, G., V.L. Moal, G. Pages, A.L. Servin, P. Hofman and V. Vouret-Craviari. 2007. Up-regulation of intestinal vascular endothelial growth factor by Afa/Dr diffusely adhering *Escherichia coli*. PLoS One. 2: e1359.

Cane, G., A. Ginouves, S. Marchetti, R. Busca, J. Pouyssegur, E. Berra, P. Hofman and V. Vouret-Craviari. 2010. HIF-1alpha mediates the induction of IL-8 and VEGF expression on infection with Afa/Dr diffusely adhering *E. coli* and promotes EMT-like behaviour. Cell Microbiol. 12: 640–653.

Canil, C., I. Rosenshine, S. Ruschkowski, M.S. Donnenberg, J.B. Kaper and B.B. Finlay. 1993. Enteropathogenic *Escherichia coli* decreases the transepithelial electrical resistance of polarized epithelial monolayers. Infect. Immun. 61: 2755–2762.

Ceponis, 2003 #1960 [mssing; no bilio details given]

Chakraborty, K., S. Ghosh, H. Koley, A.K. Mukhopadhyay, T. Ramamurthy, D.R. Saha, D. Mukhopadhyay, S. Roychowdhury, T. Hamabata, Y. Takeda and S. Das. 2008. Bacterial exotoxins downregulate cathelicidin (hCAP-18/LL-37) and human beta-defensin 1 (HBD-1) expression in the intestinal epithelial cells. Cell Microbiol. 10: 2520–2537.

Chantret, I., A. Rodolosse, A. Barbat, E. Dussaulx, E. Brot-Laroche, A. Zweibaum and M. Rousset. 1994. Differential expression of sucrase-isomaltase in clones isolated from early and late passages of the cell line Caco-2: evidence for glucose-dependent negative regulation. J. Cell Sci. 107(Pt 1): 21–225.

Chao, K.L. and L.A. Dreyfus. 1997. Interaction of *Escherichia coli* heat-stable enterotoxin B with cultured human intestinal epithelial cells. Infect. Immun. 65: 3209–3217.

Charbonneau, M.E. and M. Mourez. 2007. Functional organization of the autotransporter adhesin involved in diffuse adherence. J. Bacteriol. 189: 9020–9029.

Chassaing, B., L. Etienne-Mesmin, R. Bonnet and A. Darfeuille-Michaud. 2012. Bile salts induce long polar fimbriae expression favouring Crohn's disease-associated adherent-invasive *Escherichia coli* interaction with Peyer's patches. Environ. Microbiol.

Chassaing, B., N. Rolhion, A. de Vallee, S.Y. Salim, M. Prorok-Hamon, C. Neut, B.J. Campbell, J.D. Soderholm, J.P. Hugot, J.F. Colombel and A. Darfeuille-Michaud. 2011. Crohn disease—associated adherent-invasive *E. coli* bacteria target mouse and human Peyer's patches via long polar fimbriae. J. Clin. Invest. 121: 966–975.

Chen, H.D. and G. Frankel. 2005. Enteropathogenic *Escherichia coli*: unravelling pathogenesis. FEMS Microbiol. Rev. 29: 83–98.

Chen, M.L., Z. Ge, J.G. Fox and D.B. Schauer. 2006. Disruption of tight junctions and induction of proinflammatory cytokine responses in colonic epithelial cells by *Campylobacter jejuni*. Infect. Immun. 74: 6581–6589.

Chessa, D., M.G. Winter, M. Jakomin and A.J. Baumler. 2009. *Salmonella enterica* serotype Typhimurium Std fimbriae bind terminal alpha(1,2)fucose residues in the cecal mucosa. Mol. Microbiol. 71: 864–875.

Chessa, D., M.G. Winter, S.P. Nuccio, C. Tukel and A.J. Baumler. 2008. RosE represses Std fimbrial expression in *Salmonella enterica* serotype Typhimurium. Mol. Microbiol. 68: 573–587.

Chevance, F.F. and K.T. Hughes. 2008. Coordinating assembly of a bacterial macromolecular machine. Nat. Rev. Microbiol. 6: 455–465.

Cieplak, W., Jr., R.J. Messer, M.E. Konkel and C.C. Grant. 1995. Role of a potential endoplasmic reticulum retention sequence (RDEL) and the Golgi complex in the cytotonic activity of *Escherichia coli* heat-labile enterotoxin. Mol. Microbiol. 16: 789–800.

Clarke, D.J., R.R. Chaudhuri, H.M. Martin, B.J. Campbell, J.M. Rhodes, C. Constantinidou, M.J. Pallen, N.J. Loman, A.F. Cunningham, D.F. Browning and I.R. Henderson. 2011. Complete genome sequence of the Crohn's disease-associated adherent-invasive *Escherichia coli* strain HM605. J. Bacteriol. 193: 4540.

Cleary, J., L.C. Lai, R.K. Shaw, A. Straatman-Iwanowska, M.S. Donnenberg, G. Frankel and S. Knutton. 2004. Enteropathogenic *Escherichia coli* (EPEC) adhesion to intestinal epithelial cells: role of bundle-forming pili (BFP), EspA filaments and intimin. Microbiology. 150: 527–538.

Coburn, B., G.A. Grassl and B.B. Finlay. 2007. *Salmonella*, the host and disease: a brief review. Immunol. Cell Biol. 85: 112–118.

Cogan, T.A., A.O. Thomas, L.E. Rees, A.H. Taylor, M.A. Jepson, P.H. Williams, J. Ketley and T.J. Humphrey. 2007. Norepinephrine increases the pathogenic potential of *Campylobacter jejuni*. Gut. 56: 1060–1065.

Collaco, A., J. Marathe, H. Kohnke, D. Kravstov and N. Ameen. 2010. Syntaxin 3 is necessary for cAMP- and cGMP-regulated exocytosis of CFTR: implications for enterotoxigenic diarrhea. Am. J. Physiol. Cell Physiol. 299: C1450–1460.

Collington, G.K., I.W. Booth and S. Knutton. 1998a. Rapid modulation of electrolyte transport in Caco-2 cell monolayers by enteropathogenic *Escherichia coli* (EPEC) infection. Gut. 42: 200–207.

Collington, G.K., I.W. Booth, M.S. Donnenberg, J.B. Kaper and S. Knutton. 1998b. Enteropathogenic *Escherichia coli* virulence genes encoding secreted signalling proteins are essential for modulation of Caco-2 cell electrolyte transport. Infect. Immun. 66: 6049–6053.

Costa de Beauregard, M.A., E. Pringault, S. Robine and D. Louvard. 1995. Suppression of villin expression by antisense RNA impairs brush border assembly in polarized epithelial intestinal cells. EMBO J. 14: 409–421.

Crane, J.K. and K.L. Shanks. 1996. Phosphorylation and activation of the intestinal guanylyl cyclase receptor for *Escherichia coli* heat-stable toxin by protein kinase C. Mol. Cell Biochem. 165: 111–120.

Crane, J.K., E.L. Hewlett and C.S. Weikel. 1989. Failure of pertussis toxin to inhibit activation of guanylate cyclase by the heat-stable enterotoxin of *Escherichia coli* (STa) in the T84 cell line. Infect. Immun. 57: 1186–1191.

Crane, J.K., L.L. Burrell, C.S. Weikel and R.L. Guerrant. 1990. Carbachol mimics phorbol esters in its ability to enhance cyclic GMP production by STa, the heat-stable toxin of *Escherichia coli*. FEBS Lett. 274: 199–202.

Crane, J.K., M.S. Wehner, E.J. Bolen, J.J. Sando, J. Linden, R.L. Guerrant and C.L. Sears. 1992. Regulation of intestinal guanylate cyclase by the heat-stable enterotoxin of *Escherichia coli* (STa) and protein kinase C. Infect. Immun. 60: 5004–5012.

Czerucka, D., S. Dahan, B. Mograbi, B. Rossi and P. Rampal. 2001. Implication of mitogen-activated protein kinases in T84 cell responses to enteropathogenic *Escherichia coli* infection. Infect. Immun. 69: 1298–1305.

Darfeuille-Michaud, A., D. Aubel, G. Chauviere, C. Rich, M. Bourges, A. Servin and B. Joly. 1990. Adhesion of enterotoxigenic *Escherichia coli* to the human colon carcinoma cell line Caco-2 in culture. Infect. Immun. 58: 893–902.

Darfeuille-Michaud, A., C. Neut, N. Barnich, E. Lederman, P. Di Martino, P. Desreumaux, L. Gambiez, B. Joly, A. Cortot and J.F. Colombel. 1998. Presence of adherent *Escherichia coli* strains in ileal mucosa of patients with Crohn's disease. Gastroenterology. 115: 1405–1413.

Darfeuille-Michaud, A., J. Boudeau, P. Bulois, C. Neut, A.L. Glasser, N. Barnich, M.A. Bringer, A. Swidsinski, L. Beaugerie and J.F. Colombel. 2004. High prevalence of adherent-invasive *Escherichia coli* associated with ileal mucosa in Crohn's disease. Gastroenterology. 127: 412–421.

Day, C.J., J. Tiralongo, R.D. Hartnell, C.A. Logue, J.C. Wilson, M. von Itzstein and V. Korolik. 2009. Differential carbohydrate recognition by *Campylobacter jejuni* strain 11168: influences of temperature and growth conditions. PLoS One. 4: e4927.

de Sauvage, F.J., R. Horuk, G. Bennett, C. Quan, J.P. Burnier and D.V. Goeddel. 1992. Characterization of the recombinant human receptor for *Escherichia coli* heat-stable enterotoxin. J. Biol. Chem. 267: 6479–6482.

Dean, P. and B. Kenny. 2004. Intestinal barrier dysfunction by enteropathogenic *Escherichia coli* is mediated by two effector molecules and a bacterial surface protein. Mol. Microbiol. 54: 665–675.

Dean, P. and B. Kenny. 2011. Cell-surface nucleolin is sequestered into EPEC microcolonies and may play a role during infection. Microbiology. 157: 1761–1767.

Dean, P., M. Maresca, S. Schuller, A.D. Phillips and B. Kenny. 2006. Potent diarrheagenic mechanism mediated by the cooperative action of three enteropathogenic *Escherichia coli*-injected effector proteins. Proc. Natl. Acad. Sci. USA. 103: 1876–1881.

Del Canto, F., D.J. Botkin, P. Valenzuela, V. Popov, F. Ruiz-Perez, J.P. Nataro, M.M. Levine, O.C. Stine, M. Pop, A.G. Torres and R. Vidal. 2012. Identification of coli surface antigen 23, a novel adhesin of enterotoxigenic *Escherichia coli*. Infect. Immun. 80: 2791–2801.

Dharmsathaphorn, K., J.A. McRoberts, K.G. Mandel, L.D. Tisdale and H. Masui. 1984. A human colonic tumor cell line that maintains vectorial electrolyte transport. Am. J. Physiol. 246: G204–208.

Dibb-Fuller, M.P., E. Allen-Vercoe, C.J. Thorns and M.J. Woodward. 1999. Fimbriae- and flagella-mediated association with and invasion of cultured epithelial cells by *Salmonella enteritidis*. Microbiology. 145(Pt 5): 1023–1031.

do Valle, G.R., T.A. Gomes, K. Irino and L.R. Trabulsi. 1997. The traditional enteropathogenic *Escherichia coli* (EPEC) serogroup O125 comprises serotypes which are mainly associated with the category of enteroaggregative *E. coli*. FEMS Microbiol. Lett. 152: 95–100.

Dorsey, F.C., J.F. Fischer and J.M. Fleckenstein. 2006. Directed delivery of heat-labile enterotoxin by enterotoxigenic *Escherichia coli*. Cell Microbiol. 8: 1516–1527.

Dutta, P.R., R. Cappello, F. Navarro-Garcia and J.P. Nataro. 2002. Functional comparison of serine protease autotransporters of enterobacteriaceae. Infect. Immun. 70: 7105–7113.

Eaves-Pyles, T., C.A. Allen, J. Taormina, A. Swidsinski, C.B. Tutt, G.E. Jezek, M. Islas-Islas and A.G. Torres. 2008. *Escherichia coli* isolated from a Crohn's disease patient adheres, invades, and induces inflammatory responses in polarized intestinal epithelial cells. Int. J. Med. Microbiol. 298: 397–409.

El Bahi, S., E. Caliot, M. Bens, A. Bogdanova, S. Kerneis, A. Kahn, A. Vandewalle and E. Pringault. 2002. Lymphoepithelial interactions trigger specific regulation of gene expression in the M cell-containing follicle-associated epithelium of Peyer's patches. J. Immunol. 168: 3713–3720.

Elvers, K.T. and S.F. Park. 2002. Quorum sensing in *Campylobacter jejuni*: detection of a luxS encoded signalling molecule. Microbiology. 148: 1475–1481.

Esmaili, A., S.F. Nazir, A. Borthakur, D. Yu, J.R. Turner, S. Saksena, A. Singla, G.A. Hecht, W.A. Alrefai and R.K. Gill. 2009. Enteropathogenic *Escherichia coli* infection inhibits intestinal serotonin transporter function and expression. Gastroenterology. 137: 2074–2083.

Estrada-Garcia, T. and F. Navarro-Garcia. 2012. Enteroaggregative *Escherichia coli* pathotype: a genetically heterogeneous emerging foodborne enteropathogen. FEMS Immunol. Med. Microbiol.

Etienne-Mesmin, L., B. Chassaing, P. Sauvanet, J. Denizot, S. Blanquet-Diot, A. Darfeuille-Michaud, N. Pradel and V. Livrelli. 2011. Interactions with M cells and macrophages as key steps in the pathogenesis of enterohemorrhagic *Escherichia coli* infections. PLoS One. 6: e23594.

Finlay, B.B. and S. Falkow. 1990. *Salmonella* interactions with polarized human intestinal Caco-2 epithelial cells. J. Infect. Dis. 162: 1096–1106.

Fleckenstein, J.M., K. Roy, J.F. Fischer and M. Burkitt. 2006. Identification of a two-partner secretion locus of enterotoxigenic *Escherichia coli*. Infect. Immun. 74: 2245–2258.

Fleckenstein, J.M., P.R. Hardwidge, G.P. Munson, D.A. Rasko, H. Sommerfelt and H. Steinsland. 2010. Molecular mechanisms of enterotoxigenic *Escherichia coli* infection. Microbes Infect. 12: 89–98.

Flynn, A.N. and A.G. Buret. 2008. Caspases-3, -8, and -9 are required for induction of epithelial cell apoptosis by enteropathogenic *E. coli* but are dispensable for increased paracellular permeability. Microb. Pathog. 44: 311–319.

Fogh, J. and G. Trempe. 1975. New human tumor cell lines. In Human Tumor Cells *in vitro*. J. Fogh, editor. Plenum Publishing Corp., New York 115–141.

Fogh, J., J.M. Fogh and T. Orfeo. 1977. One hundred and twenty-seven cultured human tumor cell lines producing tumors in nude mice. J. Natl. Cancer Inst. 59: 221–226.

Forte, L.R., P.K. Thorne, S.L. Eber, W.J. Krause, R.H. Freeman, S.H. Francis and J.D. Corbin. 1992. Stimulation of intestinal Cl- transport by heat-stable enterotoxin: activation of cAMP-dependent protein kinase by cGMP. Am. J. Physiol. 263: C607–615.

Francis, C.L., M.N. Starnbach and S. Falkow. 1992. Morphological and cytoskeletal changes in epithelial cells occur immediately upon interaction with *Salmonella typhimurium* grown under low-oxygen conditions. Mol. Microbiol. 6: 3077–3087.

Gabastou, J.M., S. Kerneis, M.F. Bernet-Camard, A. Barbat, M.H. Coconnier, J.B. Kaper and A.L. Servin. 1995. Two stages of enteropathogenic *Escherichia coli* intestinal pathogenicity are up and down-regulated by the epithelial cell differentiation. Differentiation. 59: 127–134.

Galan, J.E. 2001. *Salmonella* interactions with host cells: type III secretion at work. Annu. Rev. Cell Dev. Biol. 17: 53–86.

Galan, J.E. and H. Wolf-Watz. 2006. Protein delivery into eukaryotic cells by type III secretion machines. Nature. 444: 567–573.

Gali, H., G.L. Sieckman, T.J. Hoffman, G.E. Kiefer, D.T. Chin, L.R. Forte and W.A. Volkert. 2001. Synthesis and *in vitro* evaluation of an 111 In-labeled ST-peptide enterotoxin (ST) analogue for specific targeting of guanylin receptors on human colonic cancers. Anticancer Res. 21: 2785–2792.

Garmendia, J., G. Frankel and V.F. Crepin. 2005. Enteropathogenic and enterohemorrhagic *Escherichia coli* infections: translocation, translocation, translocation. Infect. Immun. 73: 2573–2585.

Garrett, B.M. and S.S. Visweswariah. 1996. A conformational epitope in the N-terminus of the *Escherichia coli* heat-stable enterotoxins is involved in receptor-ligand interactions. Biochim. Biophys. Acta. 1317: 149–154.

Gerlach, R.G., D. Jackel, B. Stecher, C. Wagner, A. Lupas, W.D. Hardt and M. Hensel. 2007. *Salmonella* pathogenicity island 4 encodes a giant non-fimbrial adhesin and the cognate type 1 secretion system. Cell Microbiol. 9: 1834–1850.

Gewirtz, A.T., A.M. Siber, J.L. Madara and B.A. McCormick. 1999. Orchestration of neutrophil movement by intestinal epithelial cells in response to *Salmonella typhimurium* can be uncoupled from bacterial internalization. Infect. Immun. 67: 608–617.

Ghanekar, Y., A. Chandrashaker and S.S. Visweswariah. 2003. Cellular refractoriness to the heat-stable enterotoxin peptide is associated with alterations in levels of the differentially glycosylated forms of guanylyl cyclase C. Eur. J. Biochem. 270: 3848–3857.

Giannasca, K.T., P.J. Giannasca and M.R. Neutra. 1996. Adherence of *Salmonella typhimurium* to Caco-2 cells: identification of a glycoconjugate receptor. Infect. Immun. 64: 135–145.

Gill, R.K., A. Borthakur, K. Hodges, J.R. Turner, D.R. Clayburgh, S. Saksena, A. Zaheer, K. Ramaswamy, G. Hecht and P.K. Dudeja. 2007. Mechanism underlying inhibition of intestinal apical Cl/OH exchange following infection with enteropathogenic *E. coli*. J. Clin. Invest. 117: 428–437.

Glasser, A.L., J. Boudeau, N. Barnich, M.H. Perruchot, J.F. Colombel and A. Darfeuille-Michaud. 2001. Adherent invasive *Escherichia coli* strains from patients with Crohn's disease survive and replicate within macrophages without inducing host cell death. Infect. Immun. 69: 5529–5537.

Glotfelty, L.G. and G.A. Hecht. 2012. Enteropathogenic *E. coli* effectors EspG1/G2 disrupt tight junctions: new roles and mechanisms. Ann. N Y Acad. Sci. 1258: 149–158.

Gobert, A.P., M. Vareille, A.L. Glasser, T. Hindre, T. de Sablet and C. Martin. 2007. Shiga toxin produced by enterohemorrhagic *Escherichia coli* inhibits PI3K/NF-kappaB signaling pathway in globotriaosylceramide-3-negative human intestinal epithelial cells. J. Immunol. 178: 8168–8174.

Goncalves, A.G., L.C. Campos, T.A. Gomes, J. Rodrigues, V. Sperandio, T.S. Whittam and L.R. Trabulsi. 1997. Virulence properties and clonal structures of strains of *Escherichia coli* O119 serotypes. Infect. Immun. 65: 2034–2040.

Gotz, A. and W. Goebel. 2010. Glucose and glucose 6-phosphate as carbon sources in extra- and intracellular growth of enteroinvasive *Escherichia coli* and *Salmonella enterica*. Microbiology. 156: 1176–1187.

Gotz, A., E. Eylert, W. Eisenreich and W. Goebel. 2010. Carbon metabolism of enterobacterial human pathogens growing in epithelial colorectal adenocarcinoma (Caco-2) cells. PLoS One. 5: e10586.

Grant, C.C., R.J. Messer and W. Cieplak, Jr. 1994. Role of trypsin-like cleavage at arginine 192 in the enzymatic and cytotonic activities of *Escherichia coli* heat-labile enterotoxin. Infect. Immun. 62: 4270–4278.

Grant, C.C., M.E. Konkel, W. Cieplak, Jr. and L.S. Tompkins. 1993. Role of flagella in adherence, internalization, and translocation of *Campylobacter jejuni* in nonpolarized and polarized epithelial cell cultures. Infect. Immun. 61: 1764–1771.

Grasset, E., M. Pinto, E. Dussaulx, A. Zweibaum and J.F. Desjeux. 1984. Epithelial properties of human colonic carcinoma cell line Caco-2: electrical parameters. Am. J. Physiol. 247: C260–267.

Grassl, G.A. and B.B. Finlay. 2008. Pathogenesis of enteric *Salmonella* infections. Curr. Opin. Gastroenterol. 24: 22–26.

Grewal, H.M., H. Valvatne, M.K. Bhan, L. van Dijk, W. Gaastra and H. Sommerfelt. 1997. A new putative fimbrial colonization factor, CS19, of human enterotoxigenic *Escherichia coli*. Infect. Immun. 65: 507–513.

Guarino, A., M. Cohen, M. Thompson, K. Dharmsathaphorn and R. Giannella. 1987. T84 cell receptor binding and guanyl cyclase activation by *Escherichia coli* heat-stable toxin. Am. J. Physiol. 253: G775–780.

Guidry, J.J., L. Cardenas, E. Cheng and J.D. Clements. 1997. Role of receptor binding in toxicity, immunogenicity, and adjuvanticity of *Escherichia coli* heat-labile enterotoxin. Infect. Immun. 65: 4943–4950.

Guignot, J., C. Chaplais, M.H. Coconnier-Polter and A.L. Servin. 2007. The secreted autotransporter toxin, Sat, functions as a virulence factor in Afa/Dr diffusely adhering *Escherichia coli* by promoting lesions in tight junction of polarized epithelial cells. Cell Microbiol. 9: 204–221.

Guignot, J., S. Hudault, I. Kansau, I. Chau and A.L. Servin. 2009. Human decay-accelerating factor and CEACAM receptor-mediated internalization and intracellular lifestyle of Afa/Dr diffusely adhering *Escherichia coli* in epithelial cells. Infect. Immun. 77: 517–531.

Guignot, J., M.F. Bernet-Camard, C. Pous, L. Plancon, C. Le Bouguenec and A.L. Servin. 2001. Polarized entry of uropathogenic Afa/Dr diffusely adhering *Escherichia coli* strain IH11128 into human epithelial cells: evidence for alpha5beta1 integrin recognition and subsequent internalization through a pathway involving caveolae and dynamic unstable microtubules. Infect. Immun. 69: 1856–1868.

Guignot, J., J. Breard, M.F. Bernet-Camard, I. Peiffer, B.J. Nowicki, A.L. Servin and A.B. Blanc-Potard. 2000a. Pyelonephritogenic diffusely adhering *Escherichia coli* EC7372 harboring Dr-II adhesin carries classical uropathogenic virulence genes and promotes cell lysis and apoptosis in polarized epithelial Caco-2/TC7 cells. Infect. Immun. 68: 7018–7027.

Guignot, J., I. Peiffer, M.F. Bernet-Camard, D.M. Lublin, C. Carnoy, S.L. Moseley and A.L. Servin. 2000b. Recruitment of CD55 and CD66e brush border-associated glycosylphosphatidylinositol-anchored proteins by members of the Afa/Dr diffusely adhering family of *Escherichia coli* that infect the human polarized intestinal Caco-2/TC7 cells. Infect. Immun. 68: 3554–3563.

Gullberg, E., M. Leonard, J. Karlsson, A.M. Hopkins, D. Brayden, A.W. Baird and P. Artursson. 2000. Expression of specific markers and particle transport in a new human intestinal M-cell model. Biochem. Biophys. Res. Commun. 279: 808–813.

Guttman, J.A., P. Kazemi, A.E. Lin, A.W. Vogl and B.B. Finlay. 2007. Desmosomes are unaltered during infections by attaching and effacing pathogens. Anat. Rec. (Hoboken). 290: 199–205.

Haque, A., F. Bowe, R.J. Fitzhenry, G. Frankel, M. Thomson, R. Heuschkel, S. Murch, M.P. Stevens, T.S. Wallis, A.D. Phillips and G. Dougan. 2004. Early interactions of *Salmonella enterica* serovar typhimurium with human small intestinal epithelial explants. Gut. 53: 1424–1430.

Harrington, S.M., E.G. Dudley and J.P. Nataro. 2006. Pathogenesis of enteroaggregative *Escherichia coli* infection. FEMS Microbiol. Lett. 254: 12–18.

Hautefort, I., A. Thompson, S. Eriksson-Ygberg, M.L. Parker, S. Lucchini, V. Danino, R.J. Bongaerts, N. Ahmad, M. Rhen and J.C. Hinton. 2008. During infection of epithelial cells *Salmonella enterica* serovar Typhimurium undergoes a time-dependent transcriptional adaptation that results in simultaneous expression of three type 3 secretion systems. Cell Microbiol. 10: 958–984.

Hecht, G. and A. Koutsouris. 1999. Enteropathogenic *E. coli* attenuates secretagogue-induced net intestinal ion transport but not Cl- secretion. Am. J. Physiol. 276: G781–788.

Hecht, G., J.A. Marrero, A. Danilkovich, K.A. Matkowskyj, S.D. Savkovic, A. Koutsouris and R.V. Benya. 1999. Pathogenic *Escherichia coli* increase Cl- secretion from intestinal epithelia by upregulating galanin-1 receptor expression. J. Clin. Invest. 104: 253–262.

Hecht, G., K. Hodges, R.K. Gill, F. Kear, S. Tyagi, J. Malakooti, K. Ramaswamy and P.K. Dudeja. 2004. Differential regulation of Na+/H+ exchange isoform activities by enteropathogenic *E. coli* in human intestinal epithelial cells. Am. J. Physiol. Gastrointest. Liver Physiol. 287: G370–378.

Henderson, I.R. and J.P. Nataro. 2001. Virulence functions of autotransporter proteins. Infect. Immun. 69: 1231–1243.

Henderson, I.R., F. Navarro-Garcia, M. Desvaux, R.C. Fernandez and D. Ala'Aldeen. 2004. Type V protein secretion pathway: the autotransporter story. Microbiol. Mol. Biol. Rev. 68: 692–744.

Henderson, I.R., S. Hicks, F. Navarro-Garcia, W.P. Elias, A.D. Philips and J.P. Nataro. 1999. Involvement of the enteroaggregative *Escherichia coli* plasmid-encoded toxin in causing human intestinal damage. Infect. Immun. 67: 5338–5344.

Hernandes, R.T., W.P. Elias, M.A. Vieira and T.A. Gomes. 2009. An overview of atypical enteropathogenic *Escherichia coli*. FEMS Microbiol. Lett. 297: 137–149.

Higashide, W., S. Dai, V.P. Hombs and D. Zhou. 2002. Involvement of SipA in modulating actin dynamics during *Salmonella* invasion into cultured epithelial cells. Cell Microbiol. 4: 357–365.

Ho, N.K., J.C. Ossa, U. Silphaduang, R. Johnson, K.C. Johnson-Henry and P.M. Sherman. 2012. Enterohemorrhagic *Escherichia coli* O157:H7 Shiga toxins inhibit gamma interferon-mediated cellular activation. Infect. Immun. 80: 2307–2315.

Hodges, K., R. Gill, K. Ramaswamy, P.K. Dudeja and G. Hecht. 2006. Rapid activation of Na+/H+ exchange by EPEC is PKC mediated. Am. J. Physiol. Gastrointest. Liver Physiol. 291: G959–968.

Hofman, P. and V. Vouret-Craviari. 2012. Microbes-induced EMT at the crossroad of inflammation and cancer. Gut Microbes. 3: 176–185.

Holzer, S.U. and M. Hensel. 2012. Divergent roles of *Salmonella pathogenicity* island 2 and metabolic traits during interaction of S. enterica serovar typhimurium with host cells. PLoS One. 7: e33220.

Horie, K., F. Tang and R.T. Borchardt. 2003. Isolation and characterization of Caco-2 subclones expressing high levels of multidrug resistance protein efflux transporter. Pharm. Res. 20: 161–168.

Howe, K.L., C. Reardon, A. Wang, A. Nazli and D.M. McKay. 2005. Transforming growth factor-beta regulation of epithelial tight junction proteins enhances barrier function and blocks enterohemorrhagic *Escherichia coli* O157:H7-induced increased permeability. Am. J. Pathol. 167: 1587–1597.

Hoy, B., T. Geppert, M. Boehm, F. Reisen, P. Plattner, G. Gadermaier, N. Sewald, F. Ferreira, P. Briza, G. Schneider, S. Backert and S. Wessler. 2012. Distinct roles of secreted HtrA proteases from gram-negative pathogens in cleaving the junctional protein and tumor suppressor E-cadherin. J. Biol. Chem. 287: 10115–10120.

Huebner, C., L.R. Ferguson, D.Y. Han, M. Philpott, M.L. Barclay, R.B. Gearry, A. McCulloch, P.S. Demmers and B.L. Browning. 2009. Nucleotide-binding oligomerization domain containing 1 (NOD1) haplotypes and single nucleotide polymorphisms modify susceptibility to inflammatory bowel diseases in a New Zealand caucasian population: a case-control study. BMC Res. Notes. 2: 52.

Huott, P.A., W. Liu, J.A. McRoberts, R.A. Giannella and K. Dharmsathaphorn. 1988. Mechanism of action of *Escherichia coli* heat stable enterotoxin in a human colonic cell line. J. Clin. Invest. 82: 514–523.

Iizumi, Y., H. Sagara, Y. Kabe, M. Azuma, K. Kume, M. Ogawa, T. Nagai, P.G. Gillespie, C. Sasakawa and H. Handa. 2007. The enteropathogenic *E. coli* effector EspB facilitates microvillus effacing and antiphagocytosis by inhibiting myosin function. Cell Host Microbe. 2: 383–392.

Isidean, S.D., M.S. Riddle, S.J. Savarino and C.K. Porter. 2011. A systematic review of ETEC epidemiology focusing on colonization factor and toxin expression. Vaccine. 29: 6167–6178.

Jallat, C., A. Darfeuille-Michaud, C. Rich and B. Joly. 1994a. Survey of clinical isolates of diarrhoeogenic *Escherichia coli*: diffusely adhering *E. coli* strains with multiple adhesive factors. Res. Microbiol. 145: 621–632.

Jallat, C., A. Darfeuille-Michaud, J.P. Girardeau, C. Rich and B. Joly. 1994b. Self-transmissible R plasmids encoding CS31A among human *Escherichia coli* strains isolated from diarrheal stools. Infect. Immun. 62: 2865–2873.

Jandu, N., P.J. Ceponis, S. Kato, J.D. Riff, D.M. McKay and P.M. Sherman. 2006. Conditioned medium from enterohemorrhagic *Escherichia coli*-infected T84 cells inhibits signal transducer and activator of transcription 1 activation by gamma interferon. Infect. Immun. 74: 1809–1818.

Jepson, M.A. and M.A. Clark. 1998. Studying M cells and their role in infection. Trends Microbiol. 6: 359–365.

Johnson, T.J. and L.K. Nolan. 2009. Pathogenomics of the virulence plasmids of *Escherichia coli*. Microbiol. Mol. Biol. Rev. 73: 750–774.

Kansau, I., C. Berger, M. Hospital, R. Amsellem, V. Nicolas, A.L. Servin and M.F. Bernet-Camard. 2004. Zipper-like internalization of Dr-positive *Escherichia coli* by epithelial cells is preceded by an adhesin-induced mobilization of raft-associated molecules in the initial step of adhesion. Infect. Immun. 72: 3733–3742.

Kaper, J.B., J.P. Nataro and H.L. Mobley. 2004. Pathogenic *Escherichia coli*. Nat. Rev. Microbiol. 2: 123–140.

Kerneis, S., A. Bogdanova, J.P. Kraehenbuhl and E. Pringault. 1997. Conversion by Peyer's patch lymphocytes of human enterocytes into M cells that transport bacteria. Science. 277: 949–952.

Kerneis, S., S.S. Bilge, V. Fourel, G. Chauviere, M.H. Coconnier and A.L. Servin. 1991. Use of purified F1845 fimbrial adhesin to study localization and expression of receptors for diffusely adhering *Escherichia coli* during enterocytic differentiation of human colon carcinoma cell lines HT-29 and Caco-2 in culture. Infect. Immun. 59: 4013–4018.

Kerneis, S., J.M. Gabastou, M.F. Bernet-Camard, M.H. Coconnier, B.J. Nowicki and A.L. Servin. 1994. Human cultured intestinal cells express attachment sites for uropathogenic *Escherichia coli* bearing adhesins of the Dr adhesin family. FEMS Microbiol. Lett. 119: 27–32.

Kerneis, S., G. Chauviere, A. Darfeuille-Michaud, D. Aubel, M.H. Coconnier, B. Joly and A.L. Servin. 1992. Expression of receptors for enterotoxigenic *Escherichia coli* during enterocytic differentiation of human polarized intestinal epithelial cells in culture. Infect. Immun. 60: 2572–2580.

Klancnik, A., B. Guzej, P. Jamnik, D. Vuckovic, M. Abram and S.S. Mozina. 2009. Stress response and pathogenic potential of *Campylobacter jejuni* cells exposed to starvation. Res. Microbiol. 160: 345–352.

Klemm, P., R.M. Vejborg and O. Sherlock. 2006. Self-associating autotransporters, SAATs: functional and structural similarities. Int. J. Med. Microbiol. 296: 187–195.

Knappstein, S., T. Ide, M.A. Schmidt and G. Heusipp. 2004. Alpha 1-antitrypsin binds to and interferes with functionality of EspB from atypical and typical enteropathogenic *Escherichia coli* strains. Infect. Immun. 72: 4344–4350.

Knutton, S., T. Baldwin, P.H. Williams and A.S. McNeish. 1989. Actin accumulation at sites of bacterial adhesion to tissue culture cells: basis of a new diagnostic test for enteropathogenic and enterohemorrhagic *Escherichia coli*. Infect. Immun. 57: 1290–1298.

Konkel, M.E., D.J. Mead, S.F. Hayes and W. Cieplak, Jr. 1992. Translocation of *Campylobacter jejuni* across human polarized epithelial cell monolayer cultures. J. Infect. Dis. 166: 308–315.

Kops, S.K., D.K. Lowe, W.M. Bement and A.B. West. 1996. Migration of *Salmonella typhi* through intestinal epithelial monolayers: an *in vitro* study. Microbiol. Immunol. 40: 799–811.

Korotkova, N., Y. Yarova-Yarovaya, V. Tchesnokova, N. Yazvenko, M.A. Carl, A.E. Stapleton and S.L. Moseley. 2008. *Escherichia coli* DraE adhesin-associated bacterial internalization by epithelial cells is promoted independently by decay-accelerating factor and carcinoembryonic antigen-related cell adhesion molecule binding and does not require the DraD invasin. Infect. Immun. 76: 3869–3880.

Korotkova, N., E. Cota, Y. Lebedin, S. Monpouet, J. Guignot, A.L. Servin, S. Matthews and S.L. Moseley. 2006. A subfamily of Dr adhesins of *Escherichia coli* bind independently to decay-accelerating factor and the N-domain of carcinoembryonic antigen. J. Biol. Chem. 281: 29120–29130.

Kraehenbuhl, J.P. and M.R. Neutra. 2000. Epithelial M cells: differentiation and function. Annu. Rev. Cell Dev. Biol. 16: 301–332.

Krause, D.O., A.C. Little, S.E. Dowd and C.N. Bernstein. 2011. Complete genome sequence of adherent invasive *Escherichia coli* UM146 isolated from ileal Crohn's disease biopsy tissue. J. Bacteriol. 193: 583.

Laarmann, S. and M.A. Schmidt. 2003. The *Escherichia coli* AIDA autotransporter adhesin recognizes an integral membrane glycoprotein as receptor. Microbiology. 149: 1871–1882.

Laiko, M., R. Murtazina, I. Malyukova, C. Zhu, E.C. Boedeker, O. Gutsal, R. O'Malley, R.N. Cole, P.I. Tarr, K.F. Murray, A. Kane, M. Donowitz and O. Kovbasnjuk. 2010. Shiga toxin 1 interaction with enterocytes causes apical protein mistargeting through the depletion of intracellular galectin-3. Exp. Cell Res. 316: 657–666.

Le Bivic, A., A. Quaroni, B. Nichols and E. Rodriguez-Boulan. 1990. Biogenetic pathways of plasma membrane proteins in Caco-2, a human intestinal epithelial cell line. J. Cell Biol. 111: 1351–1361.

Le Bouguenec, C. and A.L. Servin. 2006. Diffusely adherent *Escherichia coli* strains expressing Afa/Dr adhesins (Afa/Dr DAEC): hitherto unrecognized pathogens. FEMS Microbiol. Lett. 256: 185–194.

Le Gall, T., P. Darlu, P. Escobar-Paramo, B. Picard and E. Denamur. 2005. Selection-driven transcriptome polymorphism in *Escherichia coli/Shigella* species. Genome Res. 15: 260–268.

Lee, C.A., M. Silva, A.M. Siber, A.J. Kelly, E. Galyov and B.A. McCormick. 2000. A secreted *Salmonella* protein induces a proinflammatory response in epithelial cells, which promotes neutrophil migration. Proc. Natl. Acad. Sci. USA. 97: 12283–12288.

Lencer, W.I., C. Constable, S. Moe, P.A. Rufo, A. Wolf, M.G. Jobling, S.P. Ruston, J.L. Madara, R.K. Holmes and T.R. Hirst. 1997. Proteolytic activation of cholera toxin and *Escherichia coli* labile toxin by entry into host epithelial cells. Signal transduction by a protease-resistant toxin variant. J. Biol. Chem. 272: 15562–15568.

Leonard, F., E.M. Collnot and C.M. Lehr. 2010. A three-dimensional coculture of enterocytes, monocytes and dendritic cells to model inflamed intestinal mucosa *in vitro*. Mol. Pharm. 7: 2103–2119.

Leung, K.Y. and B.B. Finlay. 1991. Intracellular replication is essential for the virulence of *Salmonella typhimurium*. Proc. Natl. Acad. Sci. USA. 88: 11470–11474.

Levine, S.A., M. Donowitz, A.J. Watson, G.W. Sharp, J.K. Crane and C.S. Weikel. 1991. Characterization of the synergistic interaction of *Escherichia coli* heat-stable toxin and carbachol. Am. J. Physiol. 261: G592–601.

Li, Q., Q. Zhang, C. Wang, N. Li and J. Li. 2008. Invasion of enteropathogenic *Escherichia coli* into host cells through epithelial tight junctions. Febs J. 275: 6022–6032.

Lievin-Le Moal, V. and A.L. Servin. 2006. The front line of enteric host defense against unwelcome intrusion of harmful microorganisms: mucins, antimicrobial peptides, and microbiota. Clin. Microbiol. Rev. 19: 315–337.

Lim, J.S., H.S. Na, H.C. Lee, H.E. Choy, S.C. Park, J.M. Han and K.A. Cho. 2009. Caveolae-mediated entry of *Salmonella typhimurium* in a human M-cell model. Biochem. Biophys. Res. Commun. 390: 1322–1327.

Lin, A.E., A. Benmerah and J.A. Guttman. 2011. Eps15 and Epsin1 are crucial for enteropathogenic *Escherichia coli* pedestal formation despite the absence of adaptor protein 2. J. Infect. Dis. 204: 695–703.

Lin, M., A.C. Nairn and S.E. Guggino. 1992. cGMP-dependent protein kinase regulation of a chloride channel in T84 cells. Am. J. Physiol. 262: C1304–1312.

Louvard, D., M. Kedinger and H.P. Hauri. 1992. The differentiating intestinal epithelial cell: establishment and maintenance of functions through interactions between cellular structures. Annu. Rev. Cell Biol. 8: 157–195.

Louwen, R., E.E. Nieuwenhuis, L. van Marrewijk, D. Horst-Kreft, L. de Ruiter, A.P. Heikema, W.J. van Wamel, J.A. Wagenaar, H.P. Endtz, J. Samsom, P. van Baarlen, A. Akhmanova and A. van Belkum. 2012. *Campylobacter jejuni* translocation across intestinal epithelial cells is facilitated by ganglioside-like lipooligosaccharide structures. Infect. Immun.

Luck, S.N., V. Bennett-Wood, R. Poon, R.M. Robins-Browne and E.L. Hartland. 2005. Invasion of epithelial cells by locus of enterocyte effacement-negative enterohemorrhagic *Escherichia coli*. Infect. Immun. 73: 3063–3071.

MacCallum, A., S.P. Hardy and P.H. Everest. 2005. *Campylobacter jejuni* inhibits the absorptive transport functions of Caco-2 cells and disrupts cellular tight junctions. Microbiology. 151: 2451–2458.

Madara, J.L. and K. Dharmsathaphorn. 1985. Occluding junction structure-function relationships in a cultured epithelial monolayer. J. Cell Biol. 101: 2124–2133.

Madara, J.L., J. Stafford, K. Dharmsathaphorn and S. Carlson. 1987. Structural analysis of a human intestinal epithelial cell line. Gastroenterology. 92: 1133–1145.

Malladi, V., B. Shankar, P.H. Williams and A. Balakrishnan. 2004. Enteropathogenic *Escherichia coli* outer membrane proteins induce changes in cadherin junctions of Caco-2 cells through activation of PKCalpha. Microbes Infect. 6: 38–50.

Marin-Esteban, V., I. Turbica, G. Dufour, N. Semiramoth, A. Gleizes, R. Gorges, I. Beau, A.L. Servin, V. Lievin-Le Moal, C. Sandre and S. Chollet-Martin. 2012. Afa/Dr diffusely adhering *Escherichia coli* strain C1845 induces neutrophil extracellular traps that kill bacteria and damage human enterocyte-like cells. Infect. Immun. 80: 1891–1899.

Martinez-Argudo, I. and M.A. Jepson. 2008. *Salmonella* translocates across an *in vitro* M cell model independently of SPI-1 and SPI-2. Microbiology. 154: 3887–3894.

Martinez-Argudo, I., C. Sands and M.A. Jepson. 2007. Translocation of enteropathogenic *Escherichia coli* across an *in vitro* M cell model is regulated by its type III secretion system. Cell Microbiol. 9: 1538–1546.

Martinez-Medina, M., X. Aldeguer, M. Lopez-Siles, F. Gonzalez-Huix, C. Lopez-Oliu, G. Dahbi, J.E. Blanco, J. Blanco, L.J. Garcia-Gil and A. Darfeuille-Michaud. 2009a. Molecular diversity of *Escherichia coli* in the human gut: new ecological evidence supporting the role of adherent-invasive *E. coli* (AIEC) in Crohn's disease. Inflamm. Bowel Dis. 15: 872–882.

Martinez-Medina, M., A. Mora, M. Blanco, C. Lopez, M.P. Alonso, S. Bonacorsi, M.H. Nicolas-Chanoine, A. Darfeuille-Michaud, J. Garcia-Gil and J. Blanco. 2009b. Similarity and divergence among adherent-invasive *Escherichia coli* and extraintestinal pathogenic *E. coli* strains. J. Clin. Microbiol. 47: 3968–3979.

Masseret, E., J. Boudeau, J.F. Colombel, C. Neut, P. Desreumaux, B. Joly, A. Cortot and A. Darfeuille-Michaud. 2001. Genetically related *Escherichia coli* strains associated with Crohn's disease. Gut. 48: 320–325.

Matter, K., B. Stieger, J. Klumperman, L. Ginsel and H.P. Hauri. 1990. Endocytosis, recycling, and lysosomal delivery of brush border hydrolases in cultured human intestinal epithelial cells (Caco-2). J. Biol. Chem. 265: 3503–3512.

Matthews, J.B., C.S. Awtrey, R. Thompson, T. Hung, K.J. Tally and J.L. Madara. 1993. Na(+)-K(+)-2Cl- cotransport and Cl- secretion evoked by heat-stable enterotoxin is microfilament dependent in T84 cells. Am. J. Physiol. 265: G370–378.

Mazariego-Espinosa, K., A. Cruz, M.A. Ledesma, S.A. Ochoa and J. Xicohtencatl-Cortes. 2010. Longus, a type IV pilus of enterotoxigenic *Escherichia coli*, is involved in adherence to intestinal epithelial cells. J. Bacteriol. 192: 2791–2800.

McCormick, B.A., S.I. Miller, D. Carnes and J.L. Madara. 1995a. Transepithelial signaling to neutrophils by *Salmonellae*: a novel virulence mechanism for gastroenteritis. Infect. Immun. 63: 2302–2309.

McCormick, B.A., S.P. Colgan, C. Delp-Archer, S.I. Miller and J.L. Madara. 1993. *Salmonella typhimurium* attachment to human intestinal epithelial monolayers: transcellular signalling to subepithelial neutrophils. J. Cell Biol. 123: 895–907.

McCormick, B.A., C.A. Parkos, S.P. Colgan, D.K. Carnes and J.L. Madara. 1998. Apical secretion of a pathogen-elicited epithelial chemoattractant activity in response to surface colonization of intestinal epithelia by *Salmonella typhimurium*. J. Immunol. 160: 455–466.

McCormick, B.A., P.M. Hofman, J. Kim, D.K. Carnes, S.I. Miller and J.L. Madara. 1995b. Surface attachment of *Salmonella typhimurium* to intestinal epithelia imprints the subepithelial matrix with gradients chemotactic for neutrophils. J. Cell Biol. 131: 1599–1608.

Melton-Celsa, A., K. Mohawk, L. Teel and A. O'Brien. 2012. Pathogenesis of Shiga-toxin producing *Escherichia coli*. Curr. Top Microbiol. Immunol. 357: 67–103.

Michail, S.K., D.R. Halm and F. Abernathy. 2003. Enteropathogenic *Escherichia coli*: stimulating neutrophil migration across a cultured intestinal epithelium without altering transepithelial conductance. J. Pediatr. Gastroenterol. Nutr. 36: 253–260.

Mihaljevic, R.R., M. Sikic, A. Klancnik, G. Brumini, S.S. Mozina and M. Abram. 2007. Environmental stress factors affecting survival and virulence of *Campylobacter jejuni*. Microb. Pathog. 43: 120–125.

Mills, D.C., O. Gundogdu, A. Elmi, M. Bajaj-Elliott, P.W. Taylor, B.W. Wren and N. Dorrell. 2012. Increase in *Campylobacter jejuni* invasion of intestinal epithelial cells under low-oxygen coculture conditions that reflect the *in vivo* environment. Infect. Immun. 80: 1690–1698.

Mills, S.D. and B.B. Finlay. 1994. Comparison of *Salmonella typhi* and *Salmonella typhimurium* invasion, intracellular growth and localization in cultured human epithelial cells. Microb. Pathog. 17: 409–423.

Miquel, S., E. Peyretaillade, L. Claret, A. de Vallee, C. Dossat, B. Vacherie, H. Zineb el, B. Segurens, V. Barbe, P. Sauvanet, C. Neut, J.F. Colombel, C. Medigue, F.J. Mojica, P. Peyret, R. Bonnet and A. Darfeuille-Michaud. 2010. Complete genome sequence of Crohn's disease-associated adherent-invasive *E. coli* strain LF82. PLoS One. 5.

Miyake, M., M. Hanajima, T. Matsuzawa, C. Kobayashi, M. Minami, A. Abe and Y. Horiguchi. 2005. Binding of intimin with Tir on the bacterial surface is prerequisite for the barrier disruption induced by enteropathogenic *Escherichia coli*. Biochem. Biophys. Res. Commun. 337: 922–927.

Monteville, M.R. and M.E. Konkel. 2002. Fibronectin-facilitated invasion of T84 eukaryotic cells by *Campylobacter jejuni* occurs preferentially at the basolateral cell surface. Infect. Immun. 70: 6665–6671.

Moreno, A.C., L.G. Ferreira and M.B. Martinez. 2009. Enteroinvasive *Escherichia coli* vs. *Shigella flexneri:* how different patterns of gene expression affect virulence. FEMS Microbiol. Lett. 301: 156–163.

Mrsny, R.J., A.T. Gewirtz, D. Siccardi, T. Savidge, B.P. Hurley, J.L. Madara and B.A. McCormick. 2004. Identification of hepoxilin A3 in inflammatory events: a required role in neutrophil migration across intestinal epithelia. Proc. Natl. Acad. Sci. USA. 101: 7421–7426.

Muller, U., M. Brandsch, P.D. Prasad, Y.J. Fei, V. Ganapathy and F.H. Leibach. 1996. Inhibition of the H+/peptide cotransporter in the human intestinal cell line Caco-2 by cyclic AMP. Biochem. Biophys. Res. Commun. 218: 461–465.

Muza-Moons, M.M., E.E. Schneeberger and G.A. Hecht. 2004. Enteropathogenic *Escherichia coli* infection leads to appearance of aberrant tight junctions strands in the lateral membrane of intestinal epithelial cells. Cell Microbiol. 6: 783–793.

Nakanishi, N., H. Abe, Y. Ogura, T. Hayashi, K. Tashiro, S. Kuhara, N. Sugimoto and T. Tobe. 2006. ppGpp with DksA controls gene expression in the locus of enterocyte effacement (LEE) pathogenicity island of enterohaemorrhagic *Escherichia coli* through activation of two virulence regulatory genes. Mol. Microbiol. 61: 194–205.

Nash, J.H., A. Villegas, A.M. Kropinski, R. Aguilar-Valenzuela, P. Konczy, M. Mascarenhas, K. Ziebell, A.G. Torres, M.A. Karmali and B.K. Coombes. 2010. Genome sequence of adherent-invasive *Escherichia coli* and comparative genomic analysis with other *E. coli* pathotypes. BMC Genomics. 11: 667.

Nash, S., J. Stafford and J.L. Madara. 1987. Effects of polymorphonuclear leukocyte transmigration on the barrier function of cultured intestinal epithelial monolayers. J. Clin. Invest. 80: 1104–1113.

Nataro, J.P. and J.B. Kaper. 1998. Diarrheagenic *Escherichia coli*. Clin. Microbiol. Rev. 11: 142–201.

Nataro, J.P., S. Hicks, A.D. Phillips, P.A. Vial and C.L. Sears. 1996. T84 cells in culture as a model for enteroaggregative *Escherichia coli* pathogenesis. Infect. Immun. 64: 4761–4768.

Nataro, J.P., J.B. Kaper, R. Robins-Browne, V. Prado, P. Vial and M.M. Levine. 1987. Patterns of adherence of diarrheagenic *Escherichia coli* to HEp-2 cells. Pediatr. Infect. Dis. J. 6: 829–831.

Navarro-Garcia, F. 2010. Enteroaggregative *Escherichia coli* plasmid-encoded toxin. Future Microbiol. 5: 1005–1013.

Navarro-Garcia, F. and W.P. Elias. 2011. Autotransporters and virulence of enteroaggregative *E. coli*. Gut Microbes. 2: 13–24.

Navarro-Garcia, F., M. Sonnested and K. Teter. 2010. Host-toxin interactions involving EspC and Pet, two serine protease autotransporters of the *Enterobacteriaceae*. Toxins. (Basel). 2: 1134–1147.

Negroni, A., M. Costanzo, R. Vitali, F. Superti, L. Bertuccini, A. Tinari, F. Minelli, G. Di Nardo, F. Nuti, M. Pierdomenico, S. Cucchiara and L. Stronati. 2012. Characterization of adherent-invasive *Escherichia coli* isolated from pediatric patients with inflammatory bowel disease. Inflamm. Bowel Dis. 18: 91–924.

Nicoletti, C. 2000. Unsolved mysteries of intestinal M cells. Gut. 47: 735–739.

Nougayrede, J.P., G.H. Foster and M.S. Donnenberg. 2007. Enteropathogenic *Escherichia coli* effector EspF interacts with host protein Abcf2. Cell Microbiol. 9: 680–693.

Nowicki, B., J. Moulds, R. Hull and S. Hull. 1988. A hemagglutinin of uropathogenic *Escherichia coli* recognizes the Dr blood group antigen. Infect. Immun. 56: 1057–1060.

Nowicki, B., A. Hart, K.E. Coyne, D.M. Lublin and S. Nowicki. 1993. Short consensus repeat-3 domain of recombinant decay-accelerating factor is recognized by *Escherichia coli* recombinant Dr adhesin in a model of a cell-cell interaction. J. Exp. Med. 178: 2115–2121.

Nowicki, B., A. Labigne, S. Moseley, R. Hull, S. Hull and J. Moulds. 1990. The Dr hemagglutinin, afimbrial adhesins AFA-I and AFA-III, and F1845 fimbriae of uropathogenic and diarrhea-associated *Escherichia coli* belong to a family of hemagglutinins with Dr receptor recognition. Infect. Immun. 58: 279–281.

Ohl, M.E. and S.I. Miller. 2001. *Salmonella*: a model for bacterial pathogenesis. Annu. Rev. Med. 52: 259–274.

Orlandi, P.A., D.R. Critchley and P.H. Fishman. 1994. The heat-labile enterotoxin of *Escherichia coli* binds to polylactosaminoglycan-containing receptors in CaCo-2 human intestinal epithelial cells. Biochemistry. 33: 12886–12895.

Papp-Wallace, K.M., M. Nartea, D.G. Kehres, S. Porwollik, M. McClelland, S.J. Libby, F.C. Fang and M.E. Maguire. 2008. The CorA Mg2+ channel is required for the virulence of *Salmonella enterica* serovar typhimurium. J. Bacteriol. 190: 6517–6523.

Parkinson, S.J., A.E. Alekseev, L.A. Gomez, F. Wagner, A. Terzic and S.A. Waldman. 1997. Interruption of *Escherichia coli* heat-stable enterotoxin-induced guanylyl cyclase signaling and associated chloride current in human intestinal cells by 2-chloroadenosine. J. Biol. Chem. 272: 754–758.

Parsot, C. 2005. *Shigella* spp. and enteroinvasive *Escherichia coli* pathogenicity factors. FEMS Microbiol. Lett. 252: 11–18.

Peiffer, I., M.F. Bernet-Camard, M. Rousset and A.L. Servin. 2001. Impairments in enzyme activity and biosynthesis of brush border-associated hydrolases in human intestinal Caco-2/TC7 cells infected by members of the Afa/Dr family of diffusely adhering *Escherichia coli*. Cell Microbiol. 3: 341–357.

Peiffer, I., A.B. Blanc-Potard, M.F. Bernet-Camard, J. Guignot, A. Barbat and A.L. Servin. 2000a. Afa/Dr diffusely adhering *Escherichia coli* C1845 infection promotes selective injuries in the junctional domain of polarized human intestinal Caco-2/TC7 cells. Infect. Immun. 68: 3431–3442.

Peiffer, I., J. Guignot, A. Barbat, C. Carnoy, S.L. Moseley, B.J. Nowicki, A.L. Servin and M.F. Bernet-Camard. 2000b. Structural and functional lesions in brush border of human polarized intestinal Caco-2/TC7 cells infected by members of the Afa/Dr diffusely adhering family of *Escherichia coli*. Infect. Immun. 68: 5979–5990.

Petersen, A.M., E.M. Nielsen, E. Litrup, J. Brynskov, H. Mirsepasi and K.A. Krogfelt. 2009. A phylogenetic group of *Escherichia coli* associated with active left-sided inflammatory bowel disease. BMC Microbiol. 9: 171.

Peterson, M.D. and M.S. Mooseker. 1992. Characterization of the enterocyte-like brush border cytoskeleton of the C2BBe clones of the human intestinal cell line, Caco-2. J. Cell Sci. 102(Pt 3): 581–600.

Peterson, M.D. and M.S. Mooseker. 1993. An *in vitro* model for the analysis of intestinal brush border assembly. I. Ultrastructural analysis of cell contact-induced brush border assembly in Caco-2BBe cells. J. Cell Sci. 105(Pt 2): 445–460.

Pettigrew, D., K.L. Anderson, J. Billington, E. Cota, P. Simpson, P. Urvil, F. Rabuzin, P. Roversi, B. Nowicki, L. du Merle, C. Le Bouguenec, S. Matthews and S.M. Lea. 2004. High resolution studies of the Afa/Dr adhesin DraE and its interaction with chloramphenicol. J. Biol Chem. 279: 46851–46857.

Phillips, A.D., J. Giron, S. Hicks, G. Dougan and G. Frankel. 2000. Intimin from enteropathogenic *Escherichia coli* mediates remodelling of the eukaryotic cell surface. Microbiology. 146(Pt 6): 1333–1344.

Philpott, D.J., D.M. McKay, P.M. Sherman and M.H. Perdue. 1996. Infection of T84 cells with enteropathogenic *Escherichia coli* alters barrier and transport functions. Am. J. Physiol. 270: G634–645.

Philpott, D.J., D.M. McKay, W. Mak, M.H. Perdue and P.M. Sherman. 1998. Signal transduction pathways involved in enterohemorrhagic *Escherichia coli*-induced alterations in T84 epithelial permeability. Infect. Immun. 66: 1680–1687.

Pichel, M., N. Binsztein and G. Viboud. 2000. CS22, a novel human enterotoxigenic *Escherichia coli* adhesin, is related to CS15. Infect. Immun. 68: 3280–3285.

Pichel, M., N. Binsztein, G. Gutkind and G. Viboud. 2001. Identification of a cluster of strains bearing a new adhesin among genetically diverse enterotoxigenic *Escherichia coli* isolates of serogroup O20. J. Clin. Microbiol. 39: 782–786.

Pielage, J.F., C. Cichon, L. Greune, M. Hirashima, T. Kucharzik and M.A. Schmidt. 2007. Reversible differentiation of Caco-2 cells reveals galectin-9 as a surface marker molecule for human follicle-associated epithelia and M cell-like cells. Int. J. Biochem. Cell Biol. 39: 1886–1901.

Pinto, M., S. Robine-Leon, M.D. Appay, M. Kedinger, N. Triadou, E. Dussaulx, B. Lacroix, P. Simon-Assmann, K. Haffen, J. Fogh and A. Zweibaum. 1983. Enterocyte-like differentiation and polarization of the human colon carcinoma cell line Caco-2 in culture. Biol. Cell. 47: 323–330.

Qadri, F., A.M. Svennerholm, A.S. Faruque and R.B. Sack. 2005. Enterotoxigenic *Escherichia coli* in developing countries: epidemiology, microbiology, clinical features, treatment, and prevention. Clin. Microbiol. Rev. 18: 465–483.

Raffatellu, M., R.P. Wilson, D. Chessa, H. Andrews-Polymenis, Q.T. Tran, S. Lawhon, S. Khare, L.G. Adams and A.J. Baumler. 2005. SipA, SopA, SopB, SopD, and SopE2 contribute to *Salmonella enterica* serotype typhimurium invasion of epithelial cells. Infect. Immun. 73: 146–154.

Raghunathan, D., T.J. Wells, F.C. Morris, R.K. Shaw, S. Bobat, S.E. Peters, G.K. Paterson, K.T. Jensen, D.L. Leyton, J.M. Blair, D.F. Browning, J. Pravin, A. Flores-Langarica, J.R. Hitchcock, C.T. Moraes, R.M. Piazza, D.J. Maskell, M.A. Webber, R.C. May, C.A. MacLennan, L.J. Piddock, A.F. Cunningham and I.R. Henderson. 2011. SadA, a trimeric autotransporter from *Salmonella enterica* serovar Typhimurium, can promote biofilm formation and provides limited protection against infection. Infect. Immun. 79: 4342–4352.

Ramsden, A.E., D.W. Holden and L.J. Mota. 2007. Membrane dynamics and spatial distribution of *Salmonella*-containing vacuoles. Trends Microbiol. 15: 516–524.

Raso, T., S. Crivellaro, M.G. Chirillo, P. Pais, E. Gaia and D. Savoia. 2011. Analysis of *Escherichia coli* isolated from patients affected by Crohn's disease. Curr. Microbiol. 63: 131–137.

Rocha, S.P., C.M. Abe, V. Sperandio, S.Y. Bando and W.P. Elias. 2011. Atypical enteropathogenic *Escherichia coli* that contains functional locus of enterocyte effacement genes can be attaching-and-effacing negative in cultured epithelial cells. Infect. Immun. 79: 1833–1841.

Rolhion, N., F.A. Carvalho and A. Darfeuille-Michaud. 2007. OmpC and the sigma(E) regulatory pathway are involved in adhesion and invasion of the Crohn's disease-associated *Escherichia coli* strain LF82. Mol. Microbiol. 63: 1684–1700.

Rolhion, N., N. Barnich, M.A. Bringer, A.L. Glasser, J. Ranc, X. Hebuterne, P. Hofman and A. Darfeuille-Michaud. 2010. Abnormally expressed ER stress response chaperone Gp96 in CD favours adherent-invasive *Escherichia coli* invasion. Gut. 59: 1355–1362.

Russell, R.G. and D.C. Blake, Jr. 1994. Cell association and invasion of Caco-2 cells by *Campylobacter jejuni*. Infect. Immun. 62: 3773–3779.

Sakarya, S., C. Gokturk, T. Ozturk and M.B. Ertugrul. 2010. Sialic acid is required for nonspecific adherence of *Salmonella enterica* ssp. enterica serovar Typhi on Caco-2 cells. FEMS Immunol. Med. Microbiol. 58: 330–335.

Sampaio, S.C., J.R. Andrade, J.L. Sampaio, C.R. Carneiro, E. Freymuller and T.A. Gomes. 2011. Distinct interaction of two atypical enteropathogenic *Escherichia coli* strains with enterocytes *in vitro*. Open Microbiol. J. 5: 65–71.

Sansonetti, P.J. and A. Phalipon. 1999. M cells as ports of entry for enteroinvasive pathogens: mechanisms of interaction, consequences for the disease process. Semin. Immunol. 11: 193–203.

Scaletsky, I.C., M.L. Silva and L.R. Trabulsi. 1984. Distinctive patterns of adherence of enteropathogenic *Escherichia coli* to HeLa cells. Infect. Immun. 45: 534–536.

Schippa, S., V. Iebba, V. Totino, F. Santangelo, M. Lepanto, C. Alessandri, F. Nuti, F. Viola, G. Di Nardo, S. Cucchiara, C. Longhi and M.P. Conte. 2012. A potential role of *Escherichia coli* pathobionts in the pathogenesis of pediatric inflammatory bowel disease. Can. J. Microbiol. 58: 426–432.

Schuller, S., G. Frankel and A.D. Phillips. 2004. Interaction of Shiga toxin from *Escherichia coli* with human intestinal epithelial cell lines and explants: Stx2 induces epithelial damage in organ culture. Cell Microbiol. 6: 289–301.

Sepehri, S., R. Kotlowski, C.N. Bernstein and D.O. Krause. 2009. Phylogenetic analysis of inflammatory bowel disease associated *Escherichia coli* and the fimH virulence determinant. Inflamm. Bowel Dis. 15: 1737–1745.

Sepehri, S., E. Khafipour, C.N. Bernstein, B.K. Coombes, A.V. Pilar, M. Karmali, K. Ziebell and D.O. Krause. 2011. Characterization of *Escherichia coli* isolated from gut biopsies of newly diagnosed patients with inflammatory bowel disease. Inflamm. Bowel Dis. 17: 1451–1463.

Servin, A.L. 2005. Pathogenesis of Afa/Dr diffusely adhering *Escherichia coli*. Clin. Microbiol. Rev. 18: 264–292.

Silva, J., D. Leite, M. Fernandes, C. Mena, P.A. Gibbs and P. Teixeira. 2011. *Campylobacter* spp. as a foodborne pathogen: A review. Front Microbiol. 2: 200.

Silva, M., C. Song, W.J. Nadeau, J.B. Matthews and B.A. McCormick. 2004. *Salmonella typhimurium* SipA-induced neutrophil transepithelial migration: involvement of a

PKC-alpha-dependent signal transduction pathway. Am. J. Physiol. Gastrointest. Liver Physiol. 286: G1024–1031.

Simpson, N., R. Shaw, V.F. Crepin, R. Mundy, A.J. FitzGerald, N. Cummings, A. Straatman-Iwanowska, I. Connerton, S. Knutton and G. Frankel. 2006. The enteropathogenic *Escherichia coli* type III secretion system effector Map binds EBP50/NHERF1: implication for cell signalling and diarrhoea. Mol. Microbiol. 60: 349–363.

Smollett, K., R.K. Shaw, J. Garmendia, S. Knutton and G. Frankel. 2006. Function and distribution of EspG2, a type III secretion system effector of enteropathogenic *Escherichia coli*. Microbes Infect. 8: 2220–2227.

Spears, K.J., A.J. Roe and D.L. Gally. 2006. A comparison of enteropathogenic and enterohaemorrhagic *Escherichia coli* pathogenesis. FEMS Microbiol. Lett. 255: 187–202.

Spitz, J., R. Yuhan, A. Koutsouris, C. Blatt, J. Alverdy and G. Hecht. 1995. Enteropathogenic *Escherichia coli* adherence to intestinal epithelial monolayers diminishes barrier function. Am. J. Physiol. 268: G374–379.

Stein, M.A., D.A. Mathers, H. Yan, K.G. Baimbridge and B.B. Finlay. 1996. Enteropathogenic *Escherichia coli* markedly decreases the resting membrane potential of Caco-2 and HeLa human epithelial cells. Infect. Immun. 64: 4820–4825.

Strauman, M.C., J.M. Harper, S.M. Harrington, E.J. Boll and J.P. Nataro. 2010. Enteroaggregative *Escherichia coli* disrupts epithelial cell tight junctions. Infect. Immun. 78: 4958–4964.

Szymanski, C.M., M. King, M. Haardt and G.D. Armstrong. 1995. *Campylobacter jejuni* motility and invasion of Caco-2 cells. Infect. Immun. 63: 4295–4300.

Taieb, F., J.P. Nougayrede, C. Watrin, A. Samba-Louaka and E. Oswald. 2006. *Escherichia coli* cyclomodulin Cif induces G2 arrest of the host cell cycle without activation of the DNA-damage checkpoint-signalling pathway. Cell Microbiol. 8: 1910–1921.

Taniguchi, T., Y. Akeda, A. Haba, Y. Yasuda, K. Yamamoto, T. Honda and K. Tochikubo. 2001. Gene cluster for assembly of pilus colonization factor antigen III of enterotoxigenic *Escherichia coli*. Infect. Immun. 69: 5864–5873.

Tatsuno, I., K. Nagano, K. Taguchi, L. Rong, H. Mori and C. Sasakawa. 2003. Increased adherence to Caco-2 cells caused by disruption of the yhiE and yhiF genes in enterohemorrhagic *Escherichia coli* O157:H7. Infect. Immun. 71: 2598–2606.

Tatsuno, I., M. Horie, H. Abe, T. Miki, K. Makino, H. Shinagawa, H. Taguchi, S. Kamiya, T. Hayashi and C. Sasakawa. 2001. toxB gene on pO157 of enterohemorrhagic *Escherichia coli* O157:H7 is required for full epithelial cell adherence phenotype. Infect. Immun. 69: 6660–6669.

Tatsuno, I., H. Kimura, A. Okutani, K. Kanamaru, H. Abe, S. Nagai, K. Makino, H. Shinagawa, M. Yoshida, K. Sato, J. Nakamoto, T. Tobe and C. Sasakawa. 2000. Isolation and characterization of mini-Tn5Km2 insertion mutants of enterohemorrhagic *Escherichia coli* O157:H7 deficient in adherence to Caco-2 cells. Infect. Immun. 68: 5943–5952.

Thanabalasuriar, A., A. Koutsouris, G. Hecht and S. Gruenheid. 2010a. The bacterial virulence factor NleA's involvement in intestinal tight junction disruption during enteropathogenic *E. coli* infection is independent of its putative PDZ binding domain. Gut Microbes. 1: 114–118.

Thanabalasuriar, A., A. Koutsouris, A. Weflen, M. Mimee, G. Hecht and S. Gruenheid. 2010b. The bacterial virulence factor NleA is required for the disruption of intestinal tight junctions by enteropathogenic *Escherichia coli*. Cell Microbiol. 12: 31–41.

Tian, J.Q. and A. Quaroni. 1999. Dissociation between growth arrest and differentiation in Caco-2 subclone expressing high levels of sucrase. Am. J. Physiol. 276: G1094–1104.

Tobe, T. and C. Sasakawa. 2001. Role of bundle-forming pilus of enteropathogenic *Escherichia coli* in host cell adherence and in microcolony development. Cell Microbiol. 3: 579–585.

Tobe, T. and C. Sasakawa. 2002. Species-specific cell adhesion of enteropathogenic *Escherichia coli* is mediated by type IV bundle-forming pili. Cell Microbiol. 4: 29–42.

Tobe, T., H. Ando, H. Ishikawa, H. Abe, K. Tashiro, T. Hayashi, S. Kuhara and N. Sugimoto. 2005. Dual regulatory pathways integrating the RcsC-RcsD-RcsB signalling system control enterohaemorrhagic *Escherichia coli* pathogenicity. Mol. Microbiol. 58: 320–333.

Tobias, J., M. Lebens, S. Kallgard, M. Nicklasson and A.M. Svennerholm. 2008. Role of different genes in the CS6 operon for surface expression of enterotoxigenic *Escherichia coli* colonization factor CS6. Vaccine. 26: 5373–5380.

Tomson, F.L., V.K. Viswanathan, K.J. Kanack, R.P. Kanteti, K.V. Straub, M. Menet, J.B. Kaper and G. Hecht. 2005. Enteropathogenic *Escherichia coli* EspG disrupts microtubules and in conjunction with Orf3 enhances perturbation of the tight junction barrier. Mol. Microbiol. 56: 447–464.

Toriano, R., A. Kierbel, M.A. Ramirez, G. Malnic and M. Parisi. 2001. Spontaneous water secretion in T84 cells: effects of STa enterotoxin, bumetanide, VIP, forskolin, and A-23187. Am. J. Physiol. Gastrointest. Liver Physiol. 281: G816–822.

Torres, A.G. and J.B. Kaper. 2003. Multiple elements controlling adherence of enterohemorrhagic *Escherichia coli* O157:H7 to HeLa cells. Infect. Immun. 71: 4985–4995.

Torres, A.G., K.J. Kanack, C.B. Tutt, V. Popov and J.B. Kaper. 2004. Characterization of the second long polar (LP) fimbriae of *Escherichia coli* O157:H7 and distribution of LP fimbriae in other pathogenic *E. coli* strains. FEMS Microbiol. Lett. 238: 333–344.

Trabulsi, L.R., R. Keller and T.A. Tardelli Gomes. 2002. Typical and atypical enteropathogenic *Escherichia coli*. Emerg. Infect. Dis. 8: 508–513.

Urbanski, R., S.L. Carrithers and S.A. Waldman. 1995. Internalization of *E. coli* ST mediated by guanylyl cyclase C in T84 human colon carcinoma cells. Biochim. Biophys. Acta. 1245: 29–36.

Vaandrager, A.B., A.G. Bot, J. De Vente and H.R. De Jonge. 1992. Atriopeptins and *Escherichia coli* enterotoxin STa have different sites of action in mammalian intestine. Gastroenterology. 102: 1161–1169.

Valdez, Y., R.B. Ferreira and B.B. Finlay. 2009. Molecular mechanisms of *Salmonella* virulence and host resistance. Curr. Top Microbiol. Immunol. 337: 93–127.

Valvatne, H., H. Sommerfelt, W. Gaastra, M.K. Bhan and H.M. Grewal. 1996. Identification and characterization of CS20, a new putative colonization factor of enterotoxigenic *Escherichia coli*. Infect. Immun. 64: 2635–2642.

van Asten, F.J., H.G. Hendriks, J.F. Koninkx and J.E. van Dijk. 2004. Flagella-mediated bacterial motility accelerates but is not required for *Salmonella serotype* Enteritidis invasion of differentiated Caco-2 cells. Int. J. Med. Microbiol. 294: 395–399.

Van Asten, F.J., H.G. Hendriks, J.F. Koninkx, B.A. Van der Zeijst and W. Gaastra. 2000. Inactivation of the flagellin gene of *Salmonella enterica* serotype Enteritidis strongly reduces invasion into differentiated Caco-2 cells. FEMS Microbiol. Lett. 185: 175–179.

Van Loy, C.P., E.V. Sokurenko and S.L. Moseley. 2002. The major structural subunits of Dr and F1845 fimbriae are adhesins. Infect. Immun. 70: 1694–1702.

Vareille, M., F. Rannou, N. Thelier, A.L. Glasser, T. de Sablet, C. Martin and A.P. Gobert. 2008. Heme oxygenase-1 is a critical regulator of nitric oxide production in enterohemorrhagic *Escherichia coli*-infected human enterocytes. J. Immunol. 180: 5720–5726.

Veilleux, S., N. Holt, B.D. Schultz and J.D. Dubreuil. 2008. *Escherichia coli* EAST1 toxin toxicity of variants 17-2 and O 42. Comp. Immunol. Microbiol. Infect. Dis. 31: 567–578.

Vejborg, R.M., V. Hancock, A.M. Petersen, K.A. Krogfelt and P. Klemm. 2011. Comparative genomics of *Escherichia coli* isolated from patients with inflammatory bowel disease. BMC Genomics. 12: 316.

Viboud, G.I., N. Binsztein and A.M. Svennerholm. 1993. A new fimbrial putative colonization factor, PCFO20, in human enterotoxigenic *Escherichia coli*. Infect. Immun. 61: 5190–5197.

Vieira, M.A., J.R. Andrade, L.R. Trabulsi, A.C. Rosa, A.M. Dias, S.R. Ramos, G. Frankel and T.A. Gomes. 2001. Phenotypic and genotypic characteristics of *Escherichia coli* strains of non-enteropathogenic *E. coli* (EPEC) serogroups that carry EAE and lack the EPEC adherence factor and Shiga toxin DNA probe sequences. J. Infect. Dis. 183: 762–772.

Viswanathan, V.K., K. Hodges and G. Hecht. 2009. Enteric infection meets intestinal function: how bacterial pathogens cause diarrhoea. Nat. Rev. Microbiol. 7: 110–119.

Viswanathan, V.K., A. Weflen, A. Koutsouris, J.L. Roxas and G. Hecht. 2008. Enteropathogenic *E. coli*-induced barrier function alteration is not a consequence of host cell apoptosis. Am. J. Physiol. Gastrointest. Liver Physiol. 294: G1165–1170.

Viswanathan, V.K., A. Koutsouris, S. Lukic, M. Pilkinton, I. Simonovic, M. Simonovic and G. Hecht. 2004. Comparative analysis of EspF from enteropathogenic and enterohemorrhagic *Escherichia coli* in alteration of epithelial barrier function. Infect. Immun. 72: 3218–3227.

Visweswariah, S.S., G. Shanthi and T.S. Balganesh. 1992. Interaction of heat-stable enterotoxins with human colonic (T84) cells: modulation of the activation of guanylyl cyclase. Microb. Pathog. 12: 209–218.

Visweswariah, S.S., V. Ramachandran, S. Ramamohan, G. Das and J. Ramachandran. 1994. Characterization and partial purification of the human receptor for the heat-stable enterotoxin. Eur. J. Biochem. 219: 727–736.

Wagner, S., L. Konigsmaier, M. Lara-Tejero, M. Lefebre, T.C. Marlovits and J.E. Galan. 2010. Organization and coordinated assembly of the type III secretion export apparatus. Proc. Natl. Acad. Sci. USA 107: 17745–17750.

Weikel, C.S., C.L. Spann, C.P. Chambers, J.K. Crane, J. Linden and E.L. Hewlett. 1990. Phorbol esters enhance the cyclic GMP response of T84 cells to the heat-stable enterotoxin of *Escherichia coli* (STa). Infect. Immun. 58: 1402–1407.

Wick, L.M., W. Qi, D.W. Lacher and T.S. Whittam. 2005. Evolution of genomic content in the stepwise emergence of *Escherichia coli* O157:H7. J. Bacteriol. 187: 1783–1791.

Wimer-Mackin, S., R.K. Holmes, A.A. Wolf, W.I. Lencer and M.G. Jobling. 2001. Characterization of receptor-mediated signal transduction by *Escherichia coli* type IIa heat-labile enterotoxin in the polarized human intestinal cell line T84. Infect. Immun. 69: 7205–7212.

Wine, E., J.C. Ossa, S.D. Gray-Owen and P.M. Sherman. 2009. Adherent-invasive *Escherichia coli*, strain LF82 disrupts apical junctional complexes in polarized epithelia. BMC Microbiol. 9: 180.

Winsor, D.K., Jr., S. Ashkenazi, R. Chiovetti and T.G. Cleary. 1992. Adherence of enterohemorrhagic *Escherichia coli* strains to a human colonic epithelial cell line (T84). Infect. Immun. 60: 1613–1617.

Wolf, A.A., M.G. Jobling, S. Wimer-Mackin, M. Ferguson-Maltzman, J.L. Madara, R.K. Holmes and W.I. Lencer. 1998. Ganglioside structure dictates signal transduction by cholera toxin and association with caveolae-like membrane domains in polarized epithelia. J. Cell Biol. 141: 917–927.

Woodcook, S., I. Williamson, I. Hassan and M. Mackay. 1991. Isolation and characterisation of clones from the Caco-2 cell line displaying increased taurocholic acid transport. J. Cell Sci. 98(Pt 3): 323–332.

Young, K.T., L.M. Davis and V.J. Dirita. 2007. *Campylobacter jejuni*: molecular biology and pathogenesis. Nat. Rev. Microbiol. 5: 665–679.

Yuhan, R., A. Koutsouris, S.D. Savkovic and G. Hecht. 1997. Enteropathogenic *Escherichia coli*-induced myosin light chain phosphorylation alters intestinal epithelial permeability. Gastroenterology. 113: 1873–1882.

Zav'yalov, V., A. Zavialov, G. Zav'yalova and T. Korpela. 2010. Adhesive organelles of Gram-negative pathogens assembled with the classical chaperone/usher machinery: structure and function from a clinical standpoint. FEMS Microbiol. Rev. 34: 317–378.

Zavialov, A., G. Zav'yalova, T. Korpela and V. Zav'yalov. 2007. FGL chaperone-assembled fimbrial polyadhesins: anti-immune armament of Gram-negative bacterial pathogens. FEMS Microbiol. Rev. 31: 478–514.

Zheng, J., J. Meng, S. Zhao, R. Singh and W. Song. 2006. Adherence to and invasion of human intestinal epithelial cells by *Campylobacter jejuni* and *Campylobacter coli* isolates from retail meat products. J. Food Prot. 69: 768–774.

Zurawski, D.V., K.L. Mumy, L. Badea, J.A. Prentice, E.L. Hartland, B.A. McCormick and A.T. Maurelli. 2008. The NleE/OspZ family of effector proteins is required for polymorphonuclear transepithelial migration, a characteristic shared by enteropathogenic *Escherichia coli* and *Shigella flexneri* infections. Infect. Immun. 76: 369–379.

Zweibaum, A., M. Laburthe, E. Grasset and D. Louvard. 1991. Use of cultured cell lines in studies of intestinal cell differentiation and function. In *Handbook of Physiology, The Gastrointestinal System, Intestinal Absorption and Secretion*. Schultz SJ, F.M., Frizell RA (ed.) Bethesda: American Physiological Society. pp. 223–255.

5

Manipulation of the Host-Cell Pathways by Bacterial Enteropathogens

Brice Sperandio[1,2] and *Philippe J. Sansonetti*[1,2,3,*]

Introduction

The intestinal epithelium is constantly exposed to many resident symbiotic bacteria and foreign bacterial pathogens that establish a docking in the gut. To cope with this microbial challenge, intestinal epithelial cells rely on multiple networks of microbial sensing and defense systems that act synergistically against bacterial invaders. Mucosal immunity consists of an innate and adaptive immune response playing a major role as dedicated host defense mechanism against pathogen survival and dissemination in the host (Macdonald and Monteleone 2005). Both the intestinal epithelial cell layer and the mucosal surface act as defensive barriers against bacterial intruders. The intestinal epithelium is covered by protective layers consisting of cell surface and gel forming mucins, antigen-specific secretory

[1] Unité de Pathogénie Microbienne Moléculaire, Institut Pasteur, 25-28 Rue du Docteur Roux, 75724 Paris Cedex 15, France.
[2] INSERM U786, Institut Pasteur, 25-28 Rue du Docteur Roux, 75724 Paris Cedex 15, France.
E-mail: brice.sperandio@pasteur.fr
[3] Chaire de Microbiologie et Maladies Infectieuses, Collège de France, 11 Place Marcelin Berthelot, 75005 Paris, France.
* Corresponding author: philippe.sansonetti@pasteur.fr

immunoglobulins A, and many nonspecific antimicrobial molecules, including alpha- and beta-defensins, cathelicidin, lysozyme, phospholipases and lectins (McGuckin et al. 2011). To maintain epithelial integrity and tissue homeostasis, intestinal epithelial cells are constantly renewed throughout the host's lifespan by a continuous supply of progenitors from cryptic stem cells (Radtke and Clevers 2005). This epithelial turnover forms a crucial physical and functional barrier, its renewal being achieved by the proliferation of epithelial progenitor cells that migrate upwards from the bottom of the intestinal crypts. The intestinal epithelium renews itself within a few days, thereby providing a major intrinsic defense system that limits the persistence of pathogen colonization. This epithelial turnover indeed helps to eliminate colonizing pathogens, to confine bacterial spreading, and to localize inflammation (Chichlowski and Hale 2008).

The interactions between pathogens, the intestinal mucosa, and host innate defense responses are among the most critical factors that define the fate of bacterial infections and diseases. Enteric pathogens use a limited set of platforms to implement a wide range of pathogenic strategies (Fig. 1). Remarkably, not only are the macromolecular systems conserved, but also some of the same host cell targets are exploited by different pathogens, even though many of the effector molecules and the resulting diseases differ. Some of these pathogens, including *Clostridium* species, choose to remain in the intestinal lumen and release exotoxins into the environment that induce epithelial cell cytoskeleton rearrangements, which ultimately

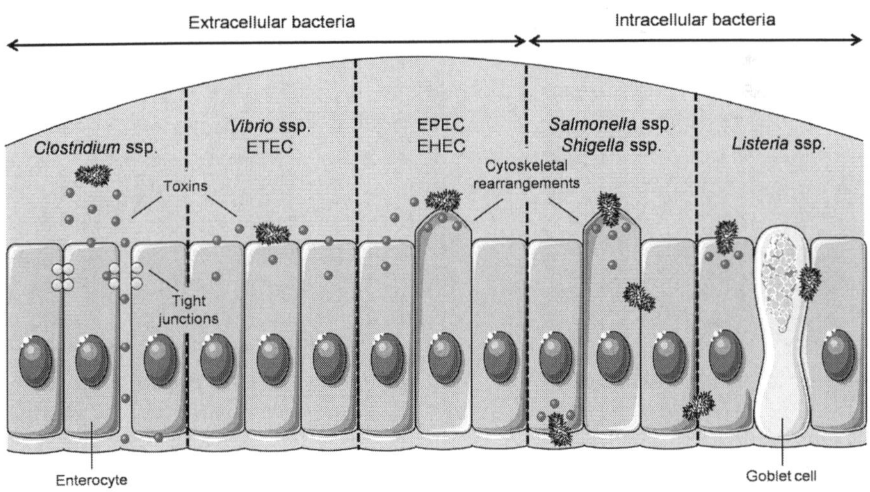

Figure 1. The cell biology of intestinal bacterial infections (Figure carried out thanks to Servier Medical Art).

Color image of this figure appears in the color plate section at the end of the book.

favor dissemination of bacteria during bacteremia (Aktories 2011). Others, such as *Vibrio cholerae* and Enterotoxic *Escherichia coli* (ETEC), adhere to the intestinal epithelial cells without inducing significant modifications of the cell morphology, and produce exotoxins that cause severe diarrhoea (Donnenberg 2000). In contrast, Enteropathogenic (EPEC) and Enterohaemorrhagic (EHEC) *Escherichia coli* intimately attach to the surface of epithelial cells with provoking cytoskeletal disruption, and can deliver toxins that enable them to cause diarrhoae or extraintestinal infections (Croxen and Finlay 2010). Some others pathogens inject toxins directly into intestinal epithelial cells, to then invade and grow in their cytoplasm. They can spread locally to adjacent cells of the epithelium, such as *Shigella* species, or spread through the host systemic route, such as *Salmonella* and *Listeria* species. Both cases are associated with a variety of disease syndromes (Donnenberg 2000). Naturally, all of these planktonic, adherent or invasive pathogens have developed subversive strategies that are essential for their persistence and dedicated to benefit from their host depending on their living characteristics. However, due to the capacity of adherent and invasive pathogens to promote actively their own survival in the intestinal mucosa, and sometimes beyond, only these ones will be used here as models to describe manipulations of the host cell pathways by bacterial enteropathogens.

To breach the epithelium lining in order to penetrate, propagate, disseminate deeper into the tissue and achieve successful infection, these pathogens can subvert and usurp host signaling cascades and defensive functions. They have developed mechanisms that modulate the host defense systems, and have evolved extra- or intracellular adaptive and survival systems. Most of them possess a type III secretion system (T3SS) dedicated to the injection of their toxins into host cells. A mechanism involving a cytoplasmic sorting platform ensures a hierarchy in the toxin secretion (Lara-Tejero et al. 2011). These injected bacterial effectors act by manipulating an array of host cell signaling pathways. The T3SS uses a needle-like system that forms a translocating apparatus through the host cell membrane to deliver the effectors into the host cytoplasm (Anderson and Schneewind 1999). Bacterial effectors subvert numerous eukaryotic pathways for the benefit of the pathogen. For intracellular bacteria, these activities include facilitating entry into the host cell and promoting survival and replication of the bacteria in the cytosol, or in intracellular compartments. Elucidating the mechanisms of action of these virulence factors provides a powerful approach to understand host cell pathways. Acquiring these virulence factors is crucial in promoting the fitness and survival of an enteropathogen while it coevolves with its host (Shames et al. 2009). The pathogenic diversity is a result of the evolution and dispersion of virulence genes that are spread between the bacterial species on transmissible genetic elements. These elements include plasmids, bacteriophages and pathogenicity islands. The evolutionary history of these genetic elements is obscured by subsequent insertions, deletions and rearrangements, but ultimately, their gain plays a central role in the degree of pathogenicity of enteric bacteria.

Recent progress has been made in our understanding of the molecular mechanisms by which bacterial enteropathogens interact with host cells and manipulate various host pathways that result in the induction of intestinal inflammation and disease. Bacterial pathogens specifically attack key intracellular signaling pathways to alter host responses in a way that benefits the pathogen. The same host pathways are often targeted by different bacterial effectors and can be targeted at several points by the same pathogen, ensuring an override of important cellular functions. Specifically, the identification of bacterial and host proteins that are involved in such processes, like the invasion of epithelial cells, induction and repression of signaling cascades, and establishment of a niche for intracellular replication, have contributed insights into the complex pathogenesis of bacterial enteropathogens and will be highlighted here in detail.

Manipulation of the Host Cell Cytoskeleton and Membranous Structures

To promote their survival and proliferation, extra- and intracellular bacterial enteropathogens can use host cells to allow their own adherence, colonization, possibly invasion and dissemination. Various intracellular microbes exploit cytoskeletal components to gain entry into host cells (Steven et al. 2006). The cytoskeleton of eukaryotic cells is composed of actin filaments, microtubules and intermediates filaments. In terms of bacterial pathogenesis, the most extensively studied are actin filaments. Bacterial enteropathogens do not usually interact directly with actin filaments themselves. Instead, they manipulate and control the polymerization of actin filaments by subverting cellular regulators of this mechanism, like host cell phosphoinositide (PI) metabolism and GTPase signalling, through the action of bacterial effectors or toxins, delivered by the pathogen to the cytoplasm of targeted cells (Hame et al. 2011). Besides manipulating the host cell cytoskeleton, bacterial enteropathogens are able to modulate host cell membrane structures to counteract the rapid and regular epithelial cell turnover, which occurs basally every four to five days in the intestine. To dampen exfoliation of infected epithelial cells, on or in which pathogens can replicate, many bacteria have evolved mechanisms to antagonize the natural epithelial detachment.

The endocytic entry pathway

Intracellular bacterial enteropathogens invade non-phagocytic host cells, such as intestinal epithelial cells, using zipper or trigger mechanisms (Cossart and Sansonetti 2004). Bacteria using the zipper mechanism, such as *Listeria monocytogenes*, express surface proteins that bind receptors on the host cell membrane on contact, inducing a cascade that reorganize the actin cytoskeleton to internalize the bacterium. In the case of *Listeria*, the InlA protein is recognized by the human adhesion molecule E-cadherin. A second invasion protein, InlB, targets the hepatocyte growth factor receptor MET and promotes host cell invasion by favoring adhesion and internalization in synergy with InlA (Bonazzi et al. 2009). The trigger mechanism, which is best characterized in *Shigella flexneri* and *Salmonella* Typhimurium, employs the T3SS to inject protein effectors across the host plasma membrane. These proteins directly interact with the cellular components that regulate actin dynamics. In *Shigella*, several effector proteins are involved in promoting bacterial basolateral entry into polarized epithelial cells, including IpaA, IpaB, IpaC, IpgB1, IpgB2, IpgD and VirA (Nhieu and Sansonetti 1999). IpaB interacts with the cell surface hyaluronan receptor, which localizes to cholesterol-rich lipid rafts, suggesting that lipid rafts are required for *Shigella* binding and entry into host cells (Lafont et al. 2002). Lipids also play a role during the active entry of bacterial enteropathogens. Phosphatidylinositol-5-phosphate (PtdIns(5)P) accumulates in epithelial cells infected with *Shigella*. The T3SS injected bacterial effector IpgD, an inositol phosphate phosphatase, specifically dephosphorylates PtdIns(4,5)P_2 into PtdIns(5)P, which facilitates both membrane ruffling and bacterial invasion (Niebuhr et al. 2002). Another effector, VirA, is delivered into the host-cell cytoplasm near the site of bacterial entry and induces local microtubule degradation (Yoshida et al. 2002).

Fate of the endocytic vacuole

After internalization, bacteria can either form an intracellular vacuole to replicate, as is the case for *Salmonella*, or escape to the cytosol, as observed for *Shigella* and *Listeria*. Bacterial enteropathogens that remain in vacuoles often use effector proteins to manipulate vesicular trafficking, thereby providing a protective niche within the host cell (Meresse et al. 1999). The bacterium-containing vacuoles are derived from host membranes, such as endosomal vesicles and endoplasmic reticulum. Many bacterial effectors interact with host endocytic pathways to maintain the integrity of the vacuole and control its maturation. Following internalization and interaction with early endosomes, *Salmonella*-containing vacuoles (SCV) avoid fusion with lysosomes and remain in a late endosomal-like compartment (Knodler and

Steele-Mortimer 2003). *Salmonella* uses two distinct T3SS, encoded by two pathogenicity islands SPI1 and SPI2, to deliver effector proteins into host cells. Effectors from SPI1 have been shown to be essential for invasion of epithelial cells (Patel and Galan 2005), but it is likely that in macrophages SopE is the only SPI1 effector that plays a role in the maturation of SCV (Mukherjee et al. 2001). The SPI2 secretion system is required for bacterial proliferation in macrophages (Cirillo et al. 1998), and the SPI2 effector protein SpiC was shown to inhibit endosome fusion (Uchiya et al. 1999). SpiC interacts with the host protein TASSC, a member of the NIPSNAP protein family involved in the control of membrane trafficking (Lee et al. 2002). Concomitant with bacterial replication in intestinal epithelial cells is the occurrence of membrane tubules defined as *Salmonella*-induced filaments (SIF) (Garcia-del Portillo et al. 1993). The SIF membrane contains lysosomal glycoproteins (Brumell et al. 2002), and SIF appearance is under the control of the SifA effector that is essential for the formation of SIFs in epithelial cells (Stein et al. 1996). Both, SifA and PipB2 effector proteins regulate trafficking of vesicles onto microtubules by interacting with the host protein kinesin 1. These steps of the *Salmonella* pathogenesis are crucial for the maintenance of the membrane integrity of SCVs and replication in infected cells (Salcedo et al. 2001; Helaine et al. 2010).

Motility in the cytoplasm

One strategy that possibly emerged to avoid the deleterious environment resulting from fusion with endocytic compartments that contain degradative enzymes, is to rapidly escape the vacuole to gain access to the cytoplasm. Active invasion by *Shigella* produces a vacuole around the invading microorganism. However, during invasion of intestinal epithelial cells, *Shigella* occupies this vacuole only briefly. Bacteria express the IpaB effector, which, in addition to its role in host cell entry, has a lytic activity that is used to breach the vacuolar membrane (High et al. 1992). *Listeria* is also initially found in vacuoles and similarly to *Shigella*, these vacuoles are short lived. *Listeria* uses the membrane-pore-forming toxin listeriolysin O (LLO), as well as the enzymes PlcA and PlcB, to destroy the surrounding membrane, thereby allowing escape from the vacuole (Shaughnessy et al. 2006). After escape from the nascent vacuoles, bacteria invade the cytoplasm and use the actin cytoskeleton to facilitate intracellular movement and, eventually spread from cell to cell. This is performed by inducing actin nucleation at one pole of the bacterium, to gain the propulsive force that is required to move through the host cell and into neighboring cells. The bacterial effectors that are involved in this mechanism vary from species to species, nevertheless, these virulence proteins share the ability to recruit and activate the host cell ARP2/3 protein complex, which is required for bacterial motility (Pantaloni et al. 2001). In the case of *Shigella*, the IcsA effector can directly interact with the host N-WASP and CD42 proteins, allowing recruitment and activation of the ARP2/3 complex and the subsequent actin nucleation and elongation at one pole of the bacterium, thus gaining propulsive force (Suzuki et al. 1998). By contrast, the motility of *Listeria* is mediated by the ActA effector, which binds directly to both the ARP2/3 complex and the actin-associated protein VASP (Welch et al. 1997).

During bacterial movement, some bacteria impinge on the host cell membrane and cause the membrane to protrude and penetrate into neighboring cells, thereby allowing the bacteria to disseminate into adjacent cells, but inducing inflammatory stimuli emanating from infected epithelial cells. The subsequent inflammatory process is elicited through a cellular mechanism of sensing implying host protein complexes termed inflammasomes (Rathinam et al. 2010). Besides, upon preferential adherence to accessible E-cadherin on mucin-secreting goblet cells, *Listeria* can be internalized, rapidly transcytosed across the intestinal epithelium, and released in the lamina propria by exocytosis from where bacteria disseminate systemically (Nikitas et al. 2011).

Pedestal formation

Manipulation of actin-associated cytoskeletal components can also occur upon infection with extracellular bacterial enteropathogens, which adhere to host cells and secrete T3SS effectors that lead to modification of cytoplasmic membrane for effective infection. The attaching and effacing pathogens *Escherichia coli* EHEC and EPEC have evolved an elaborate actin-recruiting mechanism (Croxen and Finlay 2010). For this purpose, the Tir bacterial effector mediates extensive modification of host-cell actin filaments under the adherent pathogen. The Tir protein is injected into the target cell by the T3SS and embeds itself in the cytoplasmic membrane, where it anchors the bacterium by binding to the bacterial outer-membrane protein intimin. During EPEC infection, the Tir protein is phosphorylated on the cytoplasmic side of the host cell membrane and recruits the host adaptor protein Nck (Gruenheid et al. 2001). In the case of EHEC infection, Nck is not recruited. Instead, the TccP bacterial effector is involved (Campellone et al. 2004). The host protein N-WASP and the ARP2/3 complex are recruited downstream of Nck or TccP, and lead to the polymerization of actin filaments favoring the adherent extracellular pathogen. This mechanism results in the appearance of pedestal structures on the host cell surface membrane, where the bacterium resides (Shaner et al. 2005).

Epithelial cell detachment

Bacterial enteropathogens bind to the intestinal epithelial layer and some of them invade epithelial cells, where they spread from cell to cell to avoid any extracellular stages. This infectious process induces extensive epithelial damages and would lead to shedding of epithelial cells. However, several pathogens are able to actively stabilize the epithelial lining, by promoting cell adherence. Recently, the T3SS effector protein OspE of *Shigella* has been shown to target a host cell integrin-linked kinase (ILK) in order to reinforce epithelial adhesion to the basal lamina (Kim et al. 2009). Molecular interaction between OspE and ILK increases the levels of beta-1-integrin on the epithelial cell surface and decrease the disassembly of focal adhesions, therefore stabilizing focal adhesion complexes and favoring the adhesion of infected epithelial cells to the extracellular matrix. Several other pathogens, including *Salmonella*, EPEC and EHEC, have been found to possess OspE orthologous effectors, which probably act similarly to enhance ILK recruitment and stabilize the intestinal lining, thereby promoting bacterial infection (Van Nhieu and Guignot 2009).

Manipulation of Host Cell Immune Responses

During infection, bacterial enteropathogens interact with host cell membranes through different mechanisms. The interaction between the bacterium and the host plasma membrane and its embedded receptors results in the activation of multiple host signaling pathways involved in induction of immune responses. Most of bacterial membrane components, such as peptidoglycan and lipopolysaccharide (LPS), are recognized by Toll-like receptors (TLRs) and Nod-like receptors (NLRs), which activate the host innate defense systems, stimulate pro-inflammatory signaling cascades and induce adaptive immune responses. These different pathogen-sensor systems have the ability to signal to the nucleus to regulate genes involved in immune defenses. This occurs through activation of the NF-κB and MAPK signaling pathways in a cascade of activated kinases that ultimately leads to a transcriptional response. Therefore, each of these signaling cascades and associated regulatory factors constitute crucial communication milestones in the immune response and represent targets of interest for bacterial enteropathogen subversion.

Regulation of the innate immune response by the NF-κB pathway

The NF-κB pathway is activated by the IκB kinase kinase (IKK) complex, which phosphorylates the alpha subunit of inhibitor of NF-κB (IκB-alpha), the cytoplasmic binding partner of NF-κB. Phosphorylation of IκB-alpha leads to its poly-ubiquitination and proteolytic degradation, resulting in exposure of the nuclear localization signal on the now unbound NF-κB (Alkalay et al. 1995). This mechanism promotes nuclear translocation of NF-κB and direct binding to DNA, thereby activating the transcription of genes containing a κB site, such as pro-inflammatory genes encoding cytokines and antimicrobial peptides, and genes involved in the antiapoptotic response (Beutler 2009). Bacterial enteropathogens have evolved a multitude of strategies to interfere with NF-κB signaling, often at several levels, highlighting the importance of inhibiting this signaling pathway for their own survival. The T3SS effector AvrA of *Salmonella* targets different host proteins and plays various roles during the course of infection (Liu et al. 2010). Early on during infection, this protein inhibits NF-κB signaling (Jones et al. 2008). AvrA is a deubiquitylase that inhibits the NF-κB pathway by removing ubiquitin from IκB-alpha, the negative regulator of the NF-κB pathway in epithelial cells (Ye et al. 2007). Another *Salmonella* T3SS protein encoded in SPI2, SseL, also possesses a deubiquitylase activity that inhibits degradation of IκB-alpha (Le Negrate et al. 2008). *Salmonella* lacking SseL causes increased NF-κB activation in macrophages as a result of ubiquitin-mediated degradation of IκB-alpha. Comparably, *Shigella* also blocks the NF-κB pathway on several levels to silence immune responses. Upon invasion of epithelial cells, the NF-κB pathway would be activated after recognition of *Shigella* by the cytoplasmic NLR sensor NOD1, which causes oligomerization of NOD1 and recruitment of the RIP2/RICK kinase, and the NF-κB essential modulator NEMO, a regulatory subunit of the IKK complex (Girardin et al. 2001). However, *Shigella* injects the T3SS effector IpaH9.8 that possesses E3 ligase activity for NEMO, into the host cell. IpaH9.8 binds both NEMO and ABIN-1, an ubiquitin-binding protein, and promotes poly-ubiquitination of NEMO. Poly-ubiquitinated NEMO undergoes proteolytic degradation, therefore inhibiting NEMO-dependent activation of the IKK complex (Rohde et al. 2007; Ashida et al. 2010). Besides, the T3SS effector OspG blocks the pathway downstream by preventing poly-ubiquitination and subsequent degradation of IκB-alpha (Kim et al. 2005). OspG is a serine/threonine kinase that binds various ubiquitinylated E2 ubiquitin-conjugating enzymes, including UBCH5. This effector prevents the transfer of ubiquitin onto phosphorylated IκB-alpha by an E3 ubiquitin ligase, the degradation of phosphorylated IκB-alpha, and thus blocking NF-κB activation. Furthermore, the T3SS effector OspZ blocks nuclear translocation of the RelA subunit of the NF-κB complex upon activation of the signaling cascade (Newton et al. 2010). As a probable consequence of the combined synergistic activity of these effectors, transcriptional induction of genes from the host innate immune response usually activated upon infection is dampened, including chemokine- and antimicrobial peptide-encoding genes (Sperandio et al. 2008). The NleH1 and NleH2 effector proteins from EHEC have a high level of sequence similarity with OspG from *Shigella* but have different host cell targets than OspG. These two proteins lack the ability to block IκB-alpha degradation, but instead interact with the 40S ribosomal protein S3 (RPS3), a subunit of NF-κB complexes that regulates NF-κB-dependent transcription (Gao et al. 2009). The binding affinity of the NF-κB complex to DNA is greatly increased in the presence of RPS3 (Wan et al. 2007). This regulator of NF-κB allows specificity in expression of target genes, such as those involved in cytokine production. Mechanistically, binding of NleH1 to RPS3 inhibits RPS3 phosphorylation by IKK-beta, which sequesters RPS3 in the cytoplasm and prevents its nuclear translocation, thereby disrupting transcriptional up-regulation of RSP3-dependent genes (Gao et al. 2009). Furthermore, the existence of multiple NF-κB inhibitors is observed in EHEC serogroup O111, which encodes an orthologue of the OspG effector of *Shigella*. EHEC serogroup O111 lacking OspG are able to block NF-κB transfer to the nucleus, suggesting existence of additional T3SS effector proteins targeting this host pathway in this EHEC serogroup (Nobe et al. 2009). EPEC can either activate or suppress the NF-κB pathway, through both T3SS-dependent and T3SS-independent mechanisms, probably by activation of TLRs (Malladi et al. 2004). The T3SS effector proteins NleE and NleB can inhibit NF-κB activation by inhibiting IκB-alpha phosphorylation (Nadler et al. 2010). These two proteins block the phosphorylation and activation of IKK-beta by targeting upstream proteins from the cascade, such as TAK1. In addition, the NleE protein can specifically block nuclear translocation of the c-REL subunit of the NF-κB complex, but not of the NFKB1 subunit or the transcription factors STAT1 and STAT2 (Newton et al. 2010). Recently, the NleC and NleD proteins were identified as proteases that specifically cleave and inactivate RelA and the JNK MAPK protein, respectively, thus blocking NF-κB and the activator protein 1 (AP-1) activation, leading to inhibition of the pro-inflammatory cytokine IL-8 (Baruch et al. 2011).

Regulation of the innate immune response by the MAPK pathway

For the MAPK signaling pathways, three types of kinases termed MAP kinase kinase kinase (MKKK), MAP kinase kinase (MKK), and MAPK, are sequentially phosphorylated, ending with the phosphorylation of conserved threonine residues of MAPK proteins p38, JNK and ERK. The activated MAPK translocate to the nucleus to phosphorylate and thereby activate transcription factors and proteins involved in chromatin assembly required for transcription of target genes (Zheng and Guan 1994). MAPKs control diverse processes important for cell proliferation, differentiation, apoptosis and immune regulation. Bacterial infection can activate MAPK signaling pathways directly by stimulating pathogen-sensors and indirectly by inducing cell damage and stress. As for the NF-κB signaling pathway, these components are targeted by bacterial enteropathogens to modulate immune responses. For example, *Escherichia coli* strain K1 suppresses the production of pro-inflammatory cytokines from infected monocytes. However, in the absence of the OmpA protein, the pathogen activates the NF-κB cascade via ERK1-ERK2 and p38 MAPK pathways, leading to the production of pro-inflammatory chemokines and cytokines (Selvaraj and Prasadarao 2005). Therefore, OmpA probably targets a kinase common to the NF-κB and MAPK signaling pathways. The T3SS effector protein VopA of *Vibrio parahaemolyticus* can inhibit the MAPK signaling pathway. This virulence protein has a dual function in which it acetylates the activation loop of MKKs, thereby inhibiting activation, and acetylates a conserved lysine residue in the catalytic loop of MKKs that is required for coordination of the gamma-phosphate of ATP, thereby inhibiting the binding of ATP (Trosky et al. 2007). Thus, by inhibiting MAPK pathways, the VopA effector allows survival of *Vibrio* through activation of the NF-κB pathway and the associated anti-apoptotic host cell functions.

In the case of *Shigella*, the T3SS effector OspF, which is a protein injected into host cells by the bacterium during infection, localizes to the nucleus and is able and sufficient to prevent MAPK activation (Arbibe et al. 2007). The OspF protein inhibits the activation of MAPKs p38 and ERK but not JNK, or any upstream MKKs. Its phosphothreonine lyase activity eliminates a phosphate group from a phosphothreonine residue in the activation loop of p38 and ERK, leading to an irreversible modification of the targeted MAPKs (Li et al. 2007). The kinases can no longer be reactivated by phosphorylation. The reaction catalyzed by OspF inhibits ERK-mediated activation of the mitogen- and stress-activated kinase 1 and 2 (MSK1 and MSK2). This reaction subsequently prevents phosphorylation of histone H3 at the serine residue Ser10, which is prerequisite for chromatin remodeling and presentation of κB binding sites at NF-κB-regulated gene promoters (Brennan and Barford 2009). As a consequence, transcriptional induction of at least 46 different genes usually activated by NF-κB is blocked, including the neutrophil chemo-attractant interleukin IL-8 encoding gene (Arbibe et al. 2007). The OspF family member SpvC effector produced by *Salmonella* removes phosphates from phosphorylated ERK2, suggesting that members of this family retain lyase activity against MAPK phosphorylated threonine residues. In addition, it has been shown in yeast that expression of the OspF protein inhibits activation of the MAPK proteins FUS3, KSS1, HOG1 and SLT2 (Kramer et al. 2007). Thus, the T3SS effector proteins related to OspF of *Shigella* constitute a family of phosphothreonine lyases capable of exploiting a highly conserved motif present in MAPK proteins to manipulate activation of this signaling pathway. Besides SpvC, the *Salmonella* AvrA and SteC effectors also target the host MAPK signaling proteins. Whereas AvrA acetylates MKKs to shut off the JNK MAPK pathway during later stages of the *Salmonella*-induced infectious process (Du and Galan 2009), SteC phosphorylates MEK to regulate the host actin cytoskeleton and possibly acts to restrain bacterial growth and thereby regulate virulence (Odendall et al. 2012).

The adaptive immune response

The adaptive immune response functions together with the innate immune response but is pathogen specific. Initially, this response requires precise bacterial identification, which is followed by induction of immunity. Sequentially, bacterial enteropathogens are recognized and internalized by dedicated cells termed antigen-presenting cells (APCs). Then bacteria are degraded by proteases in specialized vesicular compartments. The peptides resulting from degraded bacterial components then bind to host proteins termed major histocompatibility complexes (MHC). These antigen complexes are exposed at the surface of APCs, where they are presented to whole circulating immune cells, including dendritic cells (DCs), B and T lymphocytes. Bacterial enteropathogens manipulate adaptive immune responses through various ways. Although the mechanisms used by the pathogens are not completely defined at the molecular level, it is clear that these subversive strategies are diverse. For example, *Salmonella* can block antigen presentation by DCs, a class of APCs, through an incompletely characterized mechanism. The process is accomplished by inducing a decrease in the number of peptide-bound MHC complexes on the surface of infected DCs, decreasing activation and proliferation of T lymphocyte cells, an important class of adaptive immune cells (Cheminay et al. 2005). Recently, the consequences of the interaction between *Shigella* and lymphocytes have been investigated (Konradt et al. 2011). The pathogen can invade activated T lymphocytes, using its T3SS apparatus. This invasion process leads to inhibition of chemokine-induced T cell migration, a mechanism mediated by the T3SS effector protein IpgD, also involved in bacterial invasion and host-cell survival mechanisms (Niebuhr et al. 2002). Through its biochemical activity, IpgD decreases the PtdIns(4,5)P_2 pool at the T cell plasma membrane, leading to dephosphorylation of the host-cell Ezrin-Radixin-Moesin (ERM) proteins and their inability to relocalize at one pole of T cells upon a chemokine stimulus. Ultimately, this subversive strategy affect the formation of the polarized edge of T cells required for their migration process.

Manipulation of the Host Cell Death Pathways and Autophagy

The host responses to infection by bacterial enteropathogens are initially accomplished by activation of inflammatory pathways, secretion of antimicrobial peptides at the mucosal surface, and recruitment of immune cells to the site of infection. However, upon persistence of the infection process and unresolved conditions, cell death signaling pathways or autophagy are activated to eliminate pathogens and/or infected cells from the mucosa. These mechanisms counteract replication of intracellular pathogens in infected cells. Apoptosis, necrosis, and pyroptosis constitute the main mechanisms by which programmed host cell death occurs and leads to the exposure of microbes to extracellular immune surveillance. Besides, autophagy is a defense mechanism by which cells take up and, by lysosome fusion, degrade damaged organelles, misfolded protein aggregates, and invading intra-cellular pathogens. Each of these cell defense programs is characterized by specific morphological changes and dedicated signaling pathways. Unsurprisingly, bacterial enteropathogens have evolved various strategies to manipulate host cell death pathways and autophagy to subvert host cell defense responses.

The apoptosis pathway

Apoptosis is one of the cell defense mechanisms upon pathogen infection. Usually, apoptosis proceeds without eliciting an inflammatory response because the cytoplasmic content of infected cells is not spilled into the extracellular environment (Savill et al. 2002). Apoptosis can be induced by two distinct pathways, the extrinsic pathway or the intrinsic pathways. The extrinsic pathway is activated after stimulation of the transmembrane death receptors, such as TNF-R1, FasL, and APO2/APO3. These receptors transduce extracellular death signaling to the apoptotic machinery, which ultimately leads to activation of the caspase-3 and -7. The intrinsic pathway is initiated through the release of apoptogenic proteins from the mitochondria (Rudel et al. 2010). Several insults activate Bcl-2 homology 3 (BH3)-only proteins, which over side the inhibitory effects of anti-apoptotic Bcl-2 proteins. The activated BH3-only proteins promote the oligomerization of pro-apoptotic proteins, including the Bax protein, in the mitochondrial outer membrane, which results in the release of several proteins onto the cell cytoplasm, including cytochrome *c*. These proteins induce formation of the apoptosome, a machinery which proteolytically activates pro-caspase-9. Then, caspase-9 cleaves and activates other caspases that in turn cleave various substrates to ultimately induce apoptosis (Arnoult et al. 2009). Avoiding cell death through the apoptotic program is a strategy used by bacterial enteropathogens to promote their intracellular survival, replication and dissemination. *Shigella* has developed such a strategy to inhibit apoptosis. Upon infection of the intestinal tissue, the epithelial cell cytoplasm is subsequently overwhelmed with bacteria that are in close proximity to the mitochondria (Mantis et al. 1996). To dampen epithelial cell death and preserve host cells, *Shigella* manipulates the cell death pathway through both T3SS-dependent and T3SS-independent mechanisms (Mantis et al. 1996; Carneiro et al. 2009). The pathogen upregulates the NOD1/RIP2/IKK-beta/NF-κB/Bcl-2 prosurvival pathway, which counteracts the epithelial cell death program. In this mechanism, prosurvival signals through the NLR sensor NOD1 are stimulated by epithelial membrane fragments that are generated when *Shigella* enters the epithelium lining (Dupont et al. 2009). Another example is *Salmonella*, which delivers different effector proteins, including AvrA and SopB, to subvert both inflammatory and apoptotic cell death responses (Valdez et al. 2009). The AvrA effector possesses acetyltransferase activity onto MKKs, which allows *Salmonella* to dampen the inflammatory response (Du and Galan 2009; Jones et al. 2008). Besides, AvrA can suppress host cell apoptosis through interferences with NF-κB activation by specifically inhibiting JNK MAPK stress signaling and the downstream apoptosis program (Jones et al. 2008). SopB protects host cells from apoptosis by maintaining the activation of AKT, an anti-apoptosis factor (Knodler et al. 2005). The ability to antagonize apoptosis appears to be dependent on its inositol phosphatase activity, which directly upregulates AKT activity. EPEC and EHEC use the T3SS effector protein EspF to manipulate the host cell apoptosis program (Holmes et al. 2010). EspF, which contains a mitochondrial targeting sequence, targets the mitochondria and stimulates cell death by disrupting the mitochondrial membrane potential (Nougayrede and Donnenberg 2004; Nagai et al. 2005). However, several findings indicate that EPEC can inhibit the host cell apoptosis program despite initially triggering it through EspF (Crane et al. 1999, 2001). Recently, the T3SS effector NleH has been shown to antagonize the apoptotic response by interacting with the Bax inhibitor-1 (BI-1) protein (Hemrajani et al. 2010). NleH targets the endoplasmic reticulum BI-1 protein to prevent elevated cytoplasmic calcium levels, nuclear condensation, caspase-3 activation and the subsequent apoptosis program. Thus, NleH would favor colonization by slowing down epithelial cell loss from the intestine.

The necrosis pathway

Necrosis refers to a form of caspase-independent cell death that can be observed upon bacterial infection (Berghe at al. 2010). Whereas apoptotic cells form intact, membrane-enclosed apoptotic bodies, plasma membrane integrity is lost early during necrosis. Cell lysis inevitably leads to spilling of intracellular contents into the extracellular environment. Consequently, necrosis is considered as a pro-inflammatory cell death program that leads to extensive tissue damage. Necrotic cells trigger inflammation through the release of endogenous danger signals, including uric acid and the DNA-binding protein high mobility group box 1 (Matzinger 1994). The serine/threonine kinases receptor interacting protein RIP1 and RIP3 are key regulators of the necrotic program (Degterev et al. 2008; Cho et al. 2009; He et al. 2009; Zhang et al. 2009). RIP1 kinase activity is essential for the induction of necrosis following death receptor engagement (Holler et al. 2000). Although the necrotic program is incompletely delineated, the RIP1/RIP3 signaling cascade results in the activation of calpains, calcium-dependent nonlysosomal cysteine proteases, lysosomal destabilization, and increased ROS production (Vanlangenakker et al. 2008). Upon host cell infection, *Shigella* induces both apoptosis and necrosis cell death programs through a caspase-1-independent manner (Carneiro et al. 2009; Faherty and Maurelli 2009). However, the pathogen manipulates the host cell prosurvival pathway by modulating cell death signaling until the late stages of the infectious process when bacteria have replicated and disseminated among epithelial cells. *Shigella* restrains the mitochondrial dysfunction-mediated necrosis program by activating the NOD1/RIP2/IKK-beta/NF-κB/Bcl-2 prosurvival pathway. However, cell death signaling becomes predominant in the later stages of the infectious process. This death signaling is dependent on the BH-3-only protein and Cyclophilin D, which are key regulators of mitochondrial permeability and cell death in response to oxidative cell stress. Therefore, *Shigella* is able to stimulate two compensatory signaling pathways, the necrosis program initiated by the BH-3-only protein-mediated mitochondrial permeability transition and the prosurvival pathway (Carneiro et al. 2009; Galluzzi and Kroemer 2009). As a result, the necrosis program characterized by chromatin condensation and plasma membrane damages appears late in the course of infection (Carneiro et al. 2009). Recently, further investigations have been performed to analyze host cell chromatin injury caused by *Shigella* (Bergounioux et al. 2012). During the infectious process, bacteria induce an early genotoxic stress in host cells, which largely precedes the beginning of the necrosis program. However, infected cells are unable to induce a DNA repair response, due to degradation of the proto-oncogene p53 protein by calpain proteases. This bacterial strategy is prosurvival for infected cells by preventing the activation of the p53/NF-κB signaling pathway, which would lead to initiation of the apoptotic cell death program. In this mechanism, the T3SS effector protein VirA has been shown to mediate calpain activation, which promotes early cytoskeleton processes required for epithelial invasion of *Shigella*, but ultimately restricts *Shigella* proliferation, by inducing the necrosis cell death program (Bergounioux et al. 2012).

The pyroptosis pathway

Unlike the apoptotic initiator caspases, caspase-1 is not involved in the induction of apoptosis program (Li et al. 1995). Instead, caspase-1 activation drives the proteolytic maturation and secretion of the pro-inflammatory cytokines interleukin IL-1 beta and IL-18 (Lamkanfi and Dixit 2009). Caspase-1 itself undergoes proximity-induced autoactivation in large cytosolic protein complexes termed inflammasomes. Most inflammasome complexes contain a member of the NLR sensor family (Kanneganti et al. 2007). NLR sensors recognize specific pathogen-associated molecular patters, and drive the assembly of the inflammasome and the subsequent caspase-1 activation, which results in the extracellular release of IL-1 beta and IL-18. Distinct inflammasomes are assembled, depending on the pathogen, including NLRC4 (Mariathasan et al. 2004; Broz et al. 2010), NLRP1 and NLRP3 (Mariathasan et al. 2006), and the AIM2 inflammasome (Fernandes-Alnemri et al. 2010). In addition to processing pro-IL-1 beta and pro-IL-18, inflammasome-activated caspase-1 induces a specialized cell death program in myeloid cells termed pyroptosis (Chen et al. 1996). The fact that caspase-1 activation is linked with the production of mature IL-1 beta and IL-18 renders pyroptosis an inherently pro-inflammatory cell death program. Although pyroptosis shares features with apoptosis and necrosis, the signaling pathways involved are distinct, and the molecular mechanism by which caspase-1 induces pyroptosis remains to be uncovered. Bacterial enteropathogens use various strategies to manipulate pyroptosis cell death program, which is crucial to avoid host cell death and the release of pro-inflammatory cytokines during infection. *Shigella* was the first pathogen reported to induce pyropotosis in infected macrophages (Chen et al. 1996; Hilbi et al. 1998). Bacteria promote caspase-1 activation and IL-1 beta processing through the NLRC4 inflammasome (Lamkanfi and Dixit 2009). Cholesterol, NLRC4 and caspase-1 have been shown to be crucial for the *Shigella*-mediated pyroptosis program (Schroeder and Hilbi 2007). For most of enteropathogens, activation of inflammasome usually happens following delivery of bacterial flagellin into the macrophage cytosol. However, because *Shigella* lacks flagellin, it remains to be established how this pathogen activates the NLRC4 inflammasome. Comparatively, *Salmonella* escapes detection of the NLRC4 inflammasome by dampening the expression of the SPI-1 protein PrgJ and flagellin after macrophage phagocytosis, whereby instead expressing SPI-2, which is not recognized by NLRC4 (Miao et al. 2006, 2010a). When macrophages are infected with *Salmonella* that expresses flagellin, the pathogen is efficiently eradicated by host cells (Miao et al. 2010b). This subversive mechanism allows the pathogen to efficiently escape from NLRC4 inflammasome activation, therefore favoring *Salmonella* replication and its infectious process.

Autophagy

Autophagy is a mechanism by which cells degrade and recycle their contents. This process is triggered to protect cells as they undergo various stress conditions (Levine and Kroemer 2008). Autophagy begins with the formation of a double membrane-bound structure termed autophagosome, which encapsulates part of the cytoplasm or organelles. The autophagosome then fuses with lysosomes to form the autolysosome, the content of which is degraded and used as nutrients for the cell. Autophagy is also used as a cellular defense mechanism against invading pathogens, as the vacuoles containing these intruders can fuse with autophagosomes and deliver the pathogens to lysosomes. Recent evidences support the existence of a functional link between NLR-mediated immune signaling and the autophagic pathway in macrophages and epithelial cells. These indications highlight the crucial role played by the IKK complex in autophagy induction through a canonical signaling pathway controlled by p53 depletion, JNK activation, and the release of the host proautophagic protein Beclin1 from its inhibitory interaction with Bcl-2 (Travassos et al. 2010; Cooney et al. 2010; Criollo et al. 2010). Among intracellular enteropathogens, most of bacteria have evolved strategies to escape from autophagosomes, using motility, but also from the autophagic recognition systems. *Listeria* uses both mechanisms to multiply within epithelial cells and to spread from cell to cell. Recognition of *Listeria* is mediated by the host cell PRRs, but also through the membrane fragments that are released upon *Listeria* invasion and LLO membrane rupture, which stimulate the autophagic program as agonist components (Yano et al. 2008; Dupont et al. 2009; Meyer-Morse et al. 2010). Besides the bacterial motility strategy to escape from autophagosomes, recent analyses of *Listeria* ActA mutants have highlighted the fact that other mechanisms are employed by the pathogen to evade autophagic recognition in infected host cells (Birmingham et al. 2007; Yoshikawa et al. 2009). In the case of *Shigella*, bacteria are detected by sensors of the autophagic pathway during multiplication within host cells. However, secretion of the T3SS effector protein IcsB allows the pathogen to escape from autophagy (Ogawa et al. 2005). Interestingly, IcsB does not directly inhibit autophagy itself. Instead, the IcsA outer membrane protein, which is required for actin-based bacterial motility, is targeted for autophagic recognition by binding to ATG5, an essential autophagic protein. However, IcsB can bind IcsA with a strong affinity and thereby inhibit the interaction with the host protein ATG5, indicating that IcsB acts as an anti-ATG5-binding protein and providing a mechanism of bacterial escape from autophagic recognition. As observed for *Listeria*, the autophagic program can also be induced by vacuolar membrane fragments that are generated during the infectious process, when epithelial cells are invaded by *Shigella*, highlighting the existence of multiple host cell pathways that are able to stimulate autophagy during an infection (Dupont et al. 2009).

Final Comments

Our knowledge of the strategies used by bacterial enteropathogens to manipulate host cell pathways and escape from host cell defenses is growing. The current investigations in microbiology, cell biology, immunology and infectious diseases contribute to the progress made to improve this fundamental knowledge. Several approaches have already revealed some of the sophisticated mechanisms evolved by pathogenic bacteria to modulate host cell pathways through the coordinate action of bacterial effectors. These ultimately allow pathogens to maintain extensive contacts with host cells by manipulating their membranes and cytoskeleton, their immune responses and cell death pathways. Altogether, these studies have highlighted the complex interplay between bacterial enteropathogens and their host. Besides, diversity of the host population is a crucial factor that we have now to address, such as genetic polymorphisms, but also the diversity of environmental conditions including the constitution of the host commensal microbiota. Indeed, when the host commensal bacteria are included in the scheme of an infectious process, the complexity of pathogen-host interactions becomes even greater.

The recent realization that the host commensal microbiota plays a crucial role in mediating the outcome of disease provides another degree of complexity. It is clear now that commensal bacteria play a crucial part in diseases dynamics and in host-pathogen, host-commensal and commensal-pathogen interactions. Therefore, we have to assume that commensal bacteria, like pathogens, can express factors that modulate host cell pathways. Given this additional variable parameter in the equation, it will be interesting to analyze the interplay of signals and responses between the host, pathogens and commensals, and to see how this interplay affects the progression of disease. Thus, the extent to which commensal bacteria apply similar strategies to pathogens represents the current question that needs to be addressed, and the upcoming studies will undoubtedly reveal further surprising details of the intriguing relationship between bacteria and host mucosal immunity.

Acknowledgments

The authors thank Natalie Fischer for critical reading of the manuscript. Philippe J. Sansonetti is a Howard Hughes Medical Institute foreign scholar and tenant of the advanced ERC Grant HOMEOEPITH.

References Cited

Aktories, K. 2011. Bacterial protein toxins that modify host regulatory GTPases. Nature. Rev. Microbiol. 9: 487–498.

Alkalay, I., A. Yaron, A. Hatzubai, A. Orian, A. Ciechanover and Y. Ben-Neriah. 1995. Stimulation-dependent I kappa B alpha phosphorylation marks the NF-kappa B inhibitor for degradation via the ubiquitin-proteasome pathway. Proc. Natl. Acad. Sci. USA. 92: 10599–10603.

Anderson, D.M. and O. Schneewind. 1999. Type III machines of Gram-negative pathogens: injecting virulence factors into host cells and more. Curr. Opin. Microbiol. 2: 18–24.

Arbibe, L., D.W. Kim, E. Batsche, T. Pedron, B. Mateescu, C. Muchardt, C. Parsot and P.J. Sansonetti. 2007. An injected bacterial effector targets chromatin access for transcription factor NF-kappaB to alter transcription of host genes involved in immune responses. Nat. Immunol. 8: 47–56.

Arnoult, D.L., L. Carneiro, I. Tattoli and S.E. Girardin. 2009. The role of mitochondria in cellular defense against microbial infection. Semin. Immunol. 21: 223–232.

Ashida, H., M. Kim, M. Schmidt-Supprian, A. Ma, M. Ogawa and C. Sasakawa. 2010. A bacterial E3 ubiquitin ligase IpaH9.8 targets NEMO/IKKgamma to dampen the host NF-kappaB-mediated inflammatory response. Nat. Cell. Biol. 12: 66–73.

Backert, S. and T.F. Meyer. 2006. Type IV secretion systems and their effectors in bacterial pathogenesis. Curr. Opin. Microbiol. 9: 207–217.

Baruch, K.L., L. Gur-Arie, C. Nadler, S. Koby, G. Yerushalmi, Y. Ben-Neriah, O. Yogev, E. Shaulian, C. Guttman, R. Zarivach and I. Rosenshine. 2011. Metalloprotease type III effectors that specifically cleave JNK and NF-κB. EMBO J. 30: 221–231.

Berghe, T.V., N. Vanlangenakker, E. Parthoens, W. Deckers, M. Devos, N. Festjens, C.J. Guerin, U.T. Brunk, W. Declercq and P. Vandenabeele. 2010. Necroptosis, necrosis and secondary necrosis converge on similar cellular disintegration features. Cell. Death. Differ. 17: 922–930.

Bergounioux, J., R. Elisee, A.L. Prunier, F. Donnadieu, B. Sperandio, P.J. Sansonetti and L. Arbibe. 2012. Calpain activation by the *Shigella flexneri* effector VirA regulates key steps in the formation and life of the bacterium's epithelial niche. Cell. Host. Microbe. 11: 240–252.

Beutler, B. 2009. Microbe sensing, positive feedback loops, and the pathogenesis of inflammatory diseases. Immunol. Rev. 227: 248–263.

Birmingham, C.L., V. Canadien, E. Gouin, E.B. Troy, T. Yoshimori, P. Cossart, D.E. Higgins and J.H. Brumell. 2007. *Listeria monocytogenes* evades killings by autophagy during colonization of host cells. Autophagy. 3: 442–451.

Brennan, D.F. and D. Barford. 2009. Eliminylation: a post-translational modification catalyzed by phosphothreonine lyases. Trends. Biochem. Sci. 34: 108–114.

Bonazzi, M., M. Lecuit and P. Cossart. 2009. *Listeria monocytogenes* internalin and E-cadherin: from structure to pathogenesis. Cell. Microbiol. 11: 693–702.

Broz, P., K. Newton, M. Lamkanfi, S. Mariathasan, V.M. Dixit and D.M. Monack. 2010. Redundant roles for inflammasome receptors NLRP3 and NLRC4 in host defense against *Salmonella*. J. Exp. Med. 207: 1745–1755.

Brumell, J.H., D.L. Goosney and B.B. Finlay. 2002. SifA, a type III secreted effector of *Salmonella typhimurium*, directs *Salmonella*-induced filament (Sif) formation along microtubules. Traffic. 3: 407–415.

Campellone, K.G., D. Robbins and J.M. Leong. 2004. $EspF_U$ is a translocated EHEC effector that interacts with Tir and N-WASP and promotes Nck-independent actin assembly. Dev. Cell. 7: 217–228.

Carneiro, L.A., L.H. Travassos, F. Soares, I. Tattoli, J.G. Magalhaes, M.T. Bozza, M.C. Plotkowski, P.J. Sansonetti, J.D. Molkentin, D.J. Philpott and S.E. Girardin. 2009. *Shigella* induces mitochondrial dysfunction and cell death in nonmyeloid cells. Cell. Host. Microbe. 5: 123–136.

Cheminay, C., A. Mohlenbrink and M. Hensel. 2005. Intracellular *Salmonella* inhibit antigen presentation by dendritic cells. J. Immunol. 174: 2892–2899.

Chen, Y., M.R. Smith, K. Thirumalai and A. Zychlinsky. 1996. A bacterial invasion induces macrophage apoptosis by binding directly to ICE. EMBO J. 15: 3853–3860.

Chichlowski, M. and L.P. Hale. 2008. Bacterial-mucosal interactions in inflammatory bowel diseases: an alliance gone bad. Am. J. Physiol. Gastrointest. Liver. Physiol. 295: 1139–1149.

Cho, Y.S., S. Challa, D. Moquin, R. Genga, T.D. Ray, M. Guildford and F.K. Chan. 2009. Phosphorylation-driven assembly of the RIP1-RIP3 complex regulates programmed necrosis and virus-induced inflammation. Cell. 137: 1112–1123.

Cirillo, D.M., R.H. Valdivia, D.M. Monack and S. Falkow. 1998. Macrophage-dependent induction of the *Salmonella* pathogenicity island 2 type III secretion system and its role in intracellular survival. Mol. Microbiol. 30: 175–188.

Cooney, R., J. Baker, O. Brain, B. Danis, T. Pichulik, P. Allan, D.J. Ferguson, B.J. Campbell, D. Jewell and A. Simmons. 2010. NOD2 stimulation induces autophagy in dendritic cells influencing bacterial handling and antigen presentation. Nat. Med. 16: 90–97.

Cossart, P. and P.J. Sansonetti. 2004. Bacterial invasion: the paradigm of enteroinvasive pathogens. Science. 304: 242–248.

Crane, J.K., S. Majumdar and D.F. Pickhardt. 1999. Host cell death due to enteropathogenic *Escherichia coli* has features of apoptosis. Infect. Immun. 67: 2575–2584.

Crane, J.K., B.P. McNamara and M.S. Donnenberg. 2001. Role of EspF in host cell death induced by enteropathogenic *Escherichia coli*. Cell. Microbiol. 3: 197–211.

Criollo, A., L. Senovilla, H. Authier, M.C. Maiuri, E. Morselli, I. Vitale, O. Kepp, E. Tasdemir, L. Galluzzi and S. Shen. 2010. The IKK complex contributes to the induction of autophagy. EMBO J. 29: 619–631.

Croxen, M.A. and B.B. Finlay. 2010. Molecular mechanisms of *Escherichia coli* pathogenicity. Nat. Rev. Microbiol. 8: 26–38.

Degterev, A., J. Hitomi, M. Germscheid, I.L. Ch'en, O. Korkina, X. Teng, D. Abbott, G.D. Cuny, C. Yuan and G. Wagner. 2008. Identification of RIP1 kinase as a specific cellular target of necrostatins. Nat. Chem. Biol. 4: 313–321.

Donnenberg, M.S. 2000. Pathogenic strategies of enteric bacteria. Nature. 406: 768–774.

Du, F. and J.E. Galan. 2009. Selective inhibition of type III secretion activated signaling by the *Salmonella* effector AvrA. PLoS Pathog. 5: e1000595.

Dupont, N., S. Lacas-Gervais, J. Bertout, I. Paz, B. Freche, G.T. Van Nhieu, F.G. van der Goot, P.J. Sansonetti and F. Lafont. 2009. *Shigella* phagocytic vacuolar membrane remnants participate in the cellular response to pathogen invasion and are regulated by autophagy. Cell. Host. Microbe. 6: 137–149.

Faherty, C.S. and A.T. Maurelli. 2009. Spa15 of *Shigella flexneri* is secreted through the type III secretion system and prevents staurosporine-induced apoptosis. Infect. Immun. 77: 5281–5290.

Fernandes-Alnemri, T., J.W. Yu, C. Juliana, L. Solorzano, S. Kang, J. Wu, P. Datta, M. McCormick, L. Huang and E. McDermott. 2010. The AIM2 inflammasome is critical for innate immunity to Francisella tularensis. Nat. Immunol. 11: 385–393.

Galluzzi, L. and G. Kroemer. 2009. *Shigella* targets the mitochondrial checkpoint of programmed necrosis. Cell. Host. Microbe. 5: 107–109.

Gao, X., F. Wan, K. Mateo, E. Callegari, D. Wang, W. Deng, J. Puente, F. Li, M.S. Chaussee and B.B. Finlay. 2009. Bacterial effector binding to ribosomal protein s3 subverts NF-κB function. PLoS Pathog. 5: e1000708.

Garcia-del Portillo, F., M.B. Zwick, K.Y. Leung and B.B. Finlay. 1993. *Salmonella* induces the formation of filamentous structures containing lysosomal membrane glycoproteins in epithelial cells. Proc. Natl. Acad. Sci. USA 90: 10544–10548.

Girardin, S.E., R. Tournebize, M. Mavris, A.L. Page, X. Li, G.R. Stark, J. Bertin, P.S. DiStefano, M. Yaniv, P.J. Sansonetti and D.J. Philpott. 2001. CARD4/Nod1 mediates NF-kappaB and JNK activation by invasive *Shigella flexneri*. EMBO Rep. 2: 736–742.

Gruenheid, S., R. DeVinney, F. Bladt, D. Goosney, S. Gelkop, G.D. Gish, T. Pawson and B.B. Finlay. 2001. Enteropathogenic *E. coli* Tir binds Nck to initiate actin pedestal formation in host cells. Nat. Cell. Biol. 3: 856–859.

Ham, H., A. Sreelatha and K. Orth. 2011. Manipulation of host membranes by bacterial effectors. Nature. Rev. Microbiol. 9: 635–646.

He, S., L. Wang, L. Miao, T. Wang, F. Du, L. Zhao and X. Wang. 2009. Receptor interacting protein kinase-3 determines cellular necrotic response to TNF-alpha. Cell. 137: 1100–1111.

Helaine, S., J.A. Thompson, K.G. Watson, M. Liu, C. Boyle and D.W. Holden. 2010. Dynamics of intracellular bacterial replication at the single cell level. Proc. Natl. Acad. Sci. USA. 107: 3746–3751.

Hemrajani, C., C.N. Berger, K.S. Robinson, O. Marches, A. Mousnier and G. Frankel. 2010. NleH effectors interact with Bax inhibitor-1 to block apoptosis during enteropathogenic *Escherichia coli* infection. Proc. Natl. Acad. Sci. USA 107: 3129–3134.

High, N., J. Mounier, M.C. Prevost and P.J. Sansonetti. 1992. IpaB from *Shigella flexneri* causes entry into epithelial cells and escape from phagocytic vacuole. EMBO J. 11: 1991–1999.

Hilbi, H., J.E. Moss, D. Hersh, Y. Chen, J. Arondel, S. Banerjee, R.A. Flavell, J. Yuan, P.J. Sansonetti and A. Zychlinsky. 1998. *Shigella*-induced apoptosis is dependent on caspase-1 which binds to IpaB. J. Biol. Cham. 273: 32895–32900.

Holler, N., R. Zaru, O. Micheau, M. Thome, A. Attinger, S. Valitutti, J.L. Bodmer, P. Schneider, B. Seed and J. Tschopp. 2000. Fas triggers an alternative, caspase-8-independent cell death pathway unsing the kinase RIP as effector molecule. Nat. Immunol. 1: 489–495.

Holmes, A., S. Muhlen, A.J. Roe and P. Dean. 2010. The EspF effector, a bacterial pathogen's Swiss army knife. Infect. Immun. 78: 4445–4453.

Jones, R.M., H. Wu, C. Wentworth, L. Luo, L. Collier-Hyams and A.S. Neish. 2008. *Salmonella* AvrA coordinates suppression of host immune and apoptotic defenses via JNK pathway blockade. Cell. Host. Microbe. 3: 233–244.

Kanneganti, T.D., M. Lamkanfi and G. Nunez. 2007. Intracellular NOD-like receptors in host defense and disease. Immunity. 27: 549–559.

Kim, D.W., G. Lenzen, A.L. Page, P. Legrain, P.J. Sansonetti and C. Parsot. 2005. The *Shigella flexneri* effector OspG interferes with innate immune responses by targeting ubiquitin-conjugating enzymes. Proc. Natl. Acad. Sci. USA 102: 14046–14051.

Kim, M., M. Ogawa, Y. Fujita, Y. Yoshikawa, T. Nagai, T. Koyama, S. Nagai, A. Lange, R. Fassler and C. Sasakawa. 2009. Bacterial hijack integrin-linked kinase to stabilize focal adhesions and block cell detachment. Nature. 459: 578–582.

Knodler, L.A. and O. Steele-Mortimer. 2003. Taking possession: biogenesis of the *Salmonella*-containing vacuole. Traffic. 4: 587–599.

Knodler, L.A., B.B. Finlay and O. Steele-Mortimer. 2005. The *Salmonella* effector protein SopB protects epithelial cells from apoptosis by sustained activation of Akt. J. Biol. Chem. 280: 9058–9064.

Konradt, C., E. Frigimelica, K. Nothelfer, A. Puhar, W. Salgado-Pabon, V. di Bartolo, D. Scott-Algara, C.D. Rodrigues, P.J. Sansonetti and A. Phalipon. 2011. The *Shigella flexneri* type three secretion system effector IpgD inhibits T cell migration by manipulating host phosphoinositide metabolism. Cell. Host. Microbe. 9: 263–272.

Kramer, R.W., N.L. Slagowski, N.A. Eze, K.S. Giddings, M.F. Morrison, K.A. Siggers, M.N. Starnbach and C.F. Lesser. 2007. Yeast functional genomic screens lead to identification of a role for a bacterial effector in innate immunity regulation. PLoS Pathog. 3: e21.

Lafont, F., G.T. Nhieu, K. Hanada and P.J. Sansonetti. 2002. Initial steps of *Shigella* infection depend on the cholesterol/sphingolipid raft-mediated CD44-IpaB interaction. EMBO J. 21: 4449–4457.

Lamkanfin, M. and V.M. Dixit. 2009. Inflammasomes: guardians of cytosolic sanctity. Immunol. Rev. 227: 95–105.

Lara-Tejero, M., J. Kato, S. Wagner, X. Liu and J.E. Galan. 2011. A sorting platform determines the order of protein secretion in bacterial type III systems. Science. 331: 1188–1191.

Le Negrate, G., B. Faustin, K. Welsh, M. Loeffler, M. Krajewska, P. Hasegawa, S. Mukherjee, K. Orth, S. Krajewski, A. Godzik, D.G. Guiney and J.C. Reed. 2008. *Salmonella* secreted factor L deubuquitinase of *Salmonella typhimurium* inhibits NF-κB, suppresses IκB-alpha ubiquitination and modulates innate immune responses. J. Immunol. 180: 5045–5056.

Lee, A.H., M.P. Zareei and S. Daefler. 2002. Identification of a NIPSNAP homologue as host cell target for *Salmonella* virulence protein SpiC. Cell. Microbiol. 4: 739–750.

Levine, B. and G. Kroemer. 2008. Autophagy in the pathogenesis of disease. Cell. 132: 27–42.

Li, H., H. Xu, Y. Zhou, J. Zhang, C. Longs, S. Li, S. Chen, J.M. Zhou and F. Shao. 2007. The phosphothreonine lyase activity of a bacterial type III effector family. Science. 315: 1000–1003.

Li, P., H. Allen, S. Banerjee, S. Franklin, L. Herzog, C. Johnston, J. McDowell, M. Paskind, L. Rodman and J. Salfeld. 1995. Mice deficient in IL-1 beta-converting enzyme are defective in production of mature IL-1 beta and resistant to endotoxic shock. Cell. 80: 401–411.

Liu, X., R. Lu, Y. Xia, S. Wu and J. Sun. 2010. Eukaryotic signaling pathways targeted by *Salmonella* effector protein AvrA in intestinal infection *in vivo*. BMC Microbiol. 10: 326.

Macdonald, T.T. and G. Monteleone. 2005. Immunity, inflammation, and allergy in the gut. Science. 307: 1920–1925.

Malladi, V., M. Puthenedam, P.H. Willimas and A. Balakrishnan. 2004. Enteropathogenic *Escherichia coli* outer membrane proteins induce iNOS by activation of NF-κB and MAP kinases. Inflammation. 28: 345–353.

Mantis, N., M.C. Prévost and P. Sansonetti. 1996. Analysis of epithelial cell stress response during infection by *Shigella flexneri*. Infect. Immun. 64: 2474–2482.

Mariathasan, S., K. Newton, D.M. Monack, D. Vucic, D.M. French, W.P. Lee, M. Roose-Girma, S. Erickson and V.M. Dixit. 2004. Differential activation of the inflammasome by caspase-1 adaptors ASC and Ipaf. Nature. 430: 213–218.

Mariathasan, S., D.S. Weiss, K. Newton, J. McBride, K. O'Rourke, M. Roose-Girma, W.P. Lee, Y. Weinrauch, D.M. Monack and V.M. Dixit. 2006. Cryopyrin activates the inflammasome in response to toxins and ATP. Nature. 440: 228–232.

Matzinger, P. 1994. Tolerance, danger, and the extended family. Annu. Rev. Immunol. 12: 991–1045.

McGuckin, M.A., S.K. Linden, P. Sutton and T.H. Florin. 2011. Mucin dynamics and enteric pathogens. Nat. Rev. Microbiol. 9: 265–278.

Miao, E.A., C.M. Alpuche-Aranda, M. Dors, A.E. Clark, M.W. Bader, S.I. Miller and A. Aderem. 2006. Cytoplasmic flagellin activates caspase-1 and secretion of interleukin 1 beta via Ipaf. Nat. Immunol. 7: 569–575.

Miao, E.A., D.P. Mao, N. Yudkovsky, R. Bonneau, C.G. Lorang, S.E. Warren, I.A. Leaf and A. Aderem. 2010a. Innate immune detection of the type III secretion apparatus through the NLRC4 inflammasome. Proc. Natl. Acad. Sci. USA. 107: 3076–3080.

Miao, E.A., I.A. Leaf, P.M. Trunting, D.P. Mao, M. Dors, A. Sarkar, S.E. Warren, M.D. Wewers and A. Aderem. 2010b. Caspase-1 induced pyroptosis is an innate immune effector mechanism against intracellular bacteria. Nat. Immunol. 11: 1136–1142.

Meresse, S., O. Steele-Mortimer, E. Moreno, M. Desjardins, B. Finlay and J.P. Gorvel. 1999. Controlling maturation of pathogen-containing vacuoles: a matter of life and death. Nat. Cell. Biol. 1: 183–188.

Meyer-Morse, N., J.R. Robbins, C.S. Rae, S.N. Mochegova, M.S. Swanson, Z. Zhao, H.W. Virgin and D. Portnoy. 2010. Listeriolysin O is necessary and sufficient to induce autophagy during *Listeria monocytogenes* infection. PLoS ONE. 5: e8610.

Mukherjee, K., S. Parashuraman, M. Raje and A. Mukhopadhyay. 2001. SopE acts as an Rab5-specific nucleotide exchange factor and recruits non-prenylated Rab5 on *Salmonella*-containing phagosomes to promote fusion with early endosomes. J. Biol. Chem. 276: 23607–23615.

Nadler, C., K. Baruch, S. Kobi, E. Mills, G. Haviv, M. Farago, I. Alkalay, S. Bartfeld, T.F. Meyer, Y. Ben-Neriah and I. Rosenshine. 2010. The type III secretion effector NleE inhibits NF-κB activation. PLoS Pathog. 6: e1000743.

Nagai, T., A. Abe and C. Sasakawa. 2005. Targeting of enteropathogenic *Escherichia coli* EspF to host mitochondria is essential for bacterial pathogenesis: critical role of the 16th leucine residue in EspF. J. Biol. Cell. 280: 2998–3011.

Newton, H.J., J.S. Pearson, L. Badea, M. Kelly, M. Lucas, G. Holloway, K.M. Wagstaff, M.A. Dunstone, J. Sloan, J.C. Whisstock, J.B. Kaper, R.M. Robins-Browne, D.A. Jans, G. Frankel, A.D. Phillips, B.S. Coulson and E.L. Hartland. 2010. The type III effectors NleE and NleB from enteropathogenic *E. coli* and OspZ from *Shigella* block nuclear translocation of the NF-κB p65. PLoS Pathog. 6: e1000898.

Niebuhr, K, S. Giuriato, T. Pedron, D.J. Philpott, F. Gaits, J. Sable, M.P. Sheetz, C. Parsot, P.J. Sanonetti and B. Payrastre. 2002. Conversion of PtdIns(4,5)P2 into PtdIns(5)P by the *S. flexneri* effector IpgD reorganizes host cell morphology. EMBO J. 21: 5069–5078.

Nikitas, G., C. Deschamps, O. Disson, T. Niault, P. Cossart and M. Lecuit. 2011. Transcytosis of *Listeria monocytogenes* across the intestinal barrier upon specific targeting of goblet cell accessible E-cadherin. J. Exp. Med. 208: 2263–2277.

Nhieu, G.T. and P.J. Sansonetti. 1999. Mechanism of *Shigella* entry into epithelial cells. Curr. Opin. Microbiol. 2: 51–55.

Nobe, R., J.P. Nougayrède, F. Taieb, M. Bardiau, D. Cassart, F. Navarro-Garcia, J. Mainil, T. Hayashi and E. Oswald. 2009. Enterohaemorrhagic *Escherichia coli* serogroup O111 inhibits NF-κB-dependent innate responses in a manner independent of a type III secreted OspG orthologue. Microbiology. 155: 3214–3225.

Nougayrede, J.P. and M.S. Donnenberg. 2004. Enteropathogenic *Escherichia coli* EspF is targeted to mitochondria and is required to initiate the mitochondrial death pathway. Cell. Microbiol. 6: 1097–1111.

Odendall, C., N. Rolhion, A. Förster, J. Poh, D.J. Lamont, M. Liu, P.S. Freemont, A.D. Catling and D.W. Holden. 2012. The *Salmonella* kinase SteC targets MEK to regulate the host actin cytoskeleton. Cell. Host. Microbe. In Press.

Ogawa, M., T. Yoshimori, T. Suzuki, H. Sagara, N. Mizushima and C. Sasakawa. 2005. Escape of intracellular *Shigella* from autophagy. Science. 307: 727–731.

Pantaloni, D., C. Le Clainche and M.F. Carlier. 2001. Mechanism of actin-based motility. Science. 292: 1502–1506.

Patel, J.C. and J.E. Galan. 2005. Manipulation of the host actin cytoskeleton by *Salmonella*—all in the name of entry. Curr. Opin. Microbiol. 8: 10–15.

Radtke, F. and H. Clevers. 2005. Self-renewal and cancer of the gut: two sides of a coin. Science. 307: 1904–1909.

Rathinam, V.A.K., Z. Jiang, S.N. Waggoner, S. Sharma, L.E. Cole, L. Waggoner, S.K. Vanaja, B.G. Monks, S. Ganesan, E. Latz, V. Hornung, S.N. Vogel, E. Szomolanyi-Tsuda and K.A. Fitzgerald. 2010. The AIM2 inflammasome is essential for host defense against cytosolic bacteria and DNA viruses. Nat. Immunol. 11: 395–402.

Rohde, J.R., A. Breitkreutz, A. Chenal, P.J. Sansonetti and C. Parsot. 2007. Type III secretion effectors of the IpaH family are ubiquitin ligases. Cell. Host. Microbe. 1: 77–83.

Rudel, T., O. Kepp and V. Kozjak-Pavlovic. 2010. Interactions between bacterial pathogens and mitochondrial cell death pathways. Nat. Rev. Microbiol. 8: 693–705.

Salcedo, S.P., M. Noursadeghi, J. Cohen and D.W. Holden. 2001. Intracellular replication of *Salmonella* Typhimurium strains in specific subsets of splenic macrophages *in vivo*. Cell. Microbiol. 3: 587–597.

Savill, J., I. Dransfield, C. Gregory and C. Haslett. 2002. A blast from the past: clearance of apoptotic cells regulates immune responses. Nat. Rev. Immunol. 2: 965–975.

Schroeder, G.N. and H. Hilbi. 2007. Cholesterol is required to trigger caspase-1 activation and macrophage apoptosis after phagosomal escape of *Shigella*. Cell. Microbiol. 9: 265–278.

Selvaraj, S.K. and N.V. Prasadarao. 2005. *Escherichia coli* K1 inhibits proinflammatory cytokines induction in monocytes by preventing NF-κB activation. J. Leukoc. Biol. 78: 544–554.

Shames, S.R., S.D. Auweter and B.B. Finlay. 2009. Co-evolution and exploitation of host cell signaling pathways by bacterial pathogens. Int. J. Biochem. Cell. Biol. 41: 380–389.

Shaner, N.C., J.W. Sanger and J.M. Sanger. 2005. Actin and alpha-actinin dynamics in the adhesion and motility of EPEC and EHEC on host cells. Cell. Motil. Cytoskeleton. 60: 104–120.

Shaughnessy, L.M., A.D. Hoppe, K.A. Christensen and J.A. Swanson. 2006. Membrane perforations inhibit lysosome fusion by altering pH and calcium in *Listeria monocytogenes* vacuoles. Cell. Microbiol. 8: 781–792.

Sperandio, B., B. Regnault, J. Guo, Z. Zhang, S.L. Stanley, P.J. Sansonetti and T. Pedron. 2008. Virulent *Shigella flexneri* subverts the host innate immune response through manipulation of antimicrobial peptide gene expression. J. Exp. Med. 205: 1121–1132.

Stein, M.A., K.Y. Leung, M. Zwick, F. Garcia-del Portillo and B.B. Finlay. 1996. Identification of a *Salmonella* virulence gene required for formation of filamentous structures containing lysosomal membrane glycoproteins within epithelial cells. Mol. Microbiol. 20: 151–164.

Steven, J.M., E.E. Galyov and M.P. Stevens. 2006. Actin-dependent movement of bacterial pathogens. Nat. Rev. Microbiol. 4: 91–101.

Suzuki, T., H. Miki, T. Takenawa and C. Sasakawa. 1998. Neural Wiskott-Aldrich syndrome protein is implicated in the actin-based motility of *Shigella flexneri*. EMBO J. 17: 2767–2776.

Travassos, L.H., L.A. Carneiro, M. Ramjeet, S. Hussey, Y.G. Kim, J.G. Magalhaes, L. Yuan, F. Soares, E. Chea and L. Le Bourhis. 2010. Nod1 and Nod2 direct autophagy by recruiting ATG16L1 to the plasma membrane at the site of bacterial entry. Nat. Immunol. 11: 55–62.

Trosky, J.E., Y. Li, S. Mukherjee, G. Keitany, H. Ball and K. Orth. 2007. VopA inhibits ATP binding by acetylating the catalytic loop of MAPK kinases. J. Biol. Chem. 282: 34299–34305.

Uchiya, K., M.A. Barbieri, K. Funato, A.H. Shah, P.D. Stahl and E.A. Groisman. 1999. *Salmonella* virulence protein that inhibits cellular trafficking. EMBO J. 18: 3924–3933.

Valdez, Y., R.B. Ferreira and B.B. Finlay. 2009. Molecular mechanisms of *Salmonella* virulence and host resistance. Curr. Top. Microbiol. Immunol. 337: 93–127.

Van Nhieu, G.T. and J. Guignot. 2009. When *Shigella* tells the cell to hang on. J. Mol. Cell. Biol. 1: 64–65.

Vanlangenakker, N., T. Vanden Berghe, D.V. Krysko, N. Festjens and P. Vandenabeele. 2008. Molecular mechanisms and pathophysiology of necrotic cell death. Curr. Mol. Med. 8: 207–220.

Wan, F., D.E. Anderson, R.A. Barnitz, A. Snow, N. Bidere, L. Zheng, V. Hegde, L.T. Lam, L.M. Staudt and D. Levens. 2007. Ribosomal protein S3: a KH domain subunit in NF-kappaB complexes that mediates selective gene regulation. Cell. 131: 927–939.

Welch, M.D., A. Iwamatsu and T.J. Mitchison. 1997. Actin polymerization induced by Arp2/3 protein complex at the surface of *Listeria monocytogenes*. Nature. 385: 265–269.

Yano, T., S. Mita, H. Ohmori, Y. Fujimoto, R. Ueda, H. Takada, W.E. Goldman, K. Fukase and N. Silverman. 2008. Autophagic control of *Listeria* through intracellular innate immune recognition in drosophila. Nat. Immunol. 9: 908–916.

Ye, Z., E.O. Petrof, D. Boone, E.C. Claud and J. Sun. 2007. *Salmonella* effector AvrA regulation of colonic epithelial cell inflammation by deubiquitination. Am. J. Pathol. 171: 882–892.

Yoshida, S., E. Katayama, A. Kuwae, H. Mimuro, T. Suzuki and C. Sasakawa. 2002. *Shigella* deliver an effector protein to trigger host microtubule destabilization, which promotes Rac1 activity and efficient bacterial internalization. EMBO J. 21: 2923–2935.

Yoshikawa, Y., M. Ogaya, T. Hain, M. Yoshida, M. Fukumatsu, M. Kim, H. Mimuro, I. Nakawaga, T. Yanagawa and T. Ishii. 2009. *Listeria monocytogenes* ActA-mediated escape from autophagic recognition. Nat. Cell. Biol. 11: 1233–1240.

Zhang, D.W., J. Shao, J. Lin, N. Zhang, B.J. Lu, S.C. Lin, M.Q. Dong and J. Han. 2009. RIP3, an energy metabolism regulator that swithches TNF-induced cell death from apoptosis to necrosis. Science. 325: 332–336.

Zheng, C.F. and K.L. Guan. 1994. Cytoplasmic localization of the mitogen-activated protein kinase activator MEK. J. Biol. Chem. 269: 19947–19952.

6

The Role of Probiotics in Prevention and Treatment of GI Infections

*Christina M. Surawicz** and *Christopher Damman*

Introduction

Probiotics are ingested orally and many of them have major effects on the gastrointestinal (GI) tract. Thus their possible efficacy for GI tract diseases is of great interest. Since GI tract infections are a major cause of morbidity and mortality worldwide, especially in children in developing countries, the use of probiotics to prevent or to treat such infections is an area of great importance. This chapter will review the use of probiotics in the prevention and treatment of infectious diarrhea as well as the prevention of traveler's diarrhea (which is usually infectious in origin).

Diarrhea is a frequent side effect of antibiotics, with an incidence of 10–20%. While uncomplicated antibiotic-associated diarrhea (AAD) is not an infection, it is likely due to alterations in the gut microbiota, and thus is included in this chapter. One severe form of AAD is *Clostridium difficile* infection (CDI), an infection with gram positive anaerobic bacteria that has had a dramatic increase in frequency and morbidity since the year 2000. Thus the role of probiotics in prevention and treatment of AAD and CDI is also reviewed.

Department of Medicine, Division of Gastroenterology, University of Washington School of Medicine.
* Corresponding author: surawicz@uw.edu

There are two conditions with intestinal inflammation in which infectious organisms may play a role in pathogenesis. Necrotizing enterocolitis (NEC) is a severe inflammatory bowel condition affecting low birth weight infants. Pouchitis is inflammation of the pouch created after ileal pouch-anal anastomosis (IPAA) surgery that often responds to antibiotic therapy; thus there may be alterations in the colon microbiota that are important in pathophysiology. For this reason studies of probiotics in these two conditions are included in this chapter.

Infectious Diarrhea of all Causes

Prevention of Infectious Diarrhea

The use of probiotics to prevent infectious diarrhea is appealing. Animal studies point to the potential of probiotics to prevent enteric infections. For example, the *Lactobacillus casei* strain Shirota decreased colonization of *E. coli* 0157:H7 in an animal model (Ogawa et al. 2001), and *Lactobacillus salivarus* competed effectively with *Salmonella enteriditis* in animal models (Pascul et al. 1999). Studies in humans have mostly evaluated children, often in child care centers where the most common cause of diarrhea is usually rotaviurs. A randomized controlled trial used *L. casei shirota* to test prevention of diarrhea in 3,758 children between the ages of one and five in an urban slum in India, and found an efficacy of 14% and an NNT of 14% (Sur et al. 2011). In another study, there was no overall efficacy of *B. lactis* HN019 and a probiotic oligosaccharide in prevention of diarrhea but there was a decrease in diarrhea in children over 2 years of age (Sazawal et al. 2010).

A meta-analysis of 34 randomized controlled trials by Sazawal, evaluating prevention of acute diarrhea included 9 studies of acute diarrhea and 6 of traveler's diarrhea, but also 19 of AAD prevention (Sazawal et al. 2006). When pooled, probiotics had a 35% decrease in the risk of diarrhea, but interpretation was limited by combined studies in adults and children, by multiple probiotics studied and by multiple causes of diarrhea so that it was difficult to confirm a true effect.

While there appears to be modest efficacy of probiotics in prevention of infectious diarrhea, in a review of studies published before 2009, Hibberd was unable to recommend their use (Hibbard 2009). A clinical report from the American Academy of Pediatrics also concluded that available data did not support the routine use of probiotics to prevent nosocomial rotavirus diarrhea in children in child care centers (Thomas 2010). Since much infectious diarrhea is due to rotavirus, the use of newly available vaccines may be more effective in prevention.

Treatment of Acute Infectious Diarrhea

There are many randomized controlled trials of probiotics for treatment of acute diarrhea in children. Rotavirus was the most common cause of diarrhea in most of these studies. In six of ten trials, probiotic use decreased the duration of diarrhea (Kokzampassi et al. 2010).

Some probiotics have been shown to decrease duration of diarrhea. *Lactobacillus* products have been shown to have preventive and therapeutic benefit in children. In some, there is efficacy with viral gastroenteritis, in others there is efficacy with bacterial diarrhea. In infants with acute diarrhea, *Lactobacillus reuteri* decreased diarrhea days from 2.5 to 1.5. (Shornikova et al. 1997). *Lactobacillus rhamnosus GG* decreased the number of days in hospital for children with acute diarrhea (Guandalini 2000). In another study, *Lactobacillus GG* decreased shedding and shortened duration of rotavirus diarrhea by 40 hours but showed no benefit in diarrhea of other etiologies (Szymański et al. 2006). The contrast is with a study of *Lactobacillus paracasei* strain ST11 which was studied in children with acute diarrhea in Bangladesh. In this study, there was no benefit in the children with severe rotavirus diarrhea, but a significant decrease in diarrhea in children with non-rotavirus diarrhea (Sarkar et al. 2010).

In meta-analysis by Allen, 63 studies of probiotics to treat acute infectious diarrhea were evaluated, most (56 of 63) in infants and children (Allen et al. 2010). Studies were limited to those with duration of diarrhea of at least 4 days. Overall there was a 24 hour decrease in the duration of diarrhea with the use of probiotics. Moreover, there were fewer stools by the second day of illness, and reduced risk of diarrhea lasting longer than 4 days. There were no adverse events reported. The authors recommend further studies of specific probiotics. The American Academy of Pediatrics concluded that evidence supported the use of probiotics early in the course of acute infectious diarrhea to reduce duration by one day (Thomas et al. 2010).

Treatment of Persistent Diarrhea

Persistent diarrhea (lasting longer than 14 days) in children is a serious illness that can cause significant morbidity and mortality. While the pathogenesis is likely multifactorial, infections may play a significant role, which is the rationale for trials of probiotic therapy. A Cohrane analysis of 4 relevant trials found that probiotics reduced duration of diarrhea by 4.12 days, and reduced number of stools compared to placebo (Bernaola et al. 2010). However, the quality of the studies was not sufficiently high enough to recommend the use of probiotics.

Small studies showed decrease in diarrhea using *S. boulardii* as an adjunct in treatment of amebiasis or as sole therapy for *Blastocystis hominis* (Savaş-Erdeve et al. 2009; Dinleyicic et al. 2011). There are fewer studies of treatment in adults.

Prevention of Traveler's Diarrhea (TD)

A meta-analysis of 12 studies showed an odds ratio in favor of probiotics of 0.85 (95% CI 0.79–0.91 $p<0.001$) (McFarland 2007). Those with efficacy were *S. boulardii* and a mixture of *L. acidophilus* and *B. bifidum*. However, given heterogeneity in the risk of TD, organisms causing TD, as well as preventive behaviors, further studies are clearly needed. Nonetheless, probiotics are more appealing than use of antibiotics for prevention of TD.

Antibiotic-Associated Diarrhea and *Clostridium difficile* Infection (AAD and CDI)

Diarrhea is a frequent side effect of antibiotic therapy, occurring in up to 20% of those taking antibiotics, and likely is due to alteration in the normal fecal microbiota by the antibiotic. It can be mild, or occasionally severe. *Clostridium difficile* infection (CDI) is a form of AAD; the disturbed microbiota allows overgrowth of this gram-positive anaerobe that produces toxins and can cause mild to severe diarrhea, as well as colitis, pseudomembranous colitis and can be fatal. Since the year 2000, there has been a dramatic increase in numbers of cases of CDI in the US, Canada and other countries; a hypervirulent strain has been associated with more severe disease, causing increased morbidity and mortality (Tschudin-Sutled et al. 2012; Cecil 2012). Strategies to prevent CDI include wise antibiotic use policies and environmental cleaning, but there is increased interest in the use of probiotics as an adjunct to antibiotics to prevent CDI.

Prevention of Antibiotic-Associated Diarrhea (AAD)

The best studied single probiotics for prevention of AAD are the bacterium *Lactobacillus GG* and the yeast *Saccharomyces boulardii*. There are also studies using other bacteria and combinations of other bacteria. Many studies evaluate probiotics in children while others use adult subjects. Probiotics appear effective in AAD prevention in both groups.

Two large studies have shown efficacy of *Saccharomyces boulardii* in preventing AAD (Surawicz et al. 1998; McFarland et al. 1995), and a meta-analysis found an overall 53% reduction in risk of developing AAD (D'Souza et al. 2002). However, a recent large single center randomized controlled trials of 275 patients did not show any decrease of AAD in the treatment group (Pozzoni et al. 2012). Rates of AAD were 12/98 (13.3%) in patients given placebo and 16/106 (15.1%) in those given *S. boulardii*. This was an older group of hospitalized patients, which is the group at highest risk; all subjects were over 50 years of age, with a mean age of 79. There was also no difference in *C. difficile* infection in the 2 groups (3 in placebo and 3 in the *S. boulardii* group).

In a recent Cochrane Analysis of probiotics for prevention of AAD in children, Johnston et al. evaluated 3432 participants from 16 studies. There were trials of single strains (9), 2 strains (4), 3 strains (1), 10 strains (1) and one with two products (two and three strains). Overall the incidence of AAD was 9% in the probiotic group and 18% in controls. Subgroup analysis suggested that high-dose probiotics were more effective in prevention, with a NNT (number needed to treat) of 7. However, the authors caution that the overall quality of the evidence was low, with bias and imprecision (Johnston et al. 20011).

Other meta-analyses combine studies of children and adults. In one meta-analysis, 9 studies were analyzed (7 were in adults) and two probiotics (*Saccharomyces boulardii* and *Lactobacillus GG rhamnosus)* were effective in prevention of AAD with an odds ratio of 0.37 (95% CI 0.25–0.53, p<.001) (D'Souza et al. 2002). A recent meta-analysis of only *S. boulardii* in prevention of AAD found similar results with odds ratio 0.47 (McFarland 2010). A recent meta-analysis of probiotics in preventing AAD in children and adults pooled data from 34 studies (all RCTs), with 4138 patients. The pooled relative risk for AAD in the probiotic group compared to placebo was 0.53 (95% CI 0.44–0.63); the NNT was 8. Even though there were different probiotic species, different ages of patients, different durations of antibiotics and probiotics, the preventive effective remained significant (Vdelock and Cremonini 2012). Overall, probiotics appear effective in prevention of AAD in children and in adults.

Prevention of *Clostridium difficile* Infection (CDI)

While there is evidence that probiotics can prevent AAD, there is only a little evidence that they prevent CDI. The products studied include *S. boulardii*, several *Lactobacillus, Clostridium, Streptococcus* and *Bifidobacterum* species. In smaller trials there was less CDI in the probiotic groups compared to placebo —one study was of *S. boulardii* (Can et al. 2005) and one of *C. butyricum* (Imase et al. 2008) and one of *Lactobacillus plantarum* 299V (Klarin et al. 2008). Other trials showed no efficacy for *S. boulardii* (Lewis et al. 1998) or *Lactobacillus GG* (Thomas et al. 2001), one of *C. butyricum* (Imase et al. 2008). These very small studies can only be considered preliminary or exploratory. Several studies are frequently cited. In one, a capsule containing a mixture of *L. acidophilus* and *B. bifidum* was given to elderly hospitalized patients receiving antibiotics (Plummer et al. 2004). Of the 150 patients enrolled, 138 completed the study, 69 were given placebo and 69 were given the probiotic. Rates of diarrhea were the same in both groups (22%), which is an expected rate. Of those who developed diarrhea, there was a positive *C. difficile* toxin in 7.25% of the placebo group compared with 2.9% in the probiotic group. Among all patients, *C. difficile* was detected in 78% of the control group compared to 45% in the probiotic group. Whether this reduced colonization results in decrease in disease will require larger randomized controlled trials.

Another widely cited study tested a liquid probiotic milkshake containing *L. casei*, *L. bulgaricus* and *S. thermophiles* to older (age over 50) patients (Hickson et al. 2007). Only 7% of 1760 patients screened were enrolled, with 135 patients studied over a 2 year period. Of those given the probiotic, 12% (7/57) developed diarrhea compared to 34% (19/56) given placebo. *C. difficile* developed in 17% (9/53) controls and none in the probiotic group (p<.0.007). Again, further randomized controlled trials are needed to determine efficacy.

A recent meta-analysis and review of this topic concludes that there is a trend towards protection in nine of eleven studies (Johnson et al. 2012). One product, Bio-K+ CL1285, a probiotic combination of *Lactobacillus acidophilus* CL1285 and *L. casei* LBC8OR) has been shown to decrease AAD and CDI in three trials (Gao et al. 2010; Pzaradelis et al. 2010; Beausoleil et al. 2007).

In summary, most studies of probiotics do not have convincing efficacy in prevention of CDI; many of these studies are underpowered.

Treatment of *Clostridium difficile* infection (CDI)

There is no evidence that probiotics have a primary or adjunct role in treatment of first episodes of CDI. There is no evidence that they reduce risk of recurrence of first episodes of CDI (McFarland at al. 1994). These conclusions are supported by a Cochrane analysis (Pillai and Nelson 2008) and a systematic review (Dendukuri et al. 2005). There is also no evidence that probiotics play a role in therapy of severe disease, with the possible exception of fecal microbiota transplant (FMT), discussed later.

Treatment of Recurrent *Clostridium difficile* infection (RCDI)

The most widely studied probiotics for treatment of RCDI are *Saccharomyces boulardii* and *Lactobacillus GG*. In one trial *S. boulardii* decreased recurrences by 50% when used as an adjunct to antibiotics (McFarland et al. 1994). In a later trial its efficacy was limited to a subgroup on high dose (2 gm/day) vancomycin (Surawicz et al. 2000). *Lactobacillus GG* showed promise in early cases series, but was not effective in two subsequent small randomized controlled trials (Popachapin 2000; Lawrence at al. 2005). *Lactobacillus plantarum* 299V showed no significant decrease in RCDI (Wult et al. 2003).

Fecal microbiota therapy (FMT) is one of the most promising bacteria-based therapies for the treatment of recurrent and severe CDI. It involves the transfer of gastrointestinal microbiota from a healthy individual to an individual with disease. CDI is marked by a decrease in colonic microbial diversity, especially within the Bacteriodetes phyla (Chang et al. 2008; Hopkins and Macfarlane 2002; Tvede and Rask-Madsen 1989), and FMT reestablishes a diverse microbiota that has been shown to be stable in CDI and other gastrointestinal disorders for months to years (Tvede ND Rask-Madsen 1989; Grehan et al. 2010; Khoruts et al. 2010). Animal models have demonstrated that an intact microbiota is important for colonization resistance (Vollaard and Clasener 1994), and FMT has been shown to reestablish a disease resistant microbiota (Endt et al. 2010; Ghosh et al. 2010; Stecher et al. 2010; Willing et al. 2011). The exact mechanism by which a healthy microbiota protects from infection is not understood.

The first report of FMT in the literature was in 1958 by Eiseman et al. (1958). The group detailed the treatment of four patients with pseudomembranous colitis that were refractory to antibiotics. All four responded rapidly and completely to fecal retention enemas. The next report in the literature wasn't until 1981 when Bowden et al. documented a second series of 16 patients with pseudomembranous colitis that also responded to fecal retention enemas (Bowden et al. 1981). The first case of CDI treated by FMT was by Schwan et al. (1983). FMT has been evaluated primarily for recurrent CDI (Mattila et al. 2012; Rohlke et al. 2010; Yoon and Brandt 2010) although some groups have trialed and advocate for its use as a first-line therapy in severe CDI (You et al. 2008; Brandt at al. 2011). FMT has been recently reviewed by several independent groups with the number of cases reported in the literature now approaching 400, an efficacy rate of around 90%, and no adverse events reported to date (Bakken et al. 2011; Borody and Khoruts 2011; Guo et al. 2012). A recent RCT of feces delivered in to the duodenum is the first controlled trial to document efficacy (van Nood et al. 2013).

The FMT literature includes a variety of stool delivery methods and donor selection criteria. While initial reports used fecal retention enemas to deliver fecal slurries of whole stool and normal saline to the colon, subsequent papers have employed nasogastric (Aas and Bakken 2003; MacConnachie et al. 2009) and colonoscopic administration (Schwan et al. 1983; Yoon and Brandt 2010). Patient-administered retention enemas have also been evaluated (Silverman et al. 2010). The method of administration does not appear to affect cure rates (Postigo and Kim 2012). Donors have most often been selected by the recipient and include intimate-partners, household-contacts, family members, or friends. One research group has recently validated using frozen stool banks with meticulously screened stool specimens to maximize availability, minimize screening costs, and minimize the risk of transmitting infectious agents (Hamilton et al. 2010).

Safety is a critical issue as FMT becomes more widely used. The FMT workgroup recently published a position paper advocating for the use of specific serological and stool tests as well a modified Blood Banks Donor History Questionnaire to help eliminate the risk of transmitting known and unknown infectious agents (Bakken et al. 2011). Laboratory tests recommendations include screens for HIV, viral hepatitis, syphilis, parasites, *C. difficile*, and enteric infections. Screening the donor for metabolic, allergic, and autoimmune conditions in addition to gastrointestinal disease is especially important given the increasing awareness of the microbiota's role in diseases not traditionally thought to be related to the gastrointestinal tract. Flexibility in the stringency of donor testing for intimate verses non-intimate contacts as well as in emergent verses non-emergent cases has been taken into consideration in the guidelines.

Barriers to wide scale implementation of FMT as a treatment for CDI include lack of understanding of the mechanism of action, safety concerns, physician-perceived attitudes of patients toward the procedure, and lack of placebo-controlled trials supporting its efficacy. One study administering a subset of the gastrointestinal microbiota has postulated a role for Bacteroidetes species (Tvede and Rask-Madisen 1989) but more detailed work is necessary for identifying bacterial subsets with efficacy and their mechanism of action. These studies could ultimately be used to engineer "artificial stool" that would help limit infectious risks. In regard to patient attitudes, two studies have recently shown that the majority of patients with CDI would be interested in undergoing FMT (Brandt 2012; Zipursky et al. 2012). Lastly, a placebo controlled trial of FMT for CDI has been completed in Amsterdam and shows efficacy for treatment of RCDI. A similar trial has recently received FDA approval in the United States (Colleen Kelly, MD, Providence, RI, personal communication).

Other Intestinal Inflammatory Conditions

Prevention of Necrotizing Enterocolitis (NEC)

Necrotizing enterocolitis (NEC) is severe intestinal damage in low birth weight infants that is presumed to be of infectious origin, at least in part. The illness has a high morbidity and mortality. Trials of several probiotics (*Bifidobacterium* spp. and *Lactobacillus acidophilus*) have shown decreased rates of NEC when given prophylactically. A Cochrane review in 2008 of 9 randomized controlled trials showed that prophylactic probiotics led to a significant reduction in NEC (Alfaleh and Ballier 2008). Another meta-analysis failed to show that probiotics reduced the risk of NEC (Ritchie and

Romanuk 2012). While the results are encouraging, further data is suggested by experts before routine probiotic therapy can be recommended for these infants (Thomas et al. 2010; Milarsch et al. 2012).

Pouchitis

Pouchitis is inflammation of the pouch created after colectomy with ileo and anal pouch surgery (IPAA). It occurs in up to 20% of patients, and is more common after surgery for ulcerative colitis than other diseases (Shen 2012). The pathophysiology is suspected to be related to changes in the microbiota, as it often responds to therapy with antibiotics. Studies of a probiotic VSL#3 showed resolution of pouchitis in one small randomized controlled trial (Gionchetti 2003) and prevention of a first episode after initial surgery in two studies of VSL#3 (Mimura et al. 2004) and in one of *Lactobacillus rhamnousus GG* (Gosselink 2004). These are all very small trials. A meta-analysis of 5 studies of probiotics for pouchitis concludes that they have an odds ratio of 0.04 in favor of treatment of pouchitis (Elahi et al. 2008).

Risks of Probiotics

Overall, there are a small number of serious adverse events related to probiotic therapy considering their wide use worldwide. However, since bacteremia and fungemia have been reported (Segarra-Newham 2007), they should not be used in immunosuppressed patients, those with central lines in hospital, and children with short gut syndrome. One very important randomized clinical trial of probiotics in patients with acute severe pancreatitis showed an increase in infectious complications in both probiotic and placebo groups, but an increased mortality in the probiotic group (Besselink et al. 2008) (15% mortality) compared to the placebo group (6%). Nine of the patients had bowel ischemia of whom eight died. While the mechanism are not clear and there have been criticisms of the study and reporting (Reid 2008), the paper is a caution that these products may not always be benign.

Summary

Probiotics may have efficacy in prevention (but probably not treatment) of AAD and TD; treatment of infectious diarrhea in children and limited role in treatment of RCDI. Many studies and thus meta-analyses are limited by small numbers, heterogeneous patient groups, heterogeneous product studies, heterogeneous outcomes and lack of robust randomized controlled trials.

For probiotics to be widely used there need to be better studies of specific organisms and products, with combinations of organisms, mechanisms of action, validation of dose, product viability and stability, duration of therapy, safety studies and identification of high risk patients who are likely to benefit.

References Cited

Aas, J., C.E. Gessert and J.S. Bakken. 2003. Recurrent *Clostridium difficile* colitis: case series involving 18 patients treated with donor stool administered via a nasogastric tube. Clin. Infect. Dis. 36(5): 580–5.

Alfaleh, K. and D. Ballier. 2008. Probiotics for prevention of necrotizing enterocolitis in preterm infants. Cochrane Database Syst. Rev. (1) CD005496.

Allen, S.J., E.G. Martinez , G.V. Gregorio et al. 2010. Probiotics for treating acute infectious diarrhoea. Cochrane Database Syst. Rev. 11: CD003048.

Bakken, J.S., T. Borody, L.J. Brandt, J.V. Brill, D.C. Demarco, M.A. Franzos, C. Kelly et al. 2011. Treating *Clostridium difficile* infection with fecal microbiota transplantation. Clin. Gastroenterol. Hep. 9(12): 1044–9.

Beausoleil, M., N. Fortier, S. Guénette, A. L'Ecuyer, M. Savoie, M. Franco et al. 2007. Effect of a fermented milk combining *Lactobacillus acidophilus* CL1285 and *Lactobacillus casei* in the prevention of antibiotic-associated diarrhea: a randomized double-blind, placebo-controlled trial. Can. J. Gastroenterol. 21: 732–6.

Bernaola Aponte, G., C.A. Bada Macilla, C. Pariasca et al. 2010. Probiotics for treating persistent diarrhoea in children. Cochrane Database Syst. Rev. 11: CD007401.

Besselink, M.G., H.C. vanSantvoort, E. Buskens, M.A. Boemeester, H. van Goor, H.M. Timmerman et al. 2008. Probiotic prophylaxis in predicted severe acute pancreatitis: a randomized, double-blind, placebo-controlled trial. Lancet. 371: 651–659.

Borody, T.J. and A. Khoruts. 2011. Fecal microbiota transplantation and emerging applications. Nat. Rev. Gastroenterology & Hepatology. 10: 1–9. doi:10.1038/nrgastro.244.

Bowden, Jr., T., A. Mansberger Jr., L. Lykins et al. 1981. Pseudomembraneous enterocolitis: mechanism of restoring floral homeostasis. American Surgeon. 47(4): 178.

Brandt, L.J., T.J. Borody and J. Campbell. 2011. Endoscopic fecal microbiota transplantation: "first-line" treatment for severe *Clostridium difficile* infection? J. Clin. Gastroenterol. 45(8): 655–7.

Brandt, L.J. 2012. Fecal Microbiota Transplantation: Patient and Physician Attitudes. Clin. Infect. Dis. 101: 1–6.

Can, M., B.A. Besirbelliglu, I.Y. Avci, C.M. Beker and A. Pahsa. 2005. Prophylactic *Saccharomyces boulardii* in the prevention of antibiotic-associated diarrhea: a prospective study. Med. Sci. Monit. 12: P119–22.

Cecil, J.A. 2012. *Clostridium difficile*: changing epidemiology, treatment and infection prevention measures. Curr. Infect. Dis. Rep. 14: 612–619.

Chang, J.Y., D.A. Antonopoulos, A. Kalra, A. Tonelli, W.T. Khalife, T.M. Schmidt and V.B. Young. 2008. Decreased diversity of the fecal Microbiome in recurrent *Clostridium difficile*-associated diarrhea. Journal Infectious Diseases. 197(3): 435–8. doi:10.1086/525047.

D'Souza, A.L., C. Rajkumar, J. Cooke and C.J. Bulpitt. 2002. Probiotics in the prevention of antibiotic associated diarrhea: meta-analysis. BMJ. 324: 1361–6.

Dendukuri, N., V. Costa, M. McGregor, J.M. Brophy et al. 2005. Probiotic therapy for the prevention and treatment of *Clostridium difficile*-associated diarrhea: a systematic review. CMAJ. 173(2). doi:10.1503/cmaj.050350.

Dinleyici, E.C., M. Eren, N. Dougan, S. Reyhanioglu, Z.A. Yargic and Y. Vandenplas. 2011. Clinical efficacy of *Saccharomyces boulardii* or metronidazole in symptomatic children with *Blastocystis hominis* infection. Parasiol. Res. 108: 541–5.

Eiseman, B., W. Silen, G.S. Bascom and A.J. Kauvar. 1958. Fecal enema as an adjunct in the treatment of pseudomembranous enterocolitis. Surgery. 44(5): 854–9.

Elahi, B., S. Nikfar, S. Derakhshani et al. 2008. On the benefit of probiotics in the management of pouchitis in patients ileal pouch anal anastomosis: a meta-analysis of controlled clinical trials. Dig. Dis. Sci. 53: 1278–84.

Endt, K., B. Stecher, S. Chaffron, E. Slack, N. Tchitchek, A. Benecke, L. Van Maele et al. 2010. The microbiota mediates pathogen clearance from the gut lumen after non-typhoidal Salmonella diarrhea. PLoS pathogens. 6(9): e1001097.

Gao, X.W., M. Mubasher, C.Y. Fang, C. Reifer and L.E. Miller. 2010. Dose-response efficacy of probiotic prophylaxis for antibiotic-associated diarrhea and *Clostridium difficile*-associated diarrhea in adult patients. Am. J. Gastroenterol. 105: 1636–41.

Ghosh, S., C. Dai, K. Brown, E. Rajendiran, S. Makarenko, J. Baker, C. Ma et al. 2010. Colonic microbiota alters host susceptibility to infectious colitis by modulating inflammation, redox status, and ion transporter gene expression. Am. J. Physiol. Gastrointestinal and Liver Physiology. 301(1): G39–49.

Gionchetti, P., F. Rizzello, U. Helwig et al. 2003. Prophylaxis of pouchitis onset with probiotic therapy: a double-blind, placebo-controlled trial. Gastroenterology. 124: 1202–9.

Gionchetti, P., F. Rizzello, A. Venturi et al. 2000. Oral bacteriotherapy as maintenance treatment in patients with chronic pouchitis: a double-blind, placebo-controlled trial. Gastroenterology. 119: 305–309.

Gosselink, M.P., W.R. Schouten, L.M. van Lieshout et al. 2004. Eradication of pathogenic bacteria and restoration of normal pouch flora: comparison of metronidazole and ciprofloxacin in the treatment of pouchitis. Dis. Colon. Rectum. 47: 1519–25.

Grehan, M.J., T.J. Borody, S.M. Leis, J. Campbell, H. Mitchell and A. Wettstein. 2010. Durable alteration of the colonic microbiota by the administration of donor fecal flora. J. Clin. Gastroenterol. 44(8): 551–61.

Guandalini, S. 2000. Treatment of acute diarrhea in the new millennium. J. Pediatr. Gastroenterol. Nutr. 30S: 486–9.

Guo, B., C. Harstall, T. Louie, S. Veldhuyzen van Zanten and L.A. Dieleman. 2012. Systematic review: faecal transplantation for the treatment of *Clostridium difficile*-associated disease. Aliment Pharmacol. Ther. 35(8): 865–75.

Hamilton, M.J., A.R. Weingarden, M.J. Sadowsky and A. Khoruts. 2012. Standardized frozen preparation for transplantation of fecal microbiota for recurrent *Clostridium difficile* infection. Am. J. Gastroenterol. 107(5): 761–7.

Hibberd, P.L. 2009. Probiotics for infectious diarrhea and traveler's diarrhea – what do we really know? *In*: D. Charalampopoulos and R.A. Rastall (eds.). Prebiotics and Probiotics, Science and Technology. Springer, New York, Vol. 2, p. 845–899.

Hickson, M., A.L. D'Souza, N. Muthu, T.R. Rogers, S. Want, C. Rajkumar et al. 2007. Use of probiotic *Lactobacillus* to prevent diarrhoea associated with antibiotics: randomized double-blind, placebo controlled study. BMJ. 20; 335: 80–3.

Hopkins, M.J. and G.T. Macfarlane. 2002. Changes in predominant bacterial populations in human faeces with age and with *Clostridium difficil*e infection. J. Med. Microbiol. 51(5): 448–54.

Imase, K., M. Takahashi, A. Tanaka, K. Tokunaga, H. Sugano, M. Tanaka et al. 2008. Efficacy of *Clostridium butyricum* preparation concomitantly with *Helicobacter pylori* eradication therapy in relation to changes in the intestinal microbiota. Microbiol. Immunol. 52: 156–61.

Johnson, S., P.-J. Maziade, L.V. McFarland, W. TGrick, C. Donskey, B. Currie et al. 2012. Is primary prevention of *Clostridium difficile* infection possible with specific probiotics? Int. J. Infect. Dis. http://dx.doi/org/10.1016/j.ijid.2012.06.005.

Johnston, B.C., J.Z. Goldenberg, P.O. Vandvik, X. Sun and G.H. Guyatt. 2011. Probiotics for the prevention of pediatric antibiotic-associated diarrhea (Review). The Cochrane Database of systematic reviews. 11: Art No: CD004827, DOI:10.1002/114651858, CD004827, pub 3.

Khoruts, A., J. Dicksved, J.K. Jansson and M.J. Sadowsky. 2010. Changes in the composition of the human fecal microbiome after bacteriotherapy for recurrent *Clostridium difficile*-associated diarrhea. J. Clin. Gastroenterol. 44(5): 354–60.

Klarin, B., M. Wullt, I. Palmquist, G. Molin, A. Larsson and B. Jeppsson. 2008. *Lactobacillus plantarum* 299v reduces colonization of *Clostridium difficile* in critically ill patients treated with antibiotics. Acta Anaesthesiol. Scand. 52: 1095–1102.

Kotzampassi, K. and J. Giamarellos-Bourboulis. 2010. Probiotics for infectious diseases: more drugs, less dietary supplementation. Int. J. Antimicro. Agents. 40: 288–296.

Lawrence, S.J., J.R. Korzenk and L.M. Mundy. 2005. Probiotics for recurrent *Clostridium difficile* disease. J. Med. Microbiol. 54: 905–6.

Lewis, S.J., L.F. Potts and R.E. Barry. 1998. The lack of therapeutic effect of *Saccharomyces boulardii* in the prevention of antibiotic-related diarrhoea in elderly patients. J. Infect. 36: 171–4.

MacConnachie, A.A., R. Fox, D.R. Kennedy and R.A. Seaton. 2009. Faecal transplant for recurrent *Clostridium difficile*-associated diarrhoea: a UK case series. QJM. 102(11): 781–4.

Mattila, E., R. Uusitalo-Seppälä, M. Wuorela, L. Lehtola, H. Nurmi, M. Ristikankare, V. Moilanen et al. 2012. Fecal Transplantation, Through Colonoscopy, Is Effective Therapy for Recurrent *Clostridium difficile* Infection. Gastroenterology. 142(3): 490–6.

McFarland, L.V., C.M. Surawicz, R.N. Greenberg, G.W. Elmer, K.A. Moyer, S.A. Melcher, K.E. Bowen and J.L. Cox. 1995. Prevention of B-lactam-associated diarrhea by *Saccharomyces boulardii* compared with placebo. Am. J. Gastroenterol. 90: 439–448.

McFarland, L.V., C.M. Surawicz, R.N. Greenberg et al. 1994. A randomized placebo-controlled trial of *Saccharomyces boulardii* in combination with standard antibiotics for *Clostridium difficile* disease. JAMA. 271: 1913–8.

McFarland, L.V., C.M. Surawicz, R.N. Greenberg, R. Fekety, G.W. Elmer, K.A. Moyer et al. 1994. A randomized placebo-controlled trial of *Saccharomyces boulardii* in combination with standard antibiotics for *Clostridium difficile* disease. JAMA. 271: 1913–1918.

McFarland, L.V. 2007. Meta-analysis of probiotics for the prevention of traveler's diarrhea. Travel Med. Infect. Dis. 5: 97–105.

McFarland, L.V. 2010. Systematic review and meta-analysis of *Saccharomyces boulardii* in adult patients. World J. Gastroenterol. 16: 2202–22.

Milatsch, W.A., C.P. Braegger, T. Decsi et al. 2012. Critical systematic review of the level of evidence for routine use of probiotics for reduction of mortality and prevention of necrotizing enterocolitis and sepsis in preterm infants. Clin. Nutr. 31: 6–15.

Mimura, T., F. Rizzello, U. Helwig et al. 2004. Once daily high dose probiotic therapy (VSL#3) for maintaining remission of recurrent or refractory pouchitis. Gut. 53: 108–14.

Ogawa, M., K. Shimizu, K. Nomoto, M. Takahashi, M. Watanuki, R. Tanaka et al. 2001. Protective effect of *Lactobacillus casei* strain shirota on Shiga toxin-producing *Escherichia coli* 0157:H7 infection in infant rabbits. Infect. Immun. 69: 1101–8.

Pascual, M., M. Hugas, J.I. Badiola, J.M. Monfort and M. Garriga. 1999. *Lactobacillus salivarius* CTC2197 prevents *Salmonella enteritidia* colonization in chickens. Appl. Eviron. Microbiol. 65: 4981–6.

Pillai, A. and R. Nelson. 2008. Probiotics for treatment of *Clostridium difficile*-associated colitis in adults. Cochrane Database Syst. Rev. Jan 23: CD004611.

Plummer, S., M.A. Weaver, J.C. Harris, P. Dee and J. Hunter. 2004. *Clostridium difficile* pilot study: effect of probiotic supplementation on the incidence of *C. difficile* diarrhoea. Int. Microbiol. 7: 59–62.

Popachapin, M. 2000. The effect of probiotics on *Clostridium difficile* diarrhea. Am. J. Gastroenterol. 95(Suppl 1): S11–3.

Postigo, R. and J.H. Kim. 2012. Colonoscopic versus nasogastric fecal transplantation for the treatment of *Clostridium difficile* infection: a review and pooled analysis. Infection. doi:10.1007/s15010-012-0307-9.

Pozzoni, P., A. Riva, A.G. Bellatorre, M. Amigoni, E. Redaelli, A. Ronchetti et al. 2012. *Saccharomyces boulardii* for the prevention of antibiotic-associated diarrhea in adult hospitalized patients: a single-center, randomized, double-blind, placebo-controlled trial. Am. J. Gastroenterol. 107: 922–931.

Psaradellis, E., J. Sampalis and E. Rampakakis. 2010. Efficacy of BioK+CL1285 in the reduction of antibiotic-associated diarrhea—a placebo controlled, double-blind randomized, multi-center study. Arch. Med. Sci. 6: 56–64.

Reid, G., G. Gibson, M.E. Sanders, F. Guarner and J. Versalovic. 2008. Probiotic prophylaxis in predicted severe acute pancreatitis. Lancet. 372: 112–113.

Ritchie, M.L. and T.N. Romanuk. 2012. A meta-analysis of probiotic efficacy for gastrointestinal diseases. PlosOne. 7: www.plosone.org e34938.

Rohlke, F., C.M. Surawicz and N. Stollman. 2010. Fecal flora reconstitution for recurrent *Clostridium difficile* infection: results and methodology. J. Clin. Gastroenterol. 44(8): 567–70.

Sarker, S.A., S. Sultana, G.J. Fuchs, N.H. Alam, T. Azim, H. russow et al. 2005. *Lactobacillus paracasei* strains ST11 has no effect on rotavirus but ameliorates the outcome of nonrotavirus diarrhea in children from Bangladesh. Pediatrics. 116: e221–8.

Savaş-Erdeve, S., S. Gőkay and Y. Dallar. 2009. Efficacy and safety of *Saccharomyces boulardii* in amebiasis-associated diarrhea in children. Turk. J. Pediatr. 51: 220–4.

Sazawal, S., U. Dhingra, G. Hiremath et al. 2010. Prebiotic probiotic fortified milk in prevention of morbidities among children: community-based, randomized, double-blind, controlled trial. PLoS One. 5: e12164.

Sazawal, S., G. Hiremath, U. Dhingra, P. Malik, S. Deb et al. 2006. Efficacy of probiotics in prevention of acute diarrhoea: a meta-analysis of masked, randomized, placebo-controlled trials. Lancet Infect. Dis. 6: 374082. Dol 10.1016/S1473-3099(06)70495–9.

Schwan, A, S. Sjölin, U. Trottestam, B. Aronsson. 1983. Relapsing *clostridium difficile* enterocolitis cured by rectal infusion of homologous faeces. Lancet. 8(2): 845.

Segarra-Newnham, M. 2007. Probiotics for *Clostridium difficile*-associated diarrhea: focus on *Lactobacillus rhamnosus GG* and *Saccharomyces boulardii*. Ann. Pharmacother. 41: 1212–21.

Shen, B. 2012. Acute and chronic pouchitis—pathogenesis, diagnosis and treatment. Nat. Rev. Gastroenterol. Hepatology. 9: 323–33.

Shornikova, A.V., I.A. Casas, H. Mykkanen, E. Salo and T. Vesikari. 1997. Bacteriotherapy with *Lactobacillus reuteri* in rotovirus gastroenteritis. Pediatr. Infect. Dis. J. 16: 1103–7.

Silverman, M.S., I. Davis and D.R. Pillai. 2010. Success of self-administered home fecal transplantation for chronic *Clostridium difficile* infection. Clin. Gastroenterol. Hepatology. 8(5): 471–3.

Stecher, B., S. Chaffron, R. Käppeli, S. Hapfelmeier, S. Freedrich, T.C. Weber, J. Kirundi et al. 2010. Like will to like: abundances of closely related species can predict susceptibility to intestinal colonization by pathogenic and commensal bacteria. PLoS pathogens. 6(1): e1000711.

Sur, D., B. Manna, S.K. Niyogi et al. 2011. Role of probiotic in preventing acute diarrhoea in children: a community-based, randomized, double-blind placebo-controlled field trial in an urban slum. Epidemiol. Infect. 139: 919–926.

Surawicz, C.M., G.W. Elmer, P. Speelman, L.V. McFarland, J. Chinn and G. van Belle. 1988. Prevention of antibiotic-associated diarrhea by *Saccharomyces boulardii*: a prospective study. Gastroenterology. 96: 981–988.

Surawicz, C.M., L.V. McFarland, R.N. Greenberg et al. 2000. The search for a better treatment for recurrent *Clostridium difficile* disease: use of high-dose vancomycin combined with *Saccharomyces boulardii*. Clin. Infect. Dis. 31: 1012–7.

Szymański, H., J. Pejcz, M. Jawień, A. Chmielarczyk, M. Strus and P.B. Heczko. 2006. Treatment of acute infectious diarrhoea in infants and children with a mixture of three *Lactobacillus rhamnosus* strains: a randomized, double-blind, placebo-controlled trial. Aliment Pharmacol. Ther. 23: 247–253.

Thomas, D., F. Greer et al. 2010. Clinical Report: Probiotics and Prebiotics in Pediatrics. Pediatrics. 126: 1271–1231.

Thomas, R.T., S.C. Litin, D.R. Osmon, A.P. Corr, A.L. Weaver and C.M. Lohse. 2001. Lack of effect of *Lactobacillus GG* on antibiotic-associated diarrhea: a randomized, placebo-controlled trial. Mayo Clin. Proc. 76: 883–9.

Tschudin-Sutter, S., A.F. Widmer and T.M. Perl. 2012. *Clostridium difficile*: novel insights on an incessantly challenging disease. Curr. Opin. Infect. Dis. 25: 405–411.

Tvede, M. and J. Rask-Madsen. 1989. Bacteriotherapy for chronic relapsing *Clostridium difficile* diarrhoea in six patients. Lancet. 1: 1156–60.

van Nood, E., A. Vrieze, M. Nieuwdorp, S. Fuentes, E.G. Zoetendal et al. 2013. Duodenal infusion of donor feces for recurrent *Clostridium difficile*. N. Engl. J. Med. 368(5): 407–15. doi: 10.1056/NEJMoa1205037.

Vdelock, E.J. and F. Cremonini. 2012. Meta-analysis: probiotics in antibiotic-associated diarrhoea. Alimentary Pharmacology & Therapeutics. 35: 1355–1369.

Vollaard, E.J. and H.A. Clasener. 1994. Colonization resistance. Antimicrobial Agents and Chemotherapy. 38(3): 409–14.

Willing, B.P., A. Vacharaksa, M. Croxen, T. Thanachayanont and B.B. Finlay. 2011. Altering Host Resistance to Infections through Microbial Transplantation (S. Bereswill, Ed.) PLoS ONE. 6(10): e26988. doi:10.1371/journal.pone.0026988.

Wult, M., M.L. Hagslatt and I. Odenholt. 2003. *Lactobacillus plantarum* 299v for the treatment of recurrent *Clostridium difficile*-associated diarrhoea: a double-blind, placebo-controlled trial. Scand. J. Infect. Dis. 35: 365–7.

Yoon, S.S. and L.J. Brandt. 2010. Treatment of refractory/recurrent *C. difficile*-associated disease by donated stool transplanted via colonoscopy: a case series of 12 patients. J. Clin. Gastroenterol. 44(8): 562–6.

You, D.M., M.A. Franzos and R.P. Holman. 2008. Successful treatment of fulminant *Clostridium difficile* infection with fecal bacteriotherapy. Ann. Int. Med. 148(8): 632–3.

Zipursky, J.S., T.I. Sidorsky, C.A. Freedman, M.N. Sidorsky and K.B. Kirkland. 2012. Patient Attitudes Towards the Use of Fecal Microbiota Transplantation in the Treatment of Recurrent *Clostridium difficile* Infection. Clin. Infectious Diseases. 1–21.

7

The Microbiota in Inflammatory Bowel Disease

Philippe Marteau,[1,*] *Marion Leclerc,*[2] *Patricia Lepage,*[2] *Philippe Seksik,*[3] *Harry Sokol*[3] and *Joel Doré*[2]

Introduction

Inflammatory Bowel Disease

Inflammatory bowel diseases (IBD) affect one person out of 1000 in Europe and are increasing in many countries. They encompass several diseases with cryptogenetic and chronic inflammation of parts of the intestinal tract, the two major being ulcerative colitis and Crohn's disease (Ordás et al. 2012; Baumgart and Sandborn 2013). Ulcerative colitis (UC) affects the rectal and colonic mucosa. One third of the patients have only rectal disease (proctitis), one third distal colitis (downstream the splenic flexure of the colon) and one third extensive (sometimes total) colitis. Some patients have proctitis associated with localized periappendicular inflammation (at the other end of the colon), and some patients with pancolitis may

[1] Université Paris 7 et AP-HP Hopital Laribboisière, service de gastroentérologie, 2 rue A. Paré 75010 Paris, France.
[2] UMR1319 Micalis, INRA, Jouy-en-Josas, France. AgroParisTech, UMR 1319 MICALIS, F-78350 Jouy-en-Josas, France.
E-mails: marion.leclerc@jouy.inra.fr; patricia.lepage@jouy.inra.fr; joel.doré@jouy.inra.fr
[3] Service de Gastroentérologie et Nutrition, Hôpital Saint-Antoine, 184 rue du faubourg Saint-Antoine 75571 Paris cedex 12.
E-mails: philippe.seksik@sat.aphp.fr; harry.sokol@sat.aphp.fr
* Corresponding author: philippe.marteau@lrb.aphp.fr

have ileal inflammation called "backwash ileitis". Crohn's disease (CD) affects not only the mucosa but also deeper layers of the intestinal wall; inflammatory lesions are usually patchy with healthy mucosa surrounding inflammatory ulcers; in half of the cases, they contain granulomas, which are also observed in other "granulomatous" diseases including tuberculosis. CD affects the ileum in 2/3 of cases, the colon in 2/3 of cases, the anus in 1/4 of cases. CD less often also affects proximal parts of the gut (1/10) or other organs such as the bronchia or the skin. Over time, CD lesions can spread to larger surfaces in the intestinal tract but more often extend in the depth of the affected segment leading to local abscesses, fistulae with adjacent organs and fibrosis which narrows the intestinal lumen leading to its obstruction. The terms "penetrating" and fibrostenosing" CD are then used. It is not always easy to differentiate UC from CD when only the colon is affected, and 10% of the patients initially diagnosed with UC are later in life reclassified as having CD. Clinicians separate phenotypic subtypes of these diseases to choose treatment. For UC, the main subtypes are proctitis, distal colitis and extensive colitis. For CD the subtypes are pure ileal CD, pure colitis (sometimes "UC like colitis" when the lesions extend from the rectum), ileocolitis, and perianal disease. Both UC and CD increase the risks of cancer of the inflamed mucosa (adenocarcinoma), especially when inflammation is chronically uncontrolled, extensive, and in patients cumulating other risk factors such as slerosing cholangitis and familial history of cancer or adenomas.

Smoking increases the risk of CD and decreases the risk of UC (Ordás et al. 2012; Baumgart and Sandborn 2013). Genetic risk factors have also been identified and many loci are associated with a higher chance of developing CD and/or UC (Ordás et al. 2012; Baumgart and Sandborn 2013). The statistical risk of developing IBD for a first degree relative of a patient is however usually below 3%.

A curative treatment that would definitively avoid recurrence does not (yet) exist (Ordás et al. 2012; Baumgart and Sandborn 2013). Even surgery removing all visible lesions does not avoid recurrence. Surgery is limited to situations which cannot be sufficiently improved with a medical treatment. In UC, lesions constantly recur if a colonic segment is left in place so that the main surgical technique is removing the whole colon (proctocolectomy with ileal-pouch-anal anastomosis). In CD, surgery is often needed and is a good way to induce remission of symptoms; however, relapse of lesion occurs often and early (in 70% of patients at 1 year), most often on the anastomosis and above it. This postoperative recurrence is favored by smoking, absence of prophylactic treatment, presence of myenteric plexitis on the proximal resection margin, and composition of the microbiota (see below). The main medical treatments to induce remission of symptoms, inflammation and lesions are aminosalicylates (first line treatment of

UC), corticosteroids, immunosuppressants and anti-tumor necrosis factor antibodies. Diet influences quality of life (symptoms) in case of acute IBD but does not modify the risk of recurrence. Usually, it consists in suppressing milk and long fibers (to decrease the risks of diarrhea and obstruction respectively).

Reasons to Imagine (and Search for) a Role of Micro-organisms in IBD

Converging arguments suggest that micro-organisms trigger IBD (Table 1). The role of pathogens was already questioned in 1932 when Crohn et al. made the first description of regional enteritis (Crohn et al. 1932). Later on, it was established that the genetic factors favouring IBD involve autophagy and sensing of micro-organisms in the intestine (Ordás et al. 2012; Baumgart and Sandborn 2013). A reduced expression of α-defensins (which exhibit antimicrobial effects at the mucosal surface) has been shown in ileal CD and an attenuated induction of β-defensins in the colon of patients with colonic IBD (Ordás et al. 2012; Baumgart and Sandborn 2013). Of note, some studies have also shown that the uptake of bacteria across the epithelium is increased in CD (Keita et al. 2012), that intestinal dendritic cell function is influenced by the composition of the commensal microbiota (Ng et al. 2011), and that aberrant mucosal recognition of commensal bacteria induce pro-inflammatory cytokines in CD tissue (Edwards et al. 2011; Hotte et al. 2012).

The role of the microbiota in IBD has been experimentally approached by description studies, and intervention experiments suppressing micro-organisms using antibiotics or gnotoxenic animals, adding micro-organisms or microbial components (e.g., probiotics, CpG-DNA, culture supernatants...) or shifting the composition of the microbiota using prebiotic substrates. An abnormal chronic inflammatory response to micro-organisms

Table 1. Strongest arguments suggesting a role of the microbiota in IBD onset and/or exacerbation.

- Role of fecal stream in post-operative recurrence of CD
- In germ-free rodents, intestinal inflammation cannot be induced in most of the colitis model
- Micro-organisms can modulate (increase or decrease) inflammation in animal models of IBD
- Genetic risk factors of IBD are connected to autophagy or recognition of microbial signals
- The microbiota of patients with IBD differs from that of controls and is unstable (both in the intestinal lumen and at the surface of the mucosa).
- Imidazole antibiotics are effective to prevent early postoperative recurrence of CD

may result from specificities of the micro-organisms themselves but also on host specificities or interaction between host and microbes.

This has been observed in animal models of IBD and humans. A microbiome (coined by Joshua Lederberg) corresponds to the totality of microbes, their genetic elements (genomes), and environmental interactions in a particular environment. Among risk factors to develop IBD, micro-organisms are usually considered as "environmental factors" and human factors are mainly of genetic background, however, the human genetic background participates to the selection of the endogenous micro-organisms and the later influence human gene expression in epithelial and immune cells. For example, Nod2-deficient mice display a dysbiosis characterized by a higher bacterial load in feces and ileum compared to their wild-type counterparts (Rehman et al. 2011). A clustering of human gene expression and microbial patterns (so-called enterotypes) has been observed, and clusters differ between healthy subjects and patients with IBD (Arumugam et al. 2011; Hotte et al. 2012). Finally it is now also known that all micro-organisms do not induce the same host response, some being pro-inflammatory but other anti-inflammatory (Sokol et al. 2008; Eeckhaut et al. 2012).

Some researchers reinfused the small bowel effluent or a sterile ultrafiltrate of it into the defunctioned distal bowel of a few patients with CD treated by split ileostomy and observed that the full chyme had a far higher pro-inflammatory property than the particle and bacteria-free filtrate (Harper et al. 1985; D'Haens et al. 1998). The theory is thus that when a patient is operated for CD, the reflux of the chyme from the colon to the neoterminal ileum brings high levels of pro-inflammatory bacteria that trigger recurrence of inflammation.

Imidazole antibiotics have been shown in double blind randomized controlled trials and in their meta-analysis to decrease the risk of recurrence after ileal resection for CD (see chapter by Ng et al.). Some antibiotic combinations seem to have some therapeutic efficacy in patients with UC, at least in Japan where the studies were performed (Ohkusa et al. 2010). A limited number of studies testing probiotics were positive, especially in UC and pouchitis and this is detailed in the chapter by Ng et al. in this book (Chapter: Probiotics in IBD: Modulation of the pathological immune or inflammatory activation. Their potential benefits in the different phases of the disease, Ng and Hart).

Questions

Several questions are posed to the researcher and the clinician. Is dysbiosis the cause or the consequence of IBD? Is it disease specific? Is it localized to the lesions or "at risk parts" of the intestine? Can it be used as a diagnostic tool? Or as a therapeutic tool or target? In this chapter we will be discussing these questions but will not detail opportunistics, *E. coli*, and probiotics which are discussed in specific chapters of this book (Opportunistic pathogens in inflammatory bowel disease, the case of the adherent-invasive E.coli Darfeuille-Michaud, this book). We discuss mainly bacteria although phages (Lepage et al. 2008), and fungus communities may also be of interest as the "mycobiome" also interacts with the immune system (Iliev et al. 2012).

Descriptive Studies of the Dysbiosis

Many studies have been published and are published each year describing the intestinal ecology in various situations of IBD and using various microbiology tools (mainly molecular techniques nowadays). We summarise below their most reproducible results, but first present potential biases which should be taken into account to avoid misinterpretation.

Confounding Factors

Interpreting the results of studies describing the microbiota must be cautious. Subtypes of IBD differ and pooling them may sometimes hide information. For example genetic risk factors, defensin biology and the microbiota significantly differ between ileal and colonic CD (Baumgart and Sandborn 2013). Several confounding factors may be present in clinical studies (before sampling of feces or the mucosa) such as recent use of antibiotics, bowel cleansing, or diet. Disease activity also affects the results and one should know that in the absence of clinical signs (i.e., in periods of "clinical remission") many patients still have significant inflammatory lesions while other are in deep remission without any detectable macroscopic or histological lesions. The data also depend on the ecological niche being studied. The luminal microbiota differs between the stomach, upper small bowel, lower small bowel, right colon and the rectum (Marteau et al. 2001). It also differs at each of these sites from the mucosal microbiota which can be studied on intestinal biopsy samples. The method used to handle biopsy samples (using detergent, fixatives, rinsing, vortexing...) affects the results. In healthy subjects, the majority of the bacteria present in mucosal biopsies are located in the upper part of the mucus layer while the deeper part of the mucus close to the intestinal and colonic epithelial cells is devoid of micro-organisms (Swidsinski et al. 2005). Tools used to describe the microbiota differ in their characteristics.

The choice of controls is important in clinical trial. Studies in twins have been used to approach the role of the human genetic background. Healthy controls are easy to find for fecal analysis but not for biopsy analysis. Controls are thus not always healthy but sometimes patients suffering from irritable bowel syndrome (IBS), polyps, or cancers, now known/suspected to also involve microbiota dysbiosis.

Cigarette smoking is one of the most important environmental risk factors in IBD pathogenesis and there is a clearly divergent effect of smoking on the disease course of CD (which is worsened) and UC (which is improved by smoking). Smoking may influence disease risk activity through many mechanisms but the microbiota seems to be involved. Indeed, healthy smokers and smokers with CD have significantly higher fecal proportions of *Bacteroides-Prevotella* than non smokers (healthy or with CD) (Benjamin et al. 2012) and a lower relative frequency of *Faecalibacterium prausnitzii* spp. (Li et al. 2012). A recent pilot controlled trial also showed that smoking cessation was followed by an increase in firmicutes (Biedermann et al. 2013). The effect of some fibres on the concentrations of bifidobacteria and lactobacilli in the colon has been studied using culture techniques and molecular tools so that the "bifidogenic effect" of some of them (usually referred as "prebiotics") is established. The effect of dietary components on the dominant microbiota is less well known. Recent studies showed that long-term diets indeed were able to influence and affect enterotype of individuals: whereas diets enriched in protein and animal fat favoured the *Bacteroides* enterotype, a carbohydrate-enriched diet supported the *Prevotella* enterotype (Wu et al. 2011; Bushman et al. 2013). Diet should thus be more studied as a confounder for some microbial results in IBD (as many subjects tend to follow low fibre diets).

Antibiotics and sulfasalazine, which contains a sulfamide moiety, can affect the microbiota and subjects with recent antibiotic treatments are usually discarded from the descriptive studies. The direct impact of 5-aminosalicylic acid (5-ASA), immunosuppressants and anti-TNF agents on the microbiota has not been properly studied in the clinical setting. To assess the effects of diarrhea on the composition of the microbiota, Gorkiewicz et al. performed a longitudinal study in four healthy adults in whom they induced osmotic diarrhea by oral administration of polyethylene glycol (Gorkiewicz et al. 2013). Diarrhea led to a prominent shift of the mucosal and fecal communities. Diarrhea was associated with a significant decrease in phylotype richness in feces while the phylotype diversity and evenness were not significantly altered. The capacity of stool microbiota to reconstitute (resilience) was assessed by comparing samples from diarrhea and post-diarrhea periods, and the authors observed that although species richness increased significantly, overall reduced species richness persisted during the one-week interval after polyethylene glycol administration. The effect of colon cleansing which is needed to perform colonoscopy and uses high doses of polyethylene glycol for a short period has not been precisely assessed.

Which Microbial End Points?

Two strategies have been and will be used to search for micro-organisms potentially associated with diseases. 'Candidate micro-organism strategies' focus on one micro-organism or group supposed to be more aggressive or protective. Aggressive candidates include *Mycobacteria*, *Clostridium difficile*, some *E. coli*, *Helicobacter* species, sulfate reducing bacteria and fusobacteria. Potentially protective microbes which have been most studied include lactobacilli, bifidobacteria, *F. prausnitzii*, and other butyrate producers such as *Roseburia* or *Butyricicoccus*. The alternative "global description strategy" seeks to describe parts of the microbiota without a *priori* hypothesis on the potential role of some micro-organisms or groups. Biodiversity, a major characteristic of any ecological niche, can be (roughly) estimated using various methods of 16S rDNA ribotyping (such as Single Strain Conformation Polymorphism, Temporal Temperature Gradient gel Electrophoresis, DGGe, TRFLP…) or 16S rDNA sequencing, all of which carry limitations. Activity of bacteria can be approached by measuring 16S RNA but results are influenced by the number of ribosomes. Biostatistical tools allow assessing the beta-diversity, i.e., by comparing complex microbiota and calculating similarity indexes. This is very useful to compare subjects (for example IBD vs. controls or active vs inactive IBD) or to assess the stability of the composition of the microbiota in a defined ecological niche over time. If one looks at very specific microbial markers, the interindividual variability becomes huge and temporal variability seems frequent; however, when one uses markers of groups of bacteria, healthy ecosystems appear reasonably stable. Getting consistent information from various methods is the best way to establish that an observation is scientifically sounded.

Dysbiosis in Crohn's Disease

A dysbiosis has been almost universally observed in CD, both at the luminal (mainly fecal) and mucosal levels and very similar findings have been observed in various countries such as France, Germany, Spain, Norway, China, Japan, Canada, and the US (Sokol et al. 2008; Ricanek et al. 2012). Usually the dysbiosis includes a decrease in firmicutes, especially the *C. leptum* group and its main micro-organism prevalent species: *F. prausnitzii*. The Firmicutes/Bacteroidetes ratio is decreased in comparison to healthy controls. The dysbiosis also frequently includes increased numbers of usually subdominant micro-organisms such as enterobacteria or specific *E. coli* (see chapters Opportunistc pathogens pathogens in IBD, and the relation with specific gene susceptibilities, Fraher and Shanahan; Correction of microbiota disturbances or antagonism against specific pathogen in IBD, Flanagan and Rhodes, this book; Opportunisitc pathogens in inflammatory bowel disease, the case of adherent-invasive *E. coli*, Darfeuille-Michaud, this book).

Mucosa Associated Microbiota

The global composition of the dominant micobiota of the mucus layer (MAM) has a high similarity index along the colon in patients with CD as well as in healthy subjects (Lepage et al. 2005). We did not find significant differences between the microbiota present just at the surface of intestinal ulceration and that of uninvolved mucosa in patients with CD (Lepage et al. 2005; Vasquez et al. 2007). However, some recent studies using high resolution sequencing techniques suggested that the predominant composition of bacteria attached to the ulcerated mucosa could be distinguishable from that in the nonulcerated area (Walker et al. 2011; Li et al. 2012). The diversity of the microbiota in the ulcerated sites was especially decreased in comparison to that in the nonulcerated mucosa, while the relative abundance of *E. coli* and *Clostridium* spp. were found to be much higher. Presently, it seems that these differences vary greatly between individuals, and that there is no obvious phylogenetic bacterial signature positively associated with the gut inflammatory lesions (Walker et al. 2011).

The *Helicobacter* genus comprises two groups, gastric *Helicobacter*, describing those that preferentially colonise the stomach, and enterohepatic *Helicobacter*, which preferentially colonise the intestinal or hepatobilliary system. Many studies have shown a negative association between *H. pylori* infection and IBD (Luther et al. 2010). *H. pylori* could be a surrogate marker of childhood infection exposure modulating gastrointestinal immune development. On the opposite, enterohepatic *Helicobacter* species (*H. pullorum* and *H. canadensis*) have been found to be more frequent in the mucosal biopsies from patients with CD than in controls (Laharie et al. 2009; Thomson et al. 2011). Mucosal biopsies from affected part of the intestine contain a higher number (up to 100 times more) of bacteria than biopsies from healthy or IBS controls. *In situ* hybridization techniques showed that especially the deeper part of the mucus layer is abnormaly colonised (Swidsinski et al. 2002, 2005, 2008) and it was seen that some bacteria may even penetrate the epithelium lining. This can result from both specific invasive properties of some *E. coli*, an increased capacity of the mucosa to "sample bacteria" (Keita et al. 2012), or a decreased secretion of antimicrobial agents especially defensins (Gersemann et al. 2012), and as those three mechanisms have been observed they do not exclude each other.

Mycobacterium avium paratuberculosis has been found in several studies by PCR or culture to be more often present in intestinal tissues of patients with CD than in controls but there is a large overlap of results between the CD and healthy population. It is also difficult to reconcile the idea of a potential role of mycobacteria in CD with the therapeutic effect immunosuppressors (which usually increase risk of severe infections with mycobacteria) and the absence of therapeutic effect of antibiotics against *M. avium paratuberculosis* (Opportunistc pathogens in IBD, and the relation with specific gene susceptibilities, Fraher and Shanahan).

Virulent adherent-invasive *E. coli* pathovar (AIEC) have been isolated from the ileal mucosa of patients with CD more often than in controls and several mechanisms for colonisation and invasion by the micro-organism have been decipherised (Opportunisitc pathogens in inflammatory bowel disease, the case of adherent-invasive *E. coli*, Darfeuille-Michaud).

Many studies have observed decreased counts of firmicutes, especially *F. prausnitzii* in the ileal mucosa of patients operated for ileal CD (Swidsinski et al. 2005; Verrna et al. 2010; Willing et al. 2010; Walker et al. 2011). Interestingly, the lower these counts, the higher the risk of early recurrence of CD after surgery (Sokol et al. 2008). Studies *in vitro* and in a model of colitis in mice showed that *F. prausnitzii* had anti-inflammatory and protective effects (Sokol et al. 2008). The mechanism (active molecule) of this strong anti-inflammatory effect is still under investigation but does not seem to be butyrate, as a group of mice treated with butyrate (at the same concentration as supernatant) had a mortality rate similar to that in the colitis control group. Other micro-organisms have also been repeatedly shown to be decreased in CD and to exhibit anti-inflammatory potential, especially *Roseburia* and *Butyricicoccus pullicaecorum* (Eeckhaut et al. 2013). Local production of butyrate is one of the main scrutinized mechanisms of action of those protective bacteria although some others are also studied (Eeckhaut et al. 2013; Marteau 2013).

Fecal Microbiota

Comparisons of the microbial composition of fecal samples from patients with CD, UC and self limited colitis showed significant differences (Sokol et al. 2009), suggesting that the dysbiosis is not (only) secondary to the ecological changes induced by colitis. Joossens et al. (2011) reported that unaffected relatives of patients with CD have a different composition of their microbiota compared with healthy controls. In patients with active CD, a significant increase in enterobacteria has repeatedly been observed using both culture-based and culture-independent techniques (Seksik et al. 2003; Sokol et al. 2009; Mondot et al. 2011).

We showed that more than 30% of the dominant microflora in patients with active or inactive CD belonged to unusual phylogenetic groups (Seksik et al. 2003). The fecal flora in CD is also unstable over time (Seksik et al. 2003) and the biodiversity of firmicutes is reduced with a special and reproducible decrease in *F. prausnitzii* (Manichanh et al. 2006; Swidsinski et al. 2008; Sokol et al. 2009; Dörffel et al. 2012; Fujimoto et al. 2013). Using the molecular inventory method on faecal samples from patients with CD, we showed that numerous clones belonged to a great variety of species uncommonly detected as prevalent in healthy subjects (Mangin et al. 2004). The role of a single pathogen common to all patients with CD is thus very unlikely.

There are sometimes significant and sometimes non significant differences in the microbiota between subject with active and inactive CD, however, this is probably largely influenced by analytic methods, numbers of subjects studied (statistical power), and the fact that clinical remission of CD does not always correspond to absence of inflammatory lesions in the intestine (Baumgart and Sandborn 2013). One should also remember that ileal CD differs from colonic CD clinically but also for some biological factors such as defensins expression or role of NOD2 polymorphism. Potential other confounders (described above) must also be considered, especially smoking and fibre intake.

Dysbiosis in Ulcerative Colitis

The dysbiosis of patients with UC differs from that observed in CD but specific features of UC-associated dysbiosis are less described. A decrease in biodiversity and of the *C. coccoides* group and an increase in enterobacteria or unusual bacteria, have been reported both in the MAM and in feces (Noor et al. 2010; Lepage et al. 2011; Verma et al. 2010). Fusobacteria and sulfate reducing bacteria received special attention.

Mucosa Associated Microbiota

Mucosal bacteria differ between controls, CD and UC (Frank et al. 2007; Verma et al. 2010). They also significantly differ between patients with UC and their healthy twins, with a decreased biodiversity and a decrease in potentially protective bacteria (Lepage et al. 2011). Noticeably, the level of *Akkermansia muciniphila* was found in one study to be 100-fold reduced (Png et al. 2010). A study using *in situ* hybridization did not find microbial signal in the crypt abcesses of patients with IBD in contrast to what was observed in patients with self limited colitis (Sokol et al. 2010).

Fecal Microbiota

The stability of the fecal microbiota seems variable in different studies (Martinez et al. 2008; Rajilić-Stojanović et al. 2013). A reduction of biodiversity has also been confirmed in a pediatric study (Michail et al. 2012) and quite interestingly, children who did not improve with steroids harbored a microbiome that was less rich than steroid responders. Rajilić-Stojanović et al. (2013) reported that the significant reduction of bacterial diversity concerned members of the *Clostridium* cluster IV and bacteria involved in butyrate and propionate metabolism, including *Ruminococcus bromii, Eubacterium rectale, Roseburia* sp., and *Akkermansia* sp. Noticeably, the fecal microbiota composition seems not to be significantly influenced by geographic location, age, or gender, but it differs significantly between the patients with UC and the control subjects. Opportunistics were occasionally found including fusobacteria. *Fusobacterium varium* may induce colitis in animals (Ohkusa et al. 2003). It seems to be more often found in stools during UC than in controls, at least in Japan (Yukawa et al. 2013), and an antibiotic association against this micro-organism had a significant efficacy to treat acute episodes of UC in a blinded randomised controlled trial (Ohkusa et al. 2010). The potential influence of sulfate reducing bacteria which may produce toxic hydrogen sulfide is debated as luminal levels of hydrogen sulfide are not always increased in UC (De Preter et al. 2012; Jia et al. 2012). This metabolite has clearly numerous biological actions including inhibition of butyrate oxidation and modulates cell cycle and inflammation. More studies should be performed on this important group of bacteria which also seem of interest in irritable bowel syndrome (Chassard et al. 2012).

Microbes as Diagnostic or Therapeutic Targets or Tools—Foreseen Perspectives

Diagnosing IBD, which largely relies presently on endoscopy and histologic examination of biopsies, is sometimes not easy. Finding biological diagnostic markers would be a progress. There is also intensive research to find markers to predict the risks of complications to be used by clinicians to decipher the best treatment strategy. The first microbial markers were serum antibodies against microbes or microbial markers but other including fecal markers are under development (Papa et al. 2012; Rajca et al. 2011).

Patients with IBD, especially CD, often have antibodies against microbial molecules that could serve as disease markers. They include anti—*Saccharomyces cerevisiae* antibodies (ASCA), antibodies to the *E. coli* outer-membrane porin C (OmpC), the *Pseudomonas fluorescens* CD-related protein, anti-chitobioside carbohydrate antibodies, and the antiflagellin markers anti-CBirl antibodies, FlaX and A4-Fla2 (Kaul et al. 2012).

Microbiota as Therapeutic Target or Therapeutics in IBD (Pharmabiotics)

The effects of antibiotics and probiotics in IBD are discussed in the chapters by other authors in this book (Probiotics in IBD: Modulation of the pathological immune or inflammatory activation: Their potential benefits in the different phases of the disease, Ng and Hart; Correction of microbiota disturbances or antagonism against specific pathogens, Flanagan and Rhodes; Opportunisitc pathogens in inflammatory bowel disease, the case of adherent-invasive *E. coli*, Darfeuille-Michaud). The best evidence of clinical relevance of any marker or treatment is obtained in randomised controlled trials. Several species among the bacteria underrepresented in IBD have protecting properties against some experimental models of colitis, and this raises the question whether they (or some of their components or metabolites) could be used as treatments. Quite interestingly many of them are butyrate-producing bacteria and maintaining or increasing butyrate levels or butyrate-producing bacteria seems thus a rational target for IBD (Eeckhaut et al. 2013; Marteau 2013). This requires understanding the ecological conditions for butyrate secretion (Marteau 2013). Some authors tried to increase the levels of bifidobacteria or butyrate producing bacteria using prebiotic susbtrates in patients with IBD or combinations of prebiotics with probiotics (so called symbiotics). Despite positive prelimiray results, fructo-oligosaccharides failed to improve the clinical symptoms in CD (a worsening of the intestinal comfort was even observed) and failed to increase faecal concentration of bifidobacteria and *F. prausnitzii* (Benjamin et al. 2011). This remains a quite rational research strategy but the tolerance of the dietary treatment has to be good.

Foreseen Perspectives

Trying to manipulate the microbiota in a favourable way is one of the most promising strategies to progress in IBD treatment and hope of alleviating or even curing the diseases at an earlier step than immunomodulators. The present difficulties are i) to identify the early situations and make research using microbiota modifiers in this setting; ii) to convince investigators to search for etiological treatments rather than on immunomodulators (which also try to establish their role at early steps of the diseases); iii) to find the right treatment without the help of animal trials which are poorly relevant to predict human situation, especially for the intestinal ecology. Understanding the role of micro-organisms in IBD should benefit from metagenomic surveys which not only capture the microbial organismal and genetic diversity, but also begin to investigate the functional contributions of micro-organisms (Hamer et al. 2012; Morgan et al. 2012; Duboc et al.

2013). It may happen that finding the right ecological treatment allow the same kind of progress that occurred when the antibiotic combination was found against *Helicobacter pylori* in chronic ulcer disease. It seems likely that in the meanwhile, fecal or blood markers of dysbiosis will be developed, validated, and used by clinicians to classify patient diseases and risks.

References Cited

Arumugam, M., J. Raes, E. Pelletier, D. Le Paslier, T. Yamada, D.R. Mende, G.R. Fernandes, J. Tap, T. Bruls, J.M. Batto, M. Bertalan, N. Borruel, F. Casellas, L. Fernandez, L. Gautier, T. Hansen, M. Hattori, T. Hayashi, M. Kleerebezem, K. Kurokawa, M. Leclerc, F. Levenez, C. Manichanh, H.B. Nielsen, T. Nielsen, N. Pons, J. Poulain, J. Qin, T. Sicheritz-Ponten, S. Tims, D. Torrents, E. Ugarte, E.G. Zoetendal, J. Wang, F. Guarner, O. Pedersen, W.M. de Vos, S. Brunak, J. Doré; MetaHIT Consortium, M. Antolín, F. rtiguenave, H.M. Blottiere, M. Almeida, C. Brechot, C. Cara, C. Chervaux, A. Cultrone, C. Delorme, G. Denariaz, R. Dervyn, K.U. Foerstner, C. Friss, van de Guchte, E. Guedon, F. Haimet, W. Huber, J. van Hylckama-Vlieg, A. Jamet, C. Juste, G. Kaci, J. Knol, O. Lakhdari, S. Layec, K. Le Roux, E. Maguin, A. Mérieux, R. Melo Minardi, C. M'rini, J. Muller, R. Oozeer, J. Parkhill, P. Renault, M. Rescigno, N. Sanchez, S. Sunagawa, A. Torrejon, K. Turner, G.Vandemeulebrouck, E. Varela, Y. Winogradsky, G. Zeller, J. Weissenbach, S.D. Ehrlich and P. Bork. 2011. Enterotypes of the human gut microbiome. Nature. 473: 174–80.

Baumgart, D.C. and W.J. Sandborn. 2012. Crohn's disease. Lancet. 380: 1590–605. Erratum in: Lancet. 2013; 381:204.

Benjamin, J.L., C.R. Hedin, A. Koutsoumpas, S.C. Ng, N.E. McCarthy, A.L. Hart, M.A. Kamm, J.D. Sanderson, S.C. Knight, A. Forbes, A.J. Stagg, K. Whelan and J.O. Lindsay. 2011. Randomised, double-blind, placebo-controlled trial of fructo-oligosaccharides in active Crohn's disease. Gut. 60: 923–9.

Benjamin, J.L., C.R. Hedin, A. Koutsoumpas, S.C. Ng, N.E. McCarthy, N.J. Prescott, P. Pessoa-Lopes, C.G. Mathew, J. Sanderson, A.L. Hart, M.A. Kamm, S.C. Knight, A. Forbes, A.J. Stagg, J.O. Lindsay and K. Whelan. 2012. Smokers with active Crohn's disease have a clinically relevant dysbiosis of the gastrointestinal microbiota. Inflamm. Bowel Dis. 18: 1092–100.

Biedermann, L., J. Zeitz, J. Mwinyi, E. Sutter-Minder, A. Rehman, S.J. Ott, C. Steurer-Stey, A. Frei, P. Frei, M. Scharl, M.J. Loessner, S.R. Vavricka, M. Fried, S. Schreiber, M. Schuppler and G. Rogler. 2013. Smoking cessation induces profound changes in the composition of the intestinal microbiota in humans. PLoS One. 8(3): e59260. doi: 10.1371/journal.pone.0059260.

Bushman, F.D., J.D. Lewis and G.D. Wu. 2013. Source Diet, the human gut microbiota, and IBD. Anaerobe. Mar 30. pii: S1075–9964(13) 00052–8. doi: 10.1016/j.anaerobe.2013.03.011.

Chassard, C., M. Dapoigny, K.P. Scott, L. Crouzet, C. Del'homme, P. Marquet, J.C. Martin, G. Pickering, D. Ardid, A. Eschalier, C. Dubray, H.J. Flint and A. Bernalier-Donadille. 2012. Functional dysbiosis within the gut microbiota of patients with constipated-irritable bowel syndrome. Aliment Pharmacol. Ther. 35: 828–38.

Crohn, B.B., L. Ginzburg and G.D. Oppenheimer. 1932. Regional ileitis: a pathologic and clinical entity. JAMA. 99: 1323–9.

Darfeuille-Michaud, A., Opportunistic pathogens in inflammatory bowel disease, the case of adherent-invasive *E. coli*. In publication.

De Preter, V., I. Arijs, K. Windey, W. Vanhove, S. Vermeire, F. Schuit, P. Rutgeerts and K. Verbeke. 2012. Decreased mucosal sulfide detoxification is related to an impaired butyrate oxidation in ulcerative colitis. Inflamm. Bowel Dis. Dec. 18(12): 2371–80.

Dörffel, Y., A. Swidsinski, V. Loening-Baucke, B. Wiedenmann and M. Pavel. 2012. Common biostructure of the colonic microbiota in neuroendocrine tumors and Crohn's disease and the effect of therapy. Inflamm. Bowel Dis. 18: 1663–71.

D'Haens, G., K. Geboes, M. Peeters, F. Baert, F. Penninckx and P. Rutgeerts. 1998. Early lesions of recurrent Crohn's disease caused by infusion of intestinal contents in excluded ileum. Gastroenterology. 114(2): 262–7.

Duboc, H., S. Rajca, D. Rainteau, D. Benarous, M.A. Maubert, E. Quervain, G. Thomas, V. Barbu, L. Humbert, G. Despras, C. Bridonneau, F. Dumetz, J.P. Grill, J. Masliah, L. Beaugerie, J. Cosnes, O. Chazouillères, R. Poupon, C. Wolf, J.M. Mallet, P. Langella, G. Trugnan, H. Sokol and P. Seksik. 2013. Connecting dysbiosis, bile-acid dysmetabolism and gut inflammation in inflammatory bowel diseases. Gut. 62(4): 531–9.

Edwards, L.A., M. Lucas, E.A. Edwards, F. Torrente, R.B. Heuschkel, N.J. Klein, S.H. Murch, M. Bajaj-Elliott and A.D. Phillips 2011. Aberrant response to commensal Bacteroides thetaiotaomicron in Crohn's disease: an *ex vivo* human organ culture study. Inflamm Bowel Dis. 17(5): 1201–8.

Eeckhaut, V., K. Machiels, C. Perrier, C. Romero, S. Maes, B. Flahou, M. Steppe, F. Haesebrouck, B. Sas, R. Ducatelle, S. Vermeire and F. Van Immerseel. 2012. *Butyricicoccus pullicaecorum* in inflammatory bowel disease. Gut. Dec 22. [Epub ahead of print] Pub. Med. PMID: 23263527.

Flanagan, P. and Johnathan M. Rodes, Correction of microbiota disturbances or antagonism against specific pathogens in IBD. In publication.

Fraher and Shanahan, Opportunistc pathogens in IBD, and the relation with specific gene susceptibilities. In publication.

Frank, D.N., A.L. St. Amand, R.A. Feldman, E.C. Boedeker, N. Harpaz and N.R. Pace. 2007. Molecular-phylogenetic characterization of microbial community imbalances in human inflammatory bowel diseases. Proc. Natl. Acad. Sci. USA 104(34): 13780–5.

Fujimoto, T., H. Imaeda, K. Takahashi, E. Kasumi, S. Bamba, Y. Fujiyama and A. Andoh. 2013. Decreased abundance of *Faecalibacterium prausnitzii* in the gut microbiota of Crohn's disease. J. Gastroenterol. Hepatol. 28: 613–9.

Gersemann, M., J. Wehkamp and E.F. Stange. 2012. Innate immune dysfunction in inflammatory bowel disease. J. Intern. Med. 271: 421–8.

Gorkiewicz, G., G.G. Thallinger, S. Trajanoski, S. Lackner, G. Stocker, T. Hinterleitner, C. Gülly and C. Högenauer. 2013. Alterations in the colonic microbiota in response to osmotic diarrhea. PLoS One. 8(2): e55817. doi: 10.1371/journal.pone.0055817.

Hamer, H.M., V. De Preter, K. Windey and K. Verbeke. 2012. Functional analysis of colonic bacterial metabolism: relevant to health? Am. J. Physiol. Gastrointest. Liver Physiol. 302: G1–9.

Harper, P.H., E.C. Lee, M.G. Kettlewell, M.K. Bennett and D.P Jewell. 1985. Role of the faecal stream in the maintenance of Crohn's colitis. 26: 279–84.

Hotte, N.S., S.Y. Salim, R.H. Tso, E.J. Albert, P. Bach, J. Walker, L.A. Dieleman, R.N. Fedorak and K.L. Madsen. 2012. Patients with inflammatory bowel disease exhibit dysregulated responses to microbial DNA. PLoS One. 7(5): e37932. doi: 10.1371/journal.pone.0037932.

Iliev, I.D., V.A. Funari, K.D. Taylor, Q. Nguyen, C.N. Reyes, S.P. Strom, J. Brown, C.A. Becker, P.R. Fleshner, M. Dubinsky, J.I. Rotter, H.L. Wang, D.P. McGovern, G.D. Brown and D.M. Underhill. 2012. Interactions between commensal fungi and the C-type lectin receptor Dectin-1 influence colitis. Science. 336: 1314–7.

Jia, W., R.N. Whitehead, L. Griffiths, C. Dawson, H. Bai, R.H. Waring, D.B. Ramsden, J.O. Hunter, M. Cauchi, C. Bessant, D.P. Fowler, C. Walton, C. Turner and J.A. Cole. 2012. Diversity and distribution of sulphate-reducing bacteria in human faeces from healthy subjects and patients with inflammatory bowel disease. FEMS Immunol. Med. Microbiol. 65: 55–68.

Joossens, M., G. Huys, M. Cnockaert, V. De Preter, K. Verbeke, P. Rutgeerts, P. Vandamme and S. Vermeire. 2011. Dysbiosis of the faecal microbiota in patients with Crohn's disease and their unaffected relatives. Gut. 60: 631–7.

Kaul, A., S. Hutfless, L. Liu, T.M. Bayless, M.R. Marohn and X. Li. 2012. Serum anti-glycan antibody biomarkers for inflammatory bowel disease diagnosis and progression: a systematic review and meta-analysis. Inflamm. Bowel Dis. 18: 1872–84.

Keita, A.V. and J.D. Söderholm. 2012. Barrier dysfunction and bacterial uptake in the follicle-associated epithelium of ileal Crohn's disease. Ann. N Y Acad. Sci. 1258: 125–34.

Laharie, D., C. Asencio, J. Asselineau, P. Bulois, A. Bourreille, J. Moreau, P. Bonjean, D. Lamarque, A. Pariente, J.C. Soulé, A. Charachon, B. Coffin, P. Perez, F. Mégraud and F. Zerbib. 2009. Association between entero-hepatic Helicobacter species and Crohn's disease: a prospective cross-sectional study. Aliment Pharmacol. Ther. 30: 283–93.

Lepage, P., P. Seksik, M. Sutren, M.F. de la Cochetiere, R. Jian, P. Marteau and J. Doré. 2005. Biodiversity of the Mucosa-associated Microbiota Is Stable Along the Distal Digestive Tract in Healthy Individuals and Patients with IBD. Inflamm. Bowel Dis. 11(5): 473–480.

Lepage, P., J. Colombet, P. Marteau, T. Sime-Ngando, J. Doré and M. Leclerc. 2008. Dysbiosis in inflammatory bowel disease: a role for bacteriophages? Gut. 57: 424–5.

Lepage, P., R. Häsler, M.E. Spehlmann, A. Rehman, A. Zvirbliene, A. Begun, S. Ott, L. Kupcinskas, J. Doré, A. Raedler and S. Schreiber. 2011. Twin study indicates loss of interaction between microbiota and mucosa of patients with ulcerative colitis. Gastroenterology. 141: 227–36.

Li, Q., C. Wang, C. Tang, N. Li and J. Li. 2012. Molecular-phylogenetic characterization of the microbiota in ulcerated and non-ulcerated regions in the patients with Crohn's disease. PLoS One. 7(4): e34939. doi: 10.1371/journal.pone.0034939.

Luther, J., M. Dave, P.D. Higgins and J.Y. Kao. 2010. Association between *Helicobacter pylori* infection and inflammatory bowel disease: a meta-analysis and systematic review of the literature. Inflamm. Bowel Dis. 16(6): 1077–84.

Mangin, I., R. Bonnet, P. Seksik, L. Rigottier-Gois, M. Sutren, Y. Bouhnik, C. Neut, M.D. Collins, J.F. Colombel, P. Marteau and J. Doré. 2004. Molecular inventory of faecal microflora in patients with Crohn's disease. FEMS Microbiol. Ecol. 50: 25–36.

Manichanh, C., L. Rigottier-Gois, E. Bonnaud, K. Gloux, E. Pelletier, L. Frangeul, R. Nalin, C. Jarrin, P. Chardon, P. Marteau, J. Roca and J. Doré. 2006. Reduced diversity of faecal microbiota in Crohn's disease revealed by a metagenomic approach. Gut. 55(2): 205–11.

Marteau, P., P. Pochart, J. Doré, C. Béra-Maillet, A. Bernalier and G. Corthier. 2001. Comparative study of bacterial groups within the human cecal and fecal microbiota. Appl. Environ. Microbiol. 67: 4939–42.

Marteau, P. 2013. Butyrate-producing bacteria as pharmabiotics for inflammatory bowel disease. Gut. Mar 5. [Epub ahead of print] Pub. Med. PMID: 23461897.

Martinez, C., M. Antolin, J. Santos, A. Torrejon, F. Casellas, N. Borruel, F. Guarner and J.R. Malagelada. 2008. Unstable composition of the fecal microbiota in ulcerative colitis during clinical remission. Am. J. Gastroenterol. 103: 643–8.

Michail, S., M. Durbin, D. Turner, A.M. Griffiths, D.R. Mack, J. Hyams, N. Leleiko, H. Kenche, A. Stolfi and E. Wine. 2012. Alterations in the gut microbiome of children with severe ulcerative colitis. Inflamm. Bowel Dis. 18: 1799–808.

Mondot, S., S. Kang, J.P. Furet et al. 2011. Highlighting new phylogenetic specificities of Crohn's disease microbiota. Inflamm. Bowel Dis. 17: 185–92.

Morgan, X.C., T.L. Tickle, H. Sokol, D. Gevers, K.L. Devaney, D.V. Ward, J.A. Reyes, S.A. Shah, N. LeLeiko, S.B. Snapper, A. Bousvaros, J. Korzenik, B.E. Sands, R.J. Xavier and C. Huttenhower. 2012. Dysfunction of the intestinal microbiome in inflammatory bowel disease and treatment. Genome Biol. 13(9): R79. doi: 10.1186/gb-2012–13–9–r79.

Ng, S.C., J.L. Benjamin, N.E. McCarthy, C.R. Hedin, A. Koutsoumpas, S. Plamondon, C.L. Price, A.L. Hart, M.A. Kamm, A. Forbes, S.C. Knight, J.O. Lindsay, K. Whelan and A.J.

Stagg. 2011. Relationship between human intestinal dendritic cells, gut microbiota, and disease activity in Crohn's disease. Inflamm. Bowel Dis. 17: 2027–37.

Ng, Siew C. and Ailsa L Hart. Probiotics in IBD: modulation of the pathological immune or inflammatory activation. Their potential benefits in the different phases of the disease. In publication.

Noor, S.O., K. Ridgway, L. Scovell, E.K. Kemsley, E.K. Lund, C. Jamieson, I.T. Johnson and A. Narbad. 2010. Ulcerative colitis and irritable bowel patients exhibit distinct abnormalities of the gut microbiota. BMC Gastroenterol. 10: 134. doi: 10.1186/147–230X-10–134.

Ohkusa, T., I. Okayasu, T. Ogihara, K. Morita, M. Ogawa and N. Sato. 2003. Induction of experimental ulcerative colitis by Fusobacterium varium isolated from colonic mucosa of patients with ulcerative colitis. Gut. 52: 79–83.

Ohkusa, T., K. Kato, S. Terao, T. Chiba, K. Mabe, K. Murakami, Y. Mizokami, T. Sugiyama, A. Yanaka, Y. Takeuchi, S. Yamato, T. Yokoyama, I. Okayasu, S. Watanabe, H. Tajiri and Sato N. Japan UC Antibiotic Therapy Study Group. 2010. Newly developed antibiotic combination therapy for ulcerative colitis: a double-blind placebo-controlled multicenter trial. Am. J. Gastroenterol. 105: 1820–9.

Ordás, I., L. Eckmann, M. Talamini, D.C. Baumgart and W.J. Sandborn. 2012. Ulcerative colitis. Lancet. 380: 1606–19.

Papa, E., M. Docktor, C. Smillie, S. Weber, S.P. Preheim, D. Gevers, G. Giannoukos, D. Ciulla, D. Tabbaa, J. Ingram, D.B. Schauer, D.V. Ward, J.R. Korzenik, R.J. Xavier, A. Bousvaros and E.J. Alm. 2012. Non-invasive mapping of the gastrointestinal microbiota identifies children with inflammatory bowel disease. PLoS One. 7(6): e39242. doi: 10.1371/journal.pone.0039242.

Png, C.W., S.K. Linden, K.S. Gilshenan,E.G. Zoetendal, C.S. McSweeney, L.I. Sly, M.A. McGuckin and T.H. Florin. 2010. Mucolytic bacteria with increased prevalence in IBD mucosa augment *in vitro* utilization of mucin by other bacteria. Am. J. Gastroenterol. 105(11): 2420–2428.

Rajca, S., V. Grondin, E. Louis et al. 2011. Crohn's Disease Associated Dysbiosis as a predictive factor of clinical relapse: a microbiological substudy of the Getaid-stori Cohort. ECCO.

Rajilić-Stojanović, M., F. Shanahan, F. Guarner and W.M. de Vos. 2013. Phylogenetic analysis of dysbiosis in ulcerative colitis during remission. Inflamm. Bowel Dis. 19: 481–8.

Rehman, A., C. Sina, O. Gavrilova, R. Häsler, S. Ott, J.F. Baines, S. Schreiber and P. Rosenstiel. 2011. Nod2 is essential for temporal development of intestinal microbial communities. Gut. 60: 1354–62.

Ricanek, P., S.M. Lothe, S.A. Frye, A. Rydning, M.H. Vatn and T. Tønjum. 2012. Gut bacterial profile in patients newly diagnosed with treatment-naïve Crohn's disease. Clin. Exp. Gastroenterol. 5: 173–86.

Seksik, P., L. Rigottier-Gois, G. Gramet, M. Sutren, P. Pochart, P. Marteau, R. Jian and J. Doré. 2003. Alterations of the dominant faecal bacterial groups in patients with Crohn's disease of the colon. Gut. 52(2): 237–42.

Sokol, H., P. Seksik, L. Rigottier-Gois, C. Lay, P. Lepage, I. Podglajen, P. Marteau and J. Dore. 2006. Specificities of the fecal microbiota in inflammatory bowel disease Inflamm. Bowel Dis. 12: 106–11.

Sokol, H., B. Pigneur, L. Watterlot, O. Lakhdari, L.G. Bermúdez-Humarán, J.J. Gratadoux, S. Blugeon, C. Bridonneau, J.P. Furet, G. Corthier, C. Grangette, N. Vasquez, P. Pochart, G. Trugnan, G. Thomas, H.M. Blottière, J. Doré, P. Marteau, P. Seksik and P. Langella. 2008. Faecalibacterium prausnitzii is an anti-inflammatory commensal bacterium identified by gut microbiota analysis of Crohn disease patients. Proc. Natl. Acad. Sci. USA 105: 16731–6.

Sokol, H., P. Seksik, J.P. Furet, O. Firmesse, I. Nion-Larmurier, L. Beaugerie, J. Cosnes, G. Corthier, P. Marteau and J. Doré. 2009. Low counts of *Faecalibacterium prausnitzii* in colitis microbiota. Inflamm. Bowel Dis. 15: 1183–9.

Sokol, H., N. Vasquez, N. Hoyeau, P. Seksik, L. Beaugerie, A. Lavergne-Slove, P. Pochart and P. Marteau. 2010. Crypt abscess associated microbiota in inflammatory bowel disease and acute self-limited colitis. World J. Gastroenterol. 16: 583–7.

Swidsinski, A., A. Ladhoff, A. Pernthaler, S. Swidsinski, V. Loening-Baucke, M. Ortner, J. Weber, U. Hoffmann, S. Schreiber, M. Dietel and H. Lochs. 2002. Mucosal flora in inflammatory bowel disease. Gastroenterology. 122(1): 44–54.

Swidsinski, A., J. Weber, V. Loening-Baucke, L.P. Hale and H. Lochs. 2005. Spatial organization and composition of the mucosal flora in patients with inflammatory bowel disease. J. Clin. Microbiol. 43: 3380–9.

Swidsinski, A., V. Loening-Baucke, M. Vaneechoutte and Y. Doerffel. 2008. Active Crohn's disease and ulcerative colitis can be specifically diagnosed and monitored based on the biostructure of the fecal flora. Inflamm. Bowel Dis. 14: 147–61.

Thomson, J.M., R. Hansen, S.H. Berry, M.E. Hope, G.I. Murray, I. Mukhopadhya, M.H. McLean, Z. Shen, J.G. Fox, E. El-Omar and G.L. Hold. 2011. Enterohepatic helicobacter in ulcerative colitis: potential pathogenic entities? PLoS One. Feb 23; 6(2): e17184. doi: 10.1371/journal.pone.0017184.

Vasquez, N., I. Mangin, P. Lepage, P. Seksik, J.P. Duong, S. Blum, E. Schiffrin, A. Suau, M. Allez, G. Vernier, X. Tréton, J. Doré, P. Marteau and P. Pochart. 2007. Patchy distribution of mucosal lesions in ileal Crohn's disease is not linked to differences in the dominant mucosa-associated bacteria: a study using fluorescence *in situ* hybridization and temporal temperature gradient gel electrophoresis. Inflamm. Bowel Dis. 13: 684–92.

Verma, R., A.K. Verma, V. Ahuja and J. Paul. 2010. Real-time analysis of mucosal flora in patients with inflammatory bowel disease in India. J. Clin. Microbiol. 48: 4279–82.

Walker, A.W., J.D. Sanderson, C. Churcher, G.C. Parkes, B.N. Hudspith, N. Rayment, J. Brostoff, J. Parkhill, G. Dougan and L. Petrovska. 2011. High-throughput clone library analysis of the mucosa-associated microbiota reveals dysbiosis and differences between inflamed and non-inflamed regions of the intestine in inflammatory bowel disease. BMC Microbiol. 11: 7.

Willing, B.P., J. Dicksved, J. Halfvarson, A.F. Andersson, M. Lucio, Z. Zheng, G. Järnerot, C. Tysk, J.K. Jansson and L. Engstrand. 2010. A pyrosequencing study in twins shows that gastrointestinal microbial profiles vary with inflammatory bowel disease phenotypes. Gastroenterology. 139(6): 1844–1854.

Wu, G.D., J. Chen, C. Hoffmann et al. 2011. Linking long-term dietary patterns with gut microbial enterotypes. Science. 334: 105–108.

Yukawa, T., T. Ohkusa, T. Shibuya, S. Tsukinaga, J. Mitobe, K. Takakura, A. Takahara, S. Odahara, H. Matsudaira, K. Nagatsuma, T. Kitahara, M. Kajihara, K. Uchiyama, H. Arakawa, S. Koido and H. Tajiri. 2013. Nested culture method improves detection of Fusobacterium from stool in patients with ulcerative colitis. Jpn. J. Infect. Dis. 66: 109–14.

8

Opportunistic Pathogens in Inflammatory Bowel Disease, and the Relation with Specific Gene Susceptibilities

*Marianne Fraher** and *Fergus Shanahan*

Introduction

The inflammatory bowel diseases, Crohn's disease and ulcerative colitis, are distinct and heterogeneous syndromes. However, a consensus of evidence suggests that most cases of inflammatory bowel disease (IBD) represent the outcome of complex interplay among genetic susceptibility risk factors, environmental triggers, and disturbed immune responses to the enteric microbiota. Genetic risk factors appear to code for proteins involved in sensing or regulating the host response to the microbiota or in maintenance of mucosal barrier function (Jostins et al. 2012). Given the underpinning disturbance of host-microbe interactions, and the degree of mucosal tissue injury in both Crohn's disease and ulcerative colitis, it is remarkable that opportunistic infections are not more common than has been evident to date. This may be about to change as increasing concern is voiced in relation to the increasing use of immunosuppressive and immunomodulatory drugs and the persistent, albeit diminishing, concomitant use of corticosteroids.

Department of Medicine and Alimentary Pharmabiotic Centre, University College Cork, Cork, Ireland.
E-mail: f.shanahan@ucc.ie
* Corresponding author: m.fraher@ucc.ie

An opportunistic infection may be defined as a serious, usually progressive infection by a micro-organism that has limited (or no) pathogenic capacity under ordinary circumstances, but which has been able to cause serious disease as a result of the predisposing effect of another disease or of its treatment (Symmers 1965). A diversity of potential opportunistic infections may occur in patients with IBD but include intracellular organisms, such as *Mycobacterium tuberculosis* and *Listeria monocytogenes*. Curiously, there is overlap between the genetic risk factors for inflammatory bowel disease and those involved in immunity to mycobacterial infections (Jostins et al. 2012). It is noteworthy that the distinction between a commensal and a pathogen is moot particularly in the context of genetic susceptibility and defective mucosal barrier function. Thus, depending on the susceptibility of the host, the gut microbiota may become opportunistically pathogenic.

Of the different genetic loci confirmed as risk factors for Crohn's disease and including genes which regulate intracellular microbial sensing, autophagy and barrier function, none has been specifically linked with a specific microbial pathogen. However, it is clear from murine studies that disturbances in innate immunity associated with inflammatory bowel disease can by turn alter the microbiota composition. Thus, inflammatory bowel disease may be due to either an appropriate immune response to a disturbed microbiota or an inappropriate immune response to a normal microbiota or both. The clinical implications of recent advances in understanding the enteric microbiota as well as details concerning the microbiota in both Crohn's disease and ulcerative colitis have been reviewed elsewhere (Shanahan 2012a,b). Our purpose here is to address real and theoretical risks of specific opportunistic infections in patients with inflammatory bowel disease.

Infection as a Disease Modifier

Clinicians have long suspected that inter-current mucosal infections can trigger either the onset or a relapse of established inflammatory bowel disease. Recently, a murine model has been used to demonstrate that sporadic infections which are associated with transient breaks in mucosal integrity can lead to the generation of long-lived memory T cells with reactivity against enteric microbial antigens (Hand et al. 2012). This could account for the latency period in susceptible individuals before the disease becomes manifest with the cumulative impact of various mucosal infections in earlier life (Hand et al. 2012). Indeed, another influential study in experimental mice demonstrated the interplay among genetic risk factors and environmental triggers and showed that the timing of onset of disease was influenced by the co-incidental exposure to two environmental agents, one of which was a norovirus (Cadwell et al. 2010). Gastrointestinal infection can cause defects in the mucosal barrier with bacterial translocation. In the healthy state, there is immunotolerance to the gut microbiota, however, this can be altered in gastrointestinal infection.

Mycobacterium avium paratuberculosis

Historically, perhaps the most widely associated infectious agent with IBD is *Mycobacterium paratuberculosis*. Johne's disease is a disease of ruminants (animals that have four compartments in their stomach; one of which is the rumen, e.g., cattle), caused by *Mycobacterium avium* subspecies *paratuberculosis* (MAP). The main signs of Johne's disease are diarrhea (non-bloody) and wasting, the clinical course is progressive with dehydration and eventual demise. Given the clinical and histological similarities, there has been much speculation as to a link between these two diseases with potential acquisition of infection through the food chain (i.e., consumption of pasteurized bovine milk (Elguezabal et al. 2012)).

This link was first postulated in 1913 and in the intervening years, there have been many studies published with conflicting results (Dalziel 1989). Defective responses to MAP have been seen in Crohn's disease patients with mutations in NOD2 and persistence of MAP within macrophages (Ferwerda et al. 2007; Mendoza et al. 2010; Rosenfeld and Bressler 2010). *Mycobacterium avium* subspecies *paratuberculosis* has also been found in tissues and blood of patients with Crohn's disease more frequently than in controls. In addition, MAP antibodies are present more frequently in Crohn's disease than in controls (Naser et al. 2000; Hulten et al. 2001; Naser et al. 2004). A systematic review and meta-analysis of studies using nucleic-acid based techniques to detect MAP concluded that MAP is detected more frequently among Crohn's disease compared to controls but the pathogenic role of MAP in Crohn's disease is unclear (Abubakar et al. 2008). This does not imply that the pathogenicity of MAP causes Crohn's disease, but may reflect altered immune responses to MAP. Peripheral blood mononuclear cells (PBMCs) incubated with MAP significantly enhanced T cell proliferation in patients with Crohn's disease compared to patients with ulcerative colitis and healthy controls (Sibartie et al. 2010). In addition, PBMCs secreted significantly higher levels of TNFα and IL-10 in patients with Crohn's disease in response to MAP than patients with ulcerative colitis and healthy controls (Sibartie et al. 2010). A study in the United Kingdom failed to demonstrate a geographical association between the prevalence of Crohn's disease and clinical cases of Johne's disease and there is a lack of evidence to suggest acquisition of MAP from consumption of contaminated food products (Jones et al. 2006; Abubakar et al. 2007). To date, the link between these two diseases is inconclusive and MAP, as a cause of Crohn's disease or an incidental association remains unanswered. If MAP is associated with Crohn's disease, it is likely that its role is not mediated by a traditional infective pathway (NACC 2009).

E. coli

Studies on the role of infectious agents in the etiology of Crohn's disease have also focused on *E. coli*. Mucosa-associated adherent-invasive *E. coli* (AIEC) is increased in Crohn's disease and it has the ability to adhere to mucosa and invade the epithelium (Darfeuille-Michaud et al. 2004). *Escherichia coli* DNA is detected more frequently in granulomas in patients with Crohn's disease than patients with non-Crohn's granulomas (Ryan et al. 2004). This topic is covered in more detail in another chapter.

Clostridium difficile a Vexed Relationship with Inflammatory Bowel Disease

The association between *C. difficile* toxin and IBD was first reported more than 30 years ago and an increased frequency of both *C. difficile* infection and colonization occurs in IBD (particularly ulcerative colitis) when compared to controls (Bolton et al. 1980; Trnka and LaMont 1981; Rodemann et al. 2007; Clayton et al. 2009). Although patients with IBD are likely to have risk factors for *C. difficile* infection (hospitalization, antimicrobial therapy and immunosuppressive therapy), both *C. difficile* infection and colonization can occur in the absence of risk factors (Bolton et al. 1980; Clayton et al. 2009; Goodhand et al. 2011). *Clostridium difficile* infection is more severe in IBD and patients have a higher mortality, longer hospital stay and are six times more likely to undergo surgery (Ananthakrishnan et al. 2008).

Colonization with toxigenic strains of *C. difficile* is significantly increased in asymptomatic IBD patients (with no risk factors for *C. difficile* infection and in remission) recruited from a specialty outpatient clinic compared to healthy controls, 8.2 versus 1 percent respectively (Clayton et al. 2009). Common local nosocomial ribotypes and a predominant clone were not identified, suggesting community acquisition of the bacterium without horizontal transfer. Further studies are needed to elucidate this link but possible reasons include defects in innate immunity and dysbiosis.

The immune response to *C. difficile* infection is strongly associated with disease recurrence and patients with a low serum IgG antibody response to toxin A are 48 times more likely to develop recurrence than patients with a high serum IgG antibody response to toxin A (Kyne et al. 2001). Another study prospectively followed elderly patients prescribed antimicrobials and monitored for subsequent *C. difficile* infection and colonization, patients who became colonized with *C. difficile* but did not develop *C. difficile* infection had a significantly higher serum IgG antibody response to toxin A than patients who developed *C. difficile* infection (Kyne et al. 2000). Both studies were conducted on hospital inpatients but it was not stated if patients with IBD were included in the study. Serum IgG antibody responses to toxin B in IBD patients have presented conflicting results with higher and lower serum IgG antibody responses to toxin B when compared to non-IBD patients (Berg 2012; Shakir et al. 2012).

The incidence and severity of *C. difficile* infection has increased in recent years both in IBD and non-IBD patients in North America, Canada and Europe; this has in part been ascribed to the emergence of the hypervirulent *C. difficile* strain characterized as restriction endonuclease (REA) group B1, North American pulsed-field type 1, and polymerase chain reaction (PCR)-ribotype 027 (B1/NAP1/027) (Rodemann et al. 2007; Issa et al. 2008; Lessa et al. 2012).

Symptoms of *C. difficile* infection and relapse of IBD may be indistinguishable and clinicians must not be overlook *C. difficile* as a potential cause of their symptoms. It is unknown why *C. difficile* infection and colonization are increased in IBD and further studies are needed to decipher if infection and colonization have a common pathophysiology.

Opportunistic Infections Linked with Treatment of IBD

The risk of developing infection depends on the therapy used and combination therapy results in a higher risk of infection as a result of synergism. Corticosteroid or thiopurine therapy alone results in a two to three fold increased risk of infection but when these agents are combined the risk is increased about 15-fold (Toruner et al. 2008). The real catalyst to increased infection risk is when corticosteroids are used in combination with immunomodulators or anti-TNF therapy (Aberra et al. 2003; Toruner et al. 2008; Fidder et al. 2009). A case-control study of patients with opportunistic infections found that corticosteroids were most commonly associated with Candida species infection and thiopurines were associated with opportunistic viral infection (Toruner et al. 2008).

The advent of biologic agents has revolutionized management of IBD since they were first introduced in 1998 with reduced surgeries, hospitalization, and improved quality of life (Peyrin-Biroulet 2010). While inhibition of TNF will modulate IBD, it may also increase susceptibility to infection or reactivation of endogenous infection as it has a critical role in host defense against infectious stimuli including intracellular pathogens (Kaur and Mahl 2007). Reactivation of latent TB is the most commonly associated opportunistic infection with anti-TNF therapy and it is also associated with infection caused by other intracellular pathogens such as *Listeria monocytogenes* and *Pneumocystis jiroveci*. The estimated annual frequency of serious infections in patients receiving anti-TNF therapy is three percent (Siegel 2011). Anti-TNF therapy should be stopped in any patient who presents with symptoms and signs of a serious infection (Smith and Kauffman 2009).

The risks associated with these therapies are not just limited to infection and these must be effectively communicated to patients prior to embarking on treatment. In an online survey of IBD patients, 81 percent wanted to be involved in the decision-making process (Baars et al. 2010). The challenge for the gastroenterologist is to ensure that the patient has a clear understanding of the risk and benefit involved. The use of vague words (e.g., rare, common) and small percentages (e.g., 0.1%) should be avoided and the discussion should be tailored to the individual patient, using pictures if possible and communicating small percentages (0.1%) that are comprehendible to the patient (e.g., one per 1,000) (Siegel 2011).

Certain pathogens are not routinely examined for in clinical specimens, e.g., endemic mycoses and *Pneumocystis jiroveci*. If an opportunistic infection is suspected, it is important that the appropriate specimen is collected and examination for the suspected pathogen is requested. This factor in addition to increased travel and the rising incidence of multi-drug resistant pathogens means that management of infections in immune compromised patients is becoming increasingly complex and early involvement of an expert opinion (infectious diseases physician or clinical microbiologist) is warranted so that timely appropriate therapy is commenced.

Mycobacterium tuberculosis

Tuberculosis (TB) is second only to HIV as the greatest killer worldwide from an infectious disease. An estimated 32 percent of the world's population has latent TB; there is an unequal global distribution with highest rates in Sub-Saharan Africa and parts of South-East Asia (Dye et al. 1999). The risk of reactivation of latent TB and development of active disease is 10 percent; this risk is further increased in patients receiving immunosuppressive therapy (CDC 2000; Pai et al. 2004). The relative risk of developing TB with anti-TNF therapy is increased 1.6–25.1 times, depending on risk factors in the group studied (Solovic et al. 2010). The risk of an individual having latent or active TB is increased in immunosuppression (HIV, chemotherapy, anti-TNF therapy, etc.), health care workers, close living conditions, poverty, malnutrition and residency in or travel to countries with high rates of disease (Sub-Saharan Africa, South-East Asia, former states of Soviet Union). With increasing worldwide travel, migration from high to low incidence areas and HIV infection; TB is not just limited to developing countries but can also occur in developed countries.

Active TB in patients receiving anti-TNF therapy is usually the result of reactivation of latent TB as infection occurs soon after therapy has been started (Keane et al. 2001; Hamdi et al. 2006). Tuberculosis in patients receiving anti-TNF therapy is more commonly extra-pulmonary (56%) and disseminated (24%) disease (Keane et al. 2001). Treatment of latent TB reduces the risk of active disease by about 60 percent and it is therefore important that patients should be screened for latent TB prior to commencing immunosuppressive therapy (CDC 2000). Any patient who is found to have latent TB should be referred to a respiratory physician for further management and treatment should be commenced (Rahier et al. 2009a). It is unclear what the optimal time to delay starting anti-TNF therapy is, when a patient has been commenced on treatment for latent TB, guidelines suggest delaying anti-TNF therapy for between 3–12 weeks after starting anti-tuberculous treatment (Rahier et al. 2009a; Mowat et al. 2011).

Screening for Latent TB

Screening for latent TB requires a chest x-ray and tuberculin skin test or interferon-gamma release assay. The tuberculin skin test (Mantoux test) is an intradermal injection of purified protein derivative, which uses antigens that are common to MTB, *Mycobacterium bovis*, *Mycobacterium bovis* BCG and nontuberculous mycobacteria. Previous TB infection results in a delayed-type hypersensitivity reaction, which can be visualized as induration at the injection site, 48–72 hours after injection.

While there are many advantages to the tuberculin skin test (low cost and laboratory testing is not required), there are many limitations. As the injected proteins use antigens that are common to *Mycobacterium bovis* BCG and nontuberculous mycobacteria, false positive results may ensue. For example, previous Bacillus Calmette-Guerin (BCG) vaccination may cause false positive results, although this generally wanes with time (especially >30 years). False positive results may also occur as a result of sensitization to non tuberculous mycobacteria, which may be present in the environment. Thus, the specificity of the tuberculin skin test is reduced, sensitivity is also reduced in patients who are immunosuppressed (due to disease or treatment) as they may have a reduced immune response to injection with purified protein derivative (Bolton et al. 1974; Andersen et al. 2000; Papay et al. 2011). A common misconception is interpretation of the tuberculin skin test result; it is not the diameter of redness that is to be measured but the induration and a positive result is defined as induration greater than or equal to five mm. Other limitations include the need for two visits and patients can be lost to follow-up.

There are two techniques of interferon-gamma release assays commercially available, with three commercially available assays; all of which are Food and Drug Administration (FDA) approved. Quantiferon®-TB Gold (Cellestis Limited, Carnegie, Victoria, Australia), Quantiferon®-TB Gold In-Tube Assay (Cellestis Limited, Carnegie, Victoria, Australia) and T-SPOT®. TB (Oxford Immunotec, Oxford, UK). The principle of this test is that T lymphocytes of patients previously infected with TB release interferon-gamma when exposed to certain antigens; early secretory antigenic target 6 (ESAT 6) and culture filtrate protein 10 (CFP 10). These proteins are encoded by genes within the *Mycobacterium tuberculosis*-specific region of difference 1 (RD1) and are more specific to MTB than purified protein derivative and most nontuberculous mycobacteria. There are no false positive results after BCG vaccination and less false positive results from previous exposure to environmental nontuberculous mycobacteria.

Advantages of interferon-gamma release assay are that they are less affected by previous nontuberculous mycobacteria exposure and BCG vaccination and diminished reader bias (less subjective). Both sample collection and processing require expertise and once collected, samples need to be processed in the laboratory within 8–16 hours. Logistically this can be challenging especially in smaller laboratories.

As there is no gold standard, it is difficult to estimate sensitivity and specificity for latent TB. Interferon-gamma release assays that use RD1 antigens have a high specificity but a lower sensitivity than the tuberculin skin test (Pai et al. 2004). The specificity for diagnosing latent TB is greater than 95 percent and the sensitivity varies between 80–90 percent depending on the assay used (highest sensitivity using Quantiferon®-TB Gold Assay) (Pai et al. 2008).

Studies on the use of interferon-gamma release assays in certain populations (young and immunosuppressed) are lacking and much needed. As treatment with immunosuppressive therapy or corticosteroids can reduce the response to the tuberculin skin test, there is currently no optimal test for patients with IBD who are about to embark on anti-TNF therapy. In practice, when anti-TNF therapy is being considered, patients are usually prescribed high dose immunosuppressant therapy (e.g., thiopurines +/− corticosteroids) and so, response to the tuberculin skin test may be decreased with a false negative result (Bocchino et al. 2008). Currently available interferon-gamma release assays can be influenced by concomitant immunosuppressive use resulting in "indeterminate" results. Due to the presence of a positive control, a lack of T lymphocyte response to the antigens (due to immunosuppression) can result in an "indeterminate" result. The presence of an "indeterminate" interferon-gamma release assay result in patients with IBD has been associated with immunosuppressive use; especially double immunosuppressive therapy (Papay et al. 2011; Helwig et al. 2012). In the case of an "indeterminate" interferon-gamma release assay result, the test should be repeated if the result is again "indeterminate" then risk factors, chest x-ray, tuberculin skin test. There are few studies on the use of interferon-gamma release assays in patients with IBD and the incidence of "indeterminate" results in this group have been reported between 2.7 and 13.7 percent (Papay et al. 2011; Qumseya et al. 2011). The study with the highest incidence of "indeterminate" results (13.7%) report that this was associated with double immunosuppressive therapy, low lymphocyte count, and low albumin. A meta-analysis of interferon-gamma release assay results showed a pooled "indeterminate" result of 2.1 percent for Quantiferon®-TB Gold In-Tube Assay and 3.8 percent for T-SPOT®; these studies included healthy and immunosuppressed patients and children. The rate of "indeterminate" results rose in immunosuppressed patients to 4.4 percent for Quantiferon®-TB Gold In-Tube Assay and 6.1 percent for T-SPOT®.

What's the Evidence to use Interferon-Gamma Release Assays in Diagnosis of Latent TB in IBD?

It is unclear which test (tuberculin skin test or interferon-gamma release assay) is the best diagnostic test for latent TB in IBD and there is no general consensus on which test to use in this population. It is important to remember that neither test can distinguish between active and latent disease. A recent systematic review and meta-analysis compared the performance of interferon-gamma release assays in diagnosing latent TB in patients with IBD (Shahidi et al. 2012). This study reported a more than 72 percent concordance between these two tests but it was noted that immunosuppressive therapy had a negative influence on both results. It is recommend that patients be investigated for latent TB prior to commencing immunosuppressive agents (i.e., early in the disease) and also prior to starting anti-TNF therapy as infection may have been acquired in the intervening time (Nordgaard-Lassen et al. 2012). They conclude that it was "difficult" to determine superiority between interferon-gamma release assays and the tuberculin skin test.

What do the Guidelines Recommend?

The Centers for Disease Control and Prevention (CDC) recommend that interferon-gamma release assays may be used instead of, but not in addition to the tuberculin skin test as an aid to diagnosing *M. tuberculosis* infection.

They also recommend caution in interpretation of interferon-gamma release assay results in certain populations (e.g., immunocompromised) because of limited data available in this group. The European Crohn's and Colitis Organization (ECCO) guidelines recommend using the tuberculin skin test (and also to follow national guidance). They acknowledge that interferon-gamma release assays do have improved sensitivity and specificity over the tuberculin skin test in this population but further studies are needed. In BCG vaccinated individuals, interferon-gamma release assays are recommended.

Listeria monocytogenes

Listeria monocytogenes is a food borne pathogen that may cause serious infection in the immunocompromised, pregnant women, the elderly and neonates. Infection in high-risk groups usually presents as febrile gastroenteritis, meningitis, septicemia, or endocarditis. Rarely, infection occurs in immunocompetent patients and usually presents as febrile gastroenteritis due to foodborne disease. Meningitis and septicemia with fatal outcomes have been reported in patients with IBD receiving anti-TNF therapy; in addition, cases of acute cholecystitis and septic arthritis have been reported in patients with rheumatoid arthritis receiving immunosuppressive therapies (Gluck et al. 2002; Aparicio et al. 2003; Slifman et al. 2003).

Pasteurization and cooking kills *Listeria monocytogenes*, but contamination of food occurs after cooking and processing. The ability of the bacterium to multiply at low temperatures (as low as 4°C) provides optimal conditions for survival and multiplication (e.g., fridge) until consumption. Foods that are typically associated with *L. monocytogenes* contamination include meat and fish products (pâtés, cooked meats, smoked salmon), dairy products (soft cheeses, unpasteurized milk), and salads (coleslaw, bean sprouts).

Like all opportunistic infections; infection is more likely in patients receiving combination immunosuppressive therapy which includes anti-TNF therapy. European Crohn's and Colitis Organization guidelines recommend avoidance of high-risk foods and the CDC recommends that patients with "weakened immune systems" avoid high-risk foods and heat ready-to-eat meats until steaming (Rahier et al. 2009a). It is not routine clinical practice to advise patients receiving anti-TNF therapy to avoid high-risk foods but Listeriosis is a serious infection with a significant mortality rate (20 percent for meningitis and 50 percent for endocarditis). Given CDC guidelines, it is reasonable to advocate avoidance of high-risk foods in patients receiving anti-TNF therapy.

Hepatitis B

Reactivation of hepatitis B infection can occur on immunosuppressive treatment and the clinical picture can vary from subtle changes in aminotransferases to fulminant hepatic failure and death (Hou et al. 2010). Mortality from fulminant liver failure as a result of hepatitis B virus reactivation is between 4 to 60 percent in patients receiving chemotherapy (Nordgaard-Lassen et al. 2012). It is prudent to screen patients for past hepatitis B virus infection, not only to identify undiagnosed hepatitis B infection but also to consider vaccination in seronegative patients.

Tumor necrosis factor plays an important role in regulation of hepatitis B virus replication, thus inhibition can result in reactivation of infection (Hou et al. 2010). The exact incidence of reactivation of hepatitis B virus infection is unknown but in the majority of cases, patients have been receiving anti-TNF therapy in addition to other immunosuppressive therapies (corticosteroids and thiopurines) (Esteve et al. 2004; Ueno et al. 2005; Colbert et al. 2007). There is no clear pattern for reactivation of disease with cases reported after one dose of infliximab and up to two years of treatment with infliximab (Ueno et al. 2005; Colbert et al. 2007). Reactivation has also been reported less frequently in patients receiving other immunosuppressive therapies (thiopurines +/– corticosteroids) alone without anti-TNF therapy (Zeitz et al. 2009; Sacco et al. 2010). Most cases of reactivation have been reported in patients who are hepatitis B surface antigen positive but it has occasionally been reported in patients who are hepatitis B virus surface antigen negative (Lok et al. 1991; Ueno et al. 2005; Colbert et al. 2007). International guidance therefore recommends prophylaxis prior to and during immunosuppressive treatment with nucleotide/nucleoside analogues for patients who are hepatitis B virus surface antigen positive (Lok and McMahon 2009; Rahier et al. 2009a; Hou et al. 2010).

An analysis of 181 clinical studies published on the efficacy of two recombinant hepatitis B vaccines (Engerix B® and Recombivax®/HB-VAX II®) in nearly 33,000 subjects found varying vaccine response rates in different populations, the vaccine response rate in healthy volunteers was 95.8 percent and 94.3 percent respectively (Coates et al. 2001). Response rates in subjects with IBD were not specifically studied but poor response rates (40–70%) were reported in other immunosuppressed states, e.g., renal impairment, liver impairment, and HIV positive, these lower response rates have been confirmed in previous studies (Bruguera et al. 1990; Bruguera et al. 1992; Rahier et al. 2010). The advisory committee on vaccination practices recommends modification of the standard 3-dose vaccination with either additional doses or doubling of the dose to improve the response rate in immunocompromised patients (Mast et al. 2005).

The vaccine response rate depends on prior immunosuppression and published studies on vaccine response rates in IBD have conflicting results. One study reported a normal vaccine response in patients treated with thiopurines while the majority of studies have reported inadequate response rates in patients receiving immunosuppressive therapy (Vida Perez et al. 2009; Altunoz et al. 2012; Dotan et al. 2012; Moses et al. 2012). The balance of evidence suggests that poor vaccine response rates occur in IBD especially in patients receiving immunosuppressive therapy (Gisbert et al. 2011). It is unclear what dosing regimen is the most effective, but one study compared standard dose and double dose vaccination with improved response rate in those patients who received double dosing (75%) compared to standard dosing (41%) (Gisbert et al. 2012). Hepatitis B vaccination is recommended for all patients who are seronegative for hepatitis B virus infection. Vaccine coverage is low in IBD, gastroenterologists need to be mindful of this and consider vaccination (with double-dosing) of patients (Melmed et al. 2006; Rahier et al. 2009a; Rahier et al. 2010).

Mycoses

Although some fungi are part of the normal flora (e.g., Candida species) or ubiquitous in the environment (e.g., *Aspergillus fumigatus*, *Pneumocystis jiroveci*); other fungi are found in specific geographical locations (e.g., *Histoplasma capsulatum, Blastomyces dermatiditis*) or associated with certain occupations, hobbies, animals, etc. Tumor necrosis factor has an important role in host defense against fungal pathogens (Filler et al. 2005).

Cryptococcus neoformans

Cryptococcus neoformans is found in high concentrations in pigeon droppings and infection can occur in patients with impaired cell-mediated immunity. Common clinical presentations include pneumonia, meningitis, and cutaneous infection. Over 35 cases of Cryptococcosis in patients with IBD have been reported, mainly in patients on anti-TNF therapy but also in patients who received corticosteroids and thiopurines (Murai et al. 2006; Hirai et al. 2011; Fraison et al. 2012).

Endemic Mycoses

The endemic mycoses (*Histoplasma capsulatum*, *Coccidioides immitis* and *Blastomyces dermatiditis*) cause fungal infections by inhalation of spores in certain geographical locations. These are dimorphic fungi that are usually molds in the environment and yeasts at body temperature. The FDA issued a safety warning in September 2008, highlighting the risk of Histoplasmosis and other fungal infections when on anti-TNF therapy (FDA 2008). They recommend consideration of Histoplasmosis infection and other fungal infections in patients who are unwell and not responding to antibiotic treatment or who have serious infections that are difficult to diagnose.

The endemic mycoses may be considered in unwell patients who live in these regions but they are less likely to be considered in patients who have returned from travel to these regions. If a patient develops an endemic mycosis, anti-fungal therapy should be commenced and anti-TNF therapy should be stopped until the infection has been treated.

Histoplasma capsulatum

Histoplasma capsulatum is endemic in the midwestern and southeastern United States (high prevalence in the Ohio and Mississippi River valleys) and is also found in Central America, Asia, and Africa. Clinical manifestations vary from asymptomatic infection to pulmonary disease and disseminated infection. Pulmonary disease may follow three clinical courses: acute, sub-acute, or chronic. Infection can occur in both immunocompetent and immunocompromised patients but is more common in the latter; Histoplasmosis is the most common endemic mycosis reported in patients receiving anti-TNF therapy (Smith and Kauffman 2009). It is found in bird and bat droppings and so, persons at high risk include cave explorers, workers in construction and farmers. Patients receiving anti-TNF therapy have a high risk of developing symptomatic infection with progression to disseminated infection and potentially fatal outcomes (Smith and Kauffman 2009).

Blastomyces dermatitidis

Blastomyces dermatitidis is endemic in southeastern and south central states that border the Mississippi and Ohio Rivers, the midwestern states, and Canadian provinces bordering the Great Lakes and the St. Lawrence River. It is also found in South America, Europe, Africa, and Asia. *Blastomyces dermatitidis* lives in decomposing organic material (e.g., wood, leaves), disruption of this material, results in spores being released and inhalation can occur; thus most cases occur in manual laborers, farmers or hunters. Manifestations of this infection include an influenza-like illness, and dissemination to other parts of the body can also occur (e.g., skin, joints). Infection occurs in immunocompetent individuals but the immunocompromised are at increased risk of infection (McCann and Smith 2012).

Coccidiodes immitis

Coccidiodes immitis is endemic in dry regions of southwestern United States, Mexico, Central, and South America. Immunocompetent individuals usually have a mild or asymptomatic clinical course with mainly influenza-like symptoms, while patients with impaired cell-mediated immunity have more severe respiratory symptoms with dissemination to skin, joint, and meninges.

Pneumocystis jirovecii

Pneumocystis (carinii) jirovecii pneumonia (PCP) is a recognized opportunistic pathogen in immunocompromised patients. Previously known as *Pneumocystis carinii*, it was renamed *Pneumocystis jirovecii* in 2001, to honor of the research done by parasitologist Otto Jírovec on this organism. In addition to being renamed, it has also undergone taxonomic reclassification; previously classified as a protozoan, it is now classified as a fungus. Initially, *P. jiroveci* infection was thought to occur in immunocompromised patients as a result of reactivation of latent infection that was acquired in early childhood (Miller and Huang 2004). There is now increasing evidence to suggest that *de novo* acquisition of infection may occur and outbreaks have been reported in renal transplant recipients with the same genotype suggesting air borne or environmental transmission (de Boer et al. 2007; Yazaki et al. 2009; Thomas et al. 2011).

Pneumocystis jirovecii pneumonia in non-HIV patients follows a different clinic course to infection to HIV positive patients (Reid et al. 2011). Non-HIV patients present with sudden onset fulminant respiratory failure, fever and a dry cough and presentation may occur after alterations in immunosuppressive medication while infection in HIV positive patients tends to more gradual in onset (Arend et al. 1995). *Pneumocystis jirovecii* pneumonia has been reported in IBD patients who are being treated with corticosteroids and thiopurines, but the majority of cases have been reported in patients receiving anti-TNF therapy in addition to other immunosuppressive therapy, e.g., thiopurines (Khatchatourian and Seaton 1997; Velayos and Sandborn 2004; Seddik et al. 2005; Itaba et al. 2007; Lee et al. 2009).

It is clear that combination immunosuppressive treatment is a risk factor for PCP but when to administer prophylaxis is less clear. International guidance recommends PCP prophylaxis with trimethoprim-sulphamethoxazole (TMP-SXT) in other immunosuppressive diseases (allogeneic hematopoietic stem cell and solid organ transplant recipients, certain autologous hematopoietic stem cell transplant recipients and HIV patients with CD4 count <200 cells per mm) (CDC 1989; Martin and Fishman 2009; Tomblyn et al. 2009; Kasiske et al. 2010). For many years there was a lack of consensus on when to administer prophylactic TMP-SXT in IBD but recent European guidelines recommend prophylaxis only for patients on triple immunosuppressive therapy, one of which being anti-TNF therapy (Kaur and Mahl 2007; Poppers and Scherl 2008; Rahier et al. 2009a; Tasaka and Tokuda 2012). There was a lack of consensus on the use of prophylaxis for patients on double immunosuppressive therapy. This consensus was reached by means of a yes or no vote made by each physician for single, double, or triple immunosuppression. Interestingly, infectious diseases physicians recommended prophylaxis unanimously for patients receiving single and double immunosuppression. The consensus document comments that this most likely reflects physicians experience with *P. jiroveci* infection. Gastroenterologists see this infection infrequently, while on the contrary; infectious diseases physicians see this infection more frequently and prescribe PCP prophylaxis on a regular basis.

Epstein-Barr Virus

Epstein-Barr virus (EBV) infection may be asymptomatic or present as infectious mononucleosis, with fatigue (which may be prolonged), lymphadenopathy, pharyngitis and fever. The worldwide seroprevalence in adults is between 90 to 95 percent. After primary infection, latency is established in B-lymphocytes. Epstein-Barr virus is also associated with malignancy, namely lymphoma, carcinoma and sarcoma especially in the immunosuppressed. Treatment with thiopurines and anti-TNF therapy increases the risk of EBV-related lymphomas (Siegel et al. 2009; Subramaniam et al. 2013). These therapies should be stopped if EBV-related lymphoma occurs (Salloum et al. 1996; Rahier et al. 2009a). Screening for latent EBV infection is not required prior to immunomodulatory treatment.

Cytomegalovirus

Between 40–100 percent of the adult population worldwide have had primary cytomegalovirus (CMV) infection, this infection is usually asymptomatic but may present as a mononucleosis-like syndrome (Krech 1973). After primary infection, latency is established in endothelial cells, macrophages or granulocyte stem cells. Cytomegalovirus does not alter the clinical course of Crohn's disease but colonic CMV reactivation has been reported in severe or refractory ulcerative colitis (Garrido et al. 2013). Cytomegalovirus reactivation may occur as a result of active IBD or due to immunosuppressive treatment; conversely CMV may also trigger active IBD (Garrido et al. 2013) Cytomegalovirus may also be an innocent bystander as CMV had a predilection for inflamed tissue. Screening for latent CMV infection is not required prior to immunomodulatory treatment.

Colonic CMV reactivation has been associated with poorer outcomes and increased surgeries (Kaufman et al. 1999; Kishore et al. 2004). Cytomegalovirus reactivation should be considered in any patient with past CMV infection (IgG positive) and severe colitis that is unresponsive to immunosuppressive therapy. Colonic biopsies should be collected for histology and CMV PCR to confirm the diagnosis (Kornbluth and Sachar 2010; Mowat et al. 2011). Immunosuppressant treatment should be stopped and CMV should be treated with Gancyclovir or Valgancyclovir for two to three weeks (Rahier et al. 2009a; Mowat et al. 2011).

Prevention of Opportunistic Infections

Prevention of some opportunistic infections is possible with recognition of risk factors, diagnostic tests for past/latent infection and prevention of infection by vaccination and education. Guidelines have been published in recent years providing recommendations on appropriate investigations and vaccinations to prevent opportunistic infections. While some guidelines incorporate their recommendations into general guidelines on the management of IBD, other guidelines are specifically aimed at prevention of opportunistic infection prior to commencing anti-TNF therapy (Rahier et al. 2009b; Mowat et al. 2011; Nordgaard-Lassen et al. 2012).

Broadly, they have similar recommendations with regards vaccinations and infections that are screened for but they differ slightly in the timing and indications for both investigations required and vaccinations. For example, Hepatitis B serology should to be checked prior to anti-TNF therapy in the Danish Society for Gastroenterology guidelines and the British Society for Gastroenterology (BSG) guidelines but the BSG guidelines in addition recommend that it be checked in high-risk patients at diagnosis. On the other hand, ECCO recommend that Hepatitis B serology be checked at diagnosis with vaccination if serology is negative. BSG guidelines notably do not recommend HIV testing, whereas ECCO and Danish guidelines recommend HIV testing. Thirty percent of newly diagnosed HIV infections in Europe appearing late in the course of the disease, the combination of long-term immunosuppressive therapy and undiagnosed HIV has potential detrimental effects (Antinori et al. 2011). Undiagnosed HIV infection has been reported in a patient with symptoms mimicking IBD and histological findings were consistent with cytomegalovirus colitis (Roskell et al. 1995).

We therefore advocate HIV testing not just at diagnosis but also prior to commencing anti-TNF therapy.

At diagnosis, we recommend baseline serology, e.g., hepatitis B, HIV and varicella zoster virus IgG (if no history of varicella infection). Ideally tuberculin skin test should also be placed early in the disease given the potential future need for anti-TNF therapy. Although it is likely that the interferon-gamma release assay will become the gold standard test for latent TB in the future, studies in IBD are needed. As the tuberculin skin test can be influenced by immunosuppressive therapy, this test should be performed early in the disease. Nevertheless, tuberculin skin test should also be performed prior to anti-TNF therapy as TB may have been acquired in the intervening period. Patients who are found to be varicella zoster virus IgG negative should have varicella zoster virus vaccination, ideally this should be done early in disease. Firstly, primary varicella zoster virus infection (chickenpox) can be severe in adulthood (even in immunocompetent adults) and require intensive care unit admission for respiratory failure. Secondly, varicella zoster virus vaccine is a live vaccine and is contraindicated in immunosuppressed patients. As IBD is a chronic disease, patients should also receive vaccinations as per local guidelines (e.g., yearly influenza vaccine, pneumococcal vaccine and human papilloma virus vaccine). Prior to commencing anti-TNF therapy, hepatitis B, and HIV serology should be repeated along with the tuberculin skin test and chest x-ray.

Travel

Patients with IBD who intend travelling should have all non-live vaccinations appropriate to the country that they are travelling to. Live vaccinations are contraindicated in patients receiving immunosuppressive therapy. It is recommended that patients with IBD who intend travelling to countries with a moderate to high incidence of TB who are receiving immunosuppressive therapy (regardless of duration of travel) and patients who are not receiving immunosuppressive therapy who travel to these areas for one month or longer have either a tuberculin skin test or interferon-gamma release assay prior to departure (Rahier et al. 2009a). If this is negative, then a repeat test is recommended about 8 to 10 weeks after arrival home.

Fever or infection in the returned traveller should include standard management of infection with attention paid to area of travel and infections that are endemic in that area. Standard advice should be given to patients with regards prevention of food-borne infection; avoid tap water and ice, boiling and peeling vegetables, etc.

Summary

- Opportunistic pathogens can be described as micro-organisms that do not cause serious illness in a healthy individual but as a result of reduced immune defenses in the host can result in serious illness.
- At diagnosis, patients should have baseline investigations for past infection: hepatitis B, HIV, varicella zoster virus IgG (if no history of varicella) and the tuberculin skin test.
- Prior to commencing anti-TNF therapy, hepatitis B and HIV serology should be repeated along with the tuberculin skin test and chest x-ray.
- Patients with IBD on immunosuppressive therapy who present with signs and symptoms of infection should be both investigated and treated diligently for opportunistic pathogens.
- Opportunistic infections in IBD are not limited to patients receiving anti-TNF therapy; they may occur with other immunosuppressive drugs
- The greatest risk for developing infection is when combination therapy includes corticosteroids.
- Gastroenterologists are likely to encounter opportunistic infections with greater frequency in the future given the increasing use of mono- and combination immunosuppressive therapy.

References Cited

Aberra, F.N., J.D. Lewis, D. Hass, J.L. Rombeau, B. Osborne and G.R. Lichtenstein. 2003. Corticosteroids and immunomodulators: postoperative infectious complication risk in inflammatory bowel disease patients. Gastroenterology. 125: 320–7.

Abubakar, I., D. Myhill, S.H. Aliyu and P.R. Hunter. 2008. Detection of *Mycobacterium avium* subspecies paratuberculosis from patients with Crohn's disease using nucleic acid-based techniques: a systematic review and meta-analysis. Inflamm. Bowel Dis. 14: 401–10.

Abubakar, I., D.J. Myhill, A.R. Hart, I.R. Lake, I. Harvey, J.M. Rhodes, R. Robinson, A.J. Lobo, C.S. Probert and P.R. Hunter. 2007. A case-control study of drinking water and dairy products in Crohn's Disease--further investigation of the possible role of Mycobacterium avium paratuberculosis. Am. J. Epidemiol. 165: 776–83.

Altunoz, M.E., E. Senates, A. Yesil, T. Calhan and A.O. Ovunc. 2012. Patients with inflammatory bowel disease have a lower response rate to HBV vaccination compared to controls. Dig. Dis. Sci. 57: 1039–44.

Ananthakrishnan, A.N., E.L. Mcginley and D.G. Binion. 2008. Excess hospitalisation burden associated with Clostridium difficile in patients with inflammatory bowel disease. Gut. 57: 205–10.

Andersen, P., M.E. Munk, J.M. Pollock and T.M. Doherty. 2000. Specific immune-based diagnosis of tuberculosis. Lancet. 356: 1099–104.

Antinori, A., T. Coenen, D. Costagiola, N. Dedes, M. Ellefson, J. Gatell, E. Girardi, M. Johnson, O. Kirk, J. Lundgren, A. Mocroft, A. D'arminio Monforte, A. Phillips, D. Raben, J.K. Rockstroh, C. Sabin, A. Sonnerborg and F. De Wolf. 2011. Late presentation of HIV infection: a consensus definition. HIV Med. 12: 61–4.

Aparicio, A.G., S. Munoz-Fernandez, G. Bonilla, A. Miralles, V. Cerdeno and E. Martin-Mola. 2003. Report of an additional case of anti-tumor necrosis factor therapy and Listeria monocytogenes infection: comment on the letter by Gluck et al. Arthritis Rheum. 48: 1764–5; author reply 1765–6.

Arend, S.M., F.P. Kroon and J.W. Van't Wout. 1995. Pneumocystis carinii pneumonia in patients without AIDS, 1980 through 1993. An analysis of 78 cases. Arch. Intern. Med. 155: 2436–41.

Baars, J.E., T. Markus, E.J. Kuipers and C.J. Van Der Woude. 2010. Patients' preferences regarding shared decision-making in the treatment of inflammatory bowel disease: results from a patient-empowerment study. Digestion. 81: 113–9.

Berg, A.M., Q.T., S.K. Wasan, A.H. Calderwood, J.A. Villafuerte-Gálvez, C.P. Kelly and F.A. Farraye. 2012. Antibodies to Clostridium difficile Toxins in Inflammatory Bowel Disease Patients. American College of Gastroenterology Annual Scientific Meeting. Las Vegas. P418.

Bocchino, M., A. Matarese, B. Bellofiore, P. Giacomelli, G. Santoro, N. Balato, F. Castiglione, R. Scarpa, F. Perna, G. Signoriello, D. Galati, A. Ponticiello and A. Sanduzzi. 2008. Performance of two commercial blood IFN-gamma release assays for the detection of Mycobacterium tuberculosis infection in patient candidates for anti-TNF-alpha treatment. Eur. J. Clin. Microbiol. Infect. Dis. 27: 907–13.

Bolton, P.M., S.L. James, R.G. Newcombe, R.H. Whitehead and L.E. Hughes. 1974. The immune competence of patients with inflammatory bowel disease. Gut. 15: 213–9.

Bolton, R.P., R.J. Sherriff and A.E. Read. 1980. Clostridium difficile associated diarrhoea: a role in inflammatory bowel disease? Lancet. 1: 383–4.

Bruguera, M., M. Cremades, R. Salinas, J. Costa, M. Grau and J. Sans. 1992. Impaired response to recombinant hepatitis B vaccine in HIV-infected persons. J. Clin. Gastroenterol. 14: 27–30.

Bruguera, M., J.L. Rodicio, J.M. Alcazar, A. Oliver, G. Del Rio and R. Esteban-Mur. 1990. Effects of different dose levels and vaccination schedules on immune response to a recombinant DNA hepatitis B vaccine in haemodialysis patients. Vaccine. 8 Suppl: S47–9; discussion S60–2.

Cadwell, K.,K.K. Patel, N.S. Maloney, T.C. Liu, A.C. Ng, C.E. Storer, R.D. Head, R. Xavier, T.S. Stappenbeck and H.W. Virgin. 2010. Virus-plus-susceptibility gene interaction determines Crohn's disease gene Atg16L1 phenotypes in intestine. Cel. 141: 1135–45.

CDC. 1989. Guidelines for prophylaxis against Pneumocystis carinii pneumonia for persons infected with human immunodeficiency virus. MMWR Morb. Mortal. Wkly. Rep. 38 Suppl. 5: 1–9.

CDC. 2000. Targeted tuberculin testing and treatment of latent tuberculosis infection. American Thoracic Society. MMWR Recomm. Rep. 49: 1–51.

Clayton, E.M., M.C. Rea, F. Shanahan, E.M. Quigley, B. Kiely, C. Hill and R.P. Ross. 2009. The vexed relationship between Clostridium difficile and inflammatory bowel disease: an assessment of carriage in an outpatient setting among patients in remission. Am. J. Gastroenterol. 104: 1162–9.

Coates, T., R. Wilson, G. Patrick, F. Andre and V. Watson. 2001. Hepatitis B vaccines: assessment of the seroprotective efficacy of two recombinant DNA vaccines. Clin. Ther. 23: 392–403.

Colbert, C., A. Chavarria and C. Berkelhammer. 2007. Fulminant hepatic failure in chronic hepatitis B on withdrawal of corticosteroids, azathioprine and infliximab for Crohn's disease. Inflamm. Bowel Dis. 13: 1453–4.

Dalziel, T.K. 1989. Thomas Kennedy Dalziel 1861–1924. Chronic interstitial enteritis. Dis. Colon Rectum. 32: 1076–8.

Darfeuille-Michaud, A., J. Boudeau, P. Bulois, C. Neut, A.L. Glasser, N. Barnich, M.A. Bringer, A. Swidsinski, L. Beaugerie and J.F. Colombel. 2004. High prevalence of adherent-invasive *Escherichia coli* associated with ileal mucosa in Crohn's disease. Gastroenterology. 127: 412–21.

de Boer, M.G., L.E. Bruijnesteijn Van Coppenraet, A. Gaasbeek, S.P. Berger, L.B. Gelinck, H.C. Van Houwelingen, P. Van Den Broek, E.J. Kuijper, F.P. Kroon and J.P. Vandenbroucke. 2007. An outbreak of Pneumocystis jiroveci pneumonia with 1 predominant genotype among renal transplant recipients: interhuman transmission or a common environmental source? Clin. Infect. Dis. 44: 1143–9.

Dotan, I., L. Werner, S. Vigodman, S. Agarwal, J. Pfeffer, N. Horowitz, L. Malter, M. Abreu, T. Ullman, H. Guzner-Gur, Z. Halpern and L. Mayer. 2012. Normal response to vaccines in inflammatory bowel disease patients treated with thiopurines. Inflamm. Bowel Dis. 18: 261–8.

Dye, C., S. Scheele, P. Dolin, V. Pathania and M.C. Raviglione. 1999. Consensus statement. Global burden of tuberculosis: estimated incidence, prevalence, and mortality by country. WHO Global Surveillance and Monitoring Project. JAMA. 282: 677–86.

Elguezabal, N., S. Chamorro, E. Molina, J.M. Garrido, A. Izeta, L. Rodrigo and R.A. Juste. 2012. Lactase persistence, NOD2 status and *Mycobacterium avium* subsp. paratuberculosis infection associations to Inflammatory Bowel Disease. Gut Pathog. 4: 6.

Esteve, M., C. Saro, F. Gonzalez-Huix, F. Suarez, M. Forne and J.M. Viver. 2004. Chronic hepatitis B reactivation following infliximab therapy in Crohn's disease patients: need for primary prophylaxis. Gut. 53: 1363–5.

FDA. 2008. FDA Requires Stronger Fungal Infection Warning for TNF Blockers.

Ferwerda, G., B.J. Kullberg, D.J. De Jong, S.E. Girardin, D.M. Langenberg, R. Van Crevel, T.H. Ottenhoff, J.W. Van Der Meer and M.G. Netea. 2007. Mycobacterium paratuberculosis is recognized by Toll-like receptors and NOD2. J. Leukoc. Biol. 82: 1011–8.

Fidder, H., F. Schnitzler, M. Ferrante, M. Noman, K. Katsanos, S. Segaert, L. Henckaerts, G. Van Assche, S. Vermeire and P. Rutgeerts. 2009. Long-term safety of infliximab for the treatment of inflammatory bowel disease: a single-centre cohort study. Gut. 58: 501–8.

Filler, S.G., M.R. Yeaman and D.C. Sheppard. 2005. Tumor necrosis factor inhibition and invasive fungal infections. Clin. Infect. Dis. 41 Suppl. 3: S208–12.

Fraison, J.B., P. Guilpain, A. Schiffmann, M. Veyrac, V. Le Moing, P. Rispail and A. Le Quellec. 2012. Pulmonary cryptococcosis in a patient with Crohn's disease treated with prednisone, azathioprine and adalimumab: Exposure to chicken manure as a source of contamination. J. Crohns Colitis.

Garrido, E., E. Carrera, R. Manzano and A. Lopez-Sanroman. 2013. Clinical significance of cytomegalo virus infection in patients with inflammatory bowel disease. World J. Gastroenterol. 19: 17–25.

Gisbert, J.P., M. Chaparro and M. Esteve. 2011. Review article: prevention and management of hepatitis B and C infection in patients with inflammatory bowel disease. Aliment Pharmacol. Ther. 33: 619–33.

Gisbert, J.P., L. Menchen, V. Garcia-Sanchez, I. Marin, J.R. Villagrasa and M. Chaparro. 2012. Comparison of the effectiveness of two protocols for vaccination (standard and double dosage) against hepatitis B virus in patients with inflammatory bowel disease. Aliment Pharmacol. Ther. 35: 1379–85.

Gluck, T., H.J. Linde, J. Scholmerich, U. Muller-Ladner, C. Fiehn and P. Bohland. 2002. Anti-tumor necrosis factor therapy and Listeria monocytogenes infection: report of two cases. Arthritis Rheum. 46: 2255–7; author reply 2257.

Goodhand, J.R., W. Alazawi and D.S. Rampton. 2011. Systematic review: Clostridium difficile and inflammatory bowel disease. Aliment Pharmacol. Ther. 33: 428–41.

Hamdi, H., X. Mariette, V. Godot, K. Weldingh, A.M. Hamid, M.V. Prejean, G. Baron, M. Lemann, X. Puechal, M. Breban, F. Berenbaum, J.C. Delchier, R.M. Flipo, B. Dautzenberg, D. Salmon, M. Humbert and D. Emilie. 2006. Inhibition of anti-tuberculosis T-lymphocyte function with tumour necrosis factor antagonists. Arthritis Res. Ther. 8: R114.

Hand, T.W., L.M. Dos Santos, N. Bouladoux, M.J. Molloy, A.J. Pagan, M. Pepper, C.L. Maynard, C.O. Elson, 3rd and Y. Belkaid. 2012. Acute gastrointestinal infection induces long-lived microbiota-specific T cell responses. Science. 337: 1553–6.

Helwig, U., M. Muller, J. Hedderich and S. Schreiber. 2012. Corticosteroids and immunosuppressive therapy influence the result of QuantiFERON TB Gold testing in inflammatory bowel disease patients. J. Crohns Colitis 6: 419–24.

Hirai, F., T. Matsui, Y. Ishibashi, D. Higashi, K. Futami, S. Haraoka and A. Iwashita. 2011. Asymptomatic pulmonary cryptococcosis in a patient with Crohn's disease on infliximab: case report. Inflamm. Bowel Dis. 17: 1637–8.

Hou, J.K., F. Velayos, N. Terrault and U. Mahadevan. 2010. Viral hepatitis and inflammatory bowel disease. Inflamm. Bowel Dis. 16: 925–32.

Hulten, K., H.M. El-Zimaity, T.J. Karttunen, A. Almashhrawi, M.R. Schwartz, D.Y. Graham and F.A. El-Zaatari. 2001. Detection of *Mycobacterium avium* subspecies paratuberculosis in Crohn's diseased tissues by *in situ* hybridization. Am. J. Gastroenterol. 96: 1529–35.

Issa, M., A.N. Ananthakrishnan and D.G. Binion. 2008. *Clostridium difficile* and inflammatory bowel disease. Inflamm. Bowel Dis. 14: 1432–42.

Itaba, S., T. Iwasa, Y. Sadamoto, T. Nasu, T. Misawa, K. Inoue, H. Shimokawa, K. Nakamura and R. Takayanagi. 2007. Pneumocystis pneumonia during combined therapy of infliximab, corticosteroid, and azathioprine in a patient with Crohn's disease. Dig. Dis. Sci. 52: 1438–41.

Jones, P.H., T.B. Farver, B. Beaman, B. Cetinkaya and K.L. Morgan. 2006. Crohn's disease in people exposed to clinical cases of bovine paratuberculosis. Epidemiol. Infect. 134: 49–56.

Jostins, L., S. Ripke, R.H. Weersma, R.H. Duerr, D.P. Mcgovern, K.Y. Hui, J.C. Lee, L.P. Schumm, Y. Sharma, C.A. Anderson, J. Essers, M. Mitrovic, K. Ning, I. Cleynen, E. Theatre, S.L. Spain, S. Raychaudhuri, P. Goyette, Z. Wei, C. Abraham, J.P. Achkar, T. Ahmad, L. Amininejad, A.N. Ananthakrishnan, V. Andersen, J.M. Andrews, L. Baidoo, T. Balschun, P.A. Bampton, A. Bitton, G. Boucher, S. Brand, C. Buning, A. Cohain, S. Cichon, M. D'amato, D. De Jong, K.L. Devaney, M. Dubinsky, C. Edwards, D. Ellinghaus, L.R. Ferguson, D. Franchimont, K. Fransen, R. Gearry, M. Georges, C. Gieger, J. Glas, T. Haritunians, A. Hart, C. Hawkey, M. Hedl, X. Hu, T.H. Karlsen, L. Kupcinskas, S. Kugathasan, A. Latiano, D. Laukens, I.C. Lawrance, C.W. Lees, E. Louis, G. Mahy, J. Mansfield, A.R. Morgan, C. Mowat, W. Newman, O. Palmieri, C.Y. Ponsioen, U. Potocnik, N.J. Prescott, M. Regueiro, J.I. Rotter, R.K. Russell, J.D. Sanderson, M. Sans, J. Satsangi, S. Schreiber, L.A. Simms, J. Sventoraityte, S.R. Targan, K.D. Taylor, M. Tremelling, H.W. Verspaget, M. De Vos, C. Wijmenga, D.C. Wilson, J. Winkelmann, R.J. Xavier, S. Zeissig, B. Zhang, C.K. Zhang, H. Zhao, M.S. Silverberg, V. Annese, H. Hakonarson, S.R. Brant, G. Radford-Smith, C.G. Mathew, J.D. Rioux, E.E. Schadt et al. 2012. Host-microbe interactions have shaped the genetic architecture of inflammatory bowel disease. Nature. 491: 119–24.

Kasiske, B.L., M.G. Zeier, J.R. Chapman, J.C. Craig, H. Ekberg, C.A. Garvey, M.D. Green, V. Jha, M.A. Josephson, B.A. Kiberd, H.A. Kreis, R.A. Mcdonald, J.M. Newmann, G.T. Obrador, F.G. Vincenti, M. Cheung, A. Earley, G. Raman, S. Abariga, M. Wagner and E.M. Balk. 2010. KDIGO clinical practice guideline for the care of kidney transplant recipients: a summary. Kidney Int. 77: 299–311.

Kaufman, H.S., A.C. Kahn, C. Iacobuzio-Donahue, M.A. Talamini, K.D. Lillemoe and S.R. Hamilton. 1999. Cytomegaloviral enterocolitis: clinical associations and outcome. Dis. Colon Rectum. 42: 24–30.

Kaur, N. and T.C. Mahl. 2007. Pneumocystis jiroveci (carinii) pneumonia after infliximab therapy: a review of 84 cases. Dig. Dis. Sci. 52: 1481–4.

Keane, J., S. Gershon, R.P. Wise, E. Mirabile-Levens, J. Kasznica, W.D. Schwieterman, J.N. Siegel and M.M. Braun. 2001. Tuberculosis associated with infliximab, a tumor necrosis factor alpha-neutralizing agent. N Engl. J. Med. 345: 1098–104.

Khatchatourian, M. and T.L. Seaton. 1997. An unusual complication of immunosuppressive therapy in inflammatory bowel disease. Am. J. Gastroenterol. 92: 1558–60.

Kishore, J., U. Ghoshal, U.C. Ghoshal, N. Krishnani, S. Kumar, M. Singh and A. Ayyagari. 2004. Infection with cytomegalovirus in patients with inflammatory bowel disease: prevalence, clinical significance and outcome. J. Med. Microbiol. 53: 1155–60.

Kornbluth, A. and D.B. Sachar. 2010. Ulcerative colitis practice guidelines in adults: American College Of Gastroenterology, Practice Parameters Committee. Am. J. Gastroenterol. 105: 501–23; quiz 524.

Krech, U. 1973. Complement-fixing antibodies against cytomegalovirus in different parts of the world. Bull. World Health Organ. 49: 103–6.

Kyne, L., M. Warny, A. Qamar and C.P. Kelly. 2000. Asymptomatic carriage of *Clostridium difficile* and serum levels of IgG antibody against toxin A. N Engl. J. Med. 342: 390–7.

Kyne, L., M.Warny, A. Qamar and C.P. Kelly. 2001. Association between antibody response to toxin A and protection against recurrent Clostridium difficile diarrhoea. Lancet. 357: 189–93.

Lee, J.C., D.C. Bell, R.M. Guinness and T. Ahmad. 2009. Pneumocystis jiroveci pneumonia and pneumomediastinum in an anti-TNFalpha naive patient with ulcerative colitis. World J. Gastroenterol. 15: 1897–900.

Lessa, F.C., C.V. Gould and L.C. Mcdonald. 2012. Current status of *Clostridium difficile* infection epidemiology. Clin. Infect. Dis. 55 Suppl 2: S65–70.

Lok, A.S., R.H. Liang, E.K. Chiu, K.L. Wong, T.K. Chan and D. Todd. 1991. Reactivation of hepatitis B virus replication in patients receiving cytotoxic therapy. Report of a prospective study. Gastroenterology. 100: 182–8.

Lok, A.S. and B.J. Mcmahon. 2009. Chronic hepatitis B: update 2009. Hepatology. 50: 661–2.

Martin, S.I. and J.A. Fishman. 2009. Pneumocystis pneumonia in solid organ transplant recipients. Am. J. Transplant 9 Suppl 4: S227–33.

Mast, E.E., H.S. Margolis, A.E. Fiore, E.W. Brink, S.T. Goldstein, S.A. Wang, L.A. Moyer, B.P. Bell and M.J. Alter. 2005. A comprehensive immunization strategy to eliminate transmission of hepatitis B virus infection in the United States: recommendations of the Advisory Committee on Immunization Practices (ACIP) part 1: immunization of infants, children, and adolescents. MMWR Recomm. Rep. 54: 1–31.

Mccann, D.A. and H.L. Smith. 2012. Infliximab associated Blastomycosis dermatitidis in treatment of ulcerative colitis. Colorectal Dis.

Melmed, G.Y., A.F. Ippoliti, K.A. Papadakis, T.T. Tran, J.L. Birt, S.K. Lee, R.W. Frenck, S.R. Targan and E.A. Vasiliauskas. 2006. Patients with inflammatory bowel disease are at risk for vaccine-preventable illnesses. Am. J. Gastroenterol. 101: 1834–40.

Mendoza, J.L., A. San-Pedro, E. Culebras, R. Cies, C. Taxonera, R. Lana, E. Urcelay, F. De La Torre, J.J. Picazo and M. Diaz-Rubio. 2010. High prevalence of viable *Mycobacterium avium* subspecies paratuberculosis in Crohn's disease. World J. Gastroenterol. 16: 4558–63.

Miller, R. and L. Huang. 2004. Pneumocystis jirovecii infection. Thorax. 59: 731–3.

Moses, J., N. Alkhouri, A. Shannon, K. Raig, R. Lopez, L. Danziger-Isakov, A.E. Feldstein, N.N. Zein, R. Wyllie and C. Carter-Kent. 2012. Hepatitis B immunity and response to booster vaccination in children with inflammatory bowel disease treated with infliximab. Am. J. Gastroenterol. 107: 133–8.

Mowat, C., A. Cole, A. Windsor, T. Ahmad, I. Arnott, R. Driscoll, S. Mitton, T. Orchard, M. Rutter, L. Younge, C. Lees, G.T. Ho, J. Satsangi and S. Bloom. 2011. Guidelines for the management of inflammatory bowel disease in adults. Gut. 60: 571–607.

Murai, H., H. Tokunaga, I. Kubo, M. Kawajiri, T. Iwaki, T. Taniwaki and J. Kira. 2006. Myeloradiculitis caused by Cryptococcus neoformans infection in a patient with ulcerative colitis: a neuropathological study. J. Neurol. Sci. 247: 236–8.

NACC. 2009. Second report on the evidence linking *Mycobacterium avium* supspecies paratuberculosis (MAP) and Crohn's disease.

Naser, S.A., G. Ghobrial, C. Romero and J.F. Valentine. 2004. Culture of *Mycobacterium avium* subspecies paratuberculosis from the blood of patients with Crohn's disease. Lancet. 364: 1039–44.

Naser, S.A., K. Hulten, I. Shafran, D.Y. Graham and F.A. El-Zaatari. 2000. Specific seroreactivity of Crohn's disease patients against p35 and p36 antigens of *M. avium* subsp. paratuberculosis. Vet Microbiol. 77: 497–504.

Nordgaard-Lassen, I., J.F. Dahlerup, E. Belard, J. Gerstoft, J. Kjeldsen, K. Kragballe, P. Ravn, I.J. Sorensen, K. Theede and L. Tjellesen. 2012. Guidelines for screening, prophylaxis and critical information prior to initiating anti-TNF-alpha treatment. Dan. Med. J. 59: C4480.

Pai, M., L.W. Riley and J.M. Colford, Jr. 2004. Interferon-gamma assays in the immunodiagnosis of tuberculosis: a systematic review. Lancet Infect. Dis. 4: 761–76.

Pai, M., A. Zwerling and D. Menzies. 2008. Systematic review: T-cell-based assays for the diagnosis of latent tuberculosis infection: an update. Ann. Intern. Med. 149: 177–84.

Papay, P., A. Eser, S. Winkler, S. Frantal, C. Primas, W. Miehsler, S. Angelberger, G. Novacek, A. Mikulits, H. Vogelsang and W. Reinisch. 2011. Predictors of indeterminate IFN-gamma release assay in screening for latent TB in inflammatory bowel diseases. Eur. J. Clin. Invest. 41: 1071–6.

Peyrin-Biroulet, L. 2010. Anti-TNF therapy in inflammatory bowel diseases: a huge review. Minerva Gastroenterol. Dietol. 56: 233–43.

Poppers, D.M. and E.J. Scherl. 2008. Prophylaxis against Pneumocystis pneumonia in patients with inflammatory bowel disease: toward a standard of care. Inflamm. Bowel Dis. 14: 106–13.

Qumseya, B.J., A.N. Ananthakrishnan, S. Skaros, M. Bonner, M. Issa, Y. Zadvornova, A. Naik, L. Perera and D.G. Binion. 2011. QuantiFERON TB gold testing for tuberculosis screening in an inflammatory bowel disease cohort in the United States. Inflamm. Bowel Dis. 17: 77–83.

Rahier, J.F., S. Ben-Horin, Y. Chowers, C. Conlon, P. De Munter, G. D'haens, E. Domenech, R. Eliakim, A. Eser, J. Frater, M. Gassull, M. Giladi, A. Kaser, M. Lemann, T. Moreels, A. Moschen, R. Pollok, W. Reinisch, M. Schunter, E.F. Stange, H. Tilg, G. Van Assche, N. Viget, B. Vucelic, A. Walsh, G. Weiss, Y. Yazdanpanah, Y. Zabana, S.P. Travis and J.F. Colombel. 2009a. European evidence-based Consensus on the prevention, diagnosis and management of opportunistic infections in inflammatory bowel disease. J. Crohns Colitis 3: 47–91.

Rahier, J.F., Y. Yazdanpanah, J.F. Colombel and S. Travis. 2009b. The European (ECCO) Consensus on infection in IBD: what does it change for the clinician? Gut. 58: 1313–5.

Rahier, J.F., M. Moutschen, A. Van Gompel, M. Van Ranst, E. Louis, S. Segaert, P. Masson and F. De Keyser. 2010. Vaccinations in patients with immune-mediated inflammatory diseases. Rheumatology (Oxford). 49: 1815–27.

Reid, A.B., S.C. Chen and L.J. Worth. 2011. Pneumocystis jirovecii pneumonia in non-HIV-infected patients: new risks and diagnostic tools. Curr. Opin. Infect. Dis. 24: 534–44.

Rodemann, J.F., E.R. Dubberke, K.A. Reske, H. Seo Da and C.D. Stone. 2007. Incidence of Clostridium difficile infection in inflammatory bowel disease. Clin. Gastroenterol. Hepatol. 5: 339–44.

Rosenfeld, G. and B. Bressler. 2010. Mycobacterium avium paratuberculosis and the etiology of Crohn's disease: a review of the controversy from the clinician's perspective. Can. J. Gastroenterol. 24: 619–24.

Roskell, D.E., G.M. Hyde, A.P. Campbell, D.P. Jewell and W. Gray. 1995. HIV associated cytomegalovirus colitis as a mimic of inflammatory bowel disease. Gut. 37: 148–50.

Ryan, P., R.G. Kelly, G. Lee, J.K. Collins, G.C. O'Sullivan, J. O'Connell and F. Shanahan. 2004. Bacterial DNA within granulomas of patients with Crohn's disease—detection by laser capture microdissection and PCR. Am. J. Gastroenterol. 99: 1539–43.

Sacco, R., M. Bertini, G. Bresci, A. Romano, E. Altomare and A. Capria. 2010. Entecavir for hepatitis B virus flare treatment in patients with Crohn's disease. Hepatogastroenterology. 57: 242–5.

Salloum, E., D.L. Cooper, G. Howe, J. Lacy, G. Tallini, J. Crouch, M. Schultz and J. Murren. 1996. Spontaneous regression of lymphoproliferative disorders in patients treated with methotrexate for rheumatoid arthritis and other rheumatic diseases. J. Clin. Oncol. 14: 1943–9.

Seddik, M., H. Melliez, D. Seguy, N. Viget, A. Cortot and J.F. Colombel. 2005. Pneumocystis jiroveci (carinii) pneumonia after initiation of infliximab and azathioprine therapy in a patient with Crohn's disease. Inflamm. Bowel Dis. 11: 618–20.

Shahidi, N., Y.T. Fu, H. Qian and B. Bressler. 2012. Performance of interferon-gamma release assays in patients with inflammatory bowel disease: A systematic review and meta-analysis. Inflamm. Bowel Dis.

Shakir, F.A., T. Ali, A.C. Bigham, J.D. Ballard, P.B. Miner, Jr. and J.R. Philpott. 2012. Determination of serum antibodies to Clostridium difficile toxin B in patients with inflammatory bowel disease. Gastroenterol. Hepatol. (N Y). 8: 313–7.

Shanahan, F. 2012a. The gut microbiota-a clinical perspective on lessons learned. Nat. Rev. Gastroenterol. Hepatol. 9: 609–14.

Shanahan, F. 2012b. The microbiota in inflammatory bowel disease: friend, bystander, and sometime-villain. Nutr. Rev. 70 Suppl. 1: S31–7.

Sibartie, S., P. Scully, J. Keohane, S. O'Neill, J. O'Mahony, D. O'Hanlon, W.O. Kirwan, L. O'Mahony and F. Shanahan. 2010. *Mycobacterium avium* subsp. Paratuberculosis (MAP) as a modifying factor in Crohn's disease. Inflamm. Bowel Dis. 16: 296–304.

Siegel, C.A. 2011. Review article: explaining risks of inflammatory bowel disease therapy to patients. Aliment Pharmacol. Ther. 33: 23–32.

Siegel, C.A., S.M. Marden, S.M. Persing, R.J. Larson and B.E. Sands. 2009. Risk of lymphoma associated with combination anti-tumor necrosis factor and immunomodulator therapy for the treatment of Crohn's disease: a meta-analysis. Clin. Gastroenterol. Hepatol. 7: 874–81.

Slifman, N.R., S.K. Gershon, J.H. Lee, E.T. Edwards and M.M. Braun. 2003. Listeria monocytogenes infection as a complication of treatment with tumor necrosis factor alpha-neutralizing agents. Arthritis Rheum. 48: 319–24.

Smith, J.A. and C.A. Kauffman. 2009. Endemic fungal infections in patients receiving tumour necrosis factor-alpha inhibitor therapy. Drugs. 69: 1403–15.

Solovic, I., M. Sester, J.J. Gomez-Reino, H.L. Rieder, S. Ehlers, H.J. Milburn, B. Kampmann, B. Hellmich, R. Groves, S. Schreiber, R.S. Wallis, G. Sotgiu, E.H. Scholvinck, D. Goletti, J.P. Zellweger, R. Diel, L. Carmona, F. Bartalesi, P. Ravn, A. Bossink, R. Duarte, C. Erkens, J. Clark, G.B. Migliori and C. Lange. 2010. The risk of tuberculosis related to tumour necrosis factor antagonist therapies: a TBNET consensus statement. Eur. Respir. J. 36: 1185–206.

Subramaniam, K., J. D'rozario and P. Pavli. 2013. Lymphoma and other lymphoproliferative disorders in inflammatory bowel disease: A review. J. Gastroenterol. Hepatol. 28: 24–30.

Symmers, W.S. 1965. Opportunistic Infections. The Concept of 'Opportunistic Infections'. Proc. R Soc. Med. 58: 341–6.

Tasaka, S. and H. Tokuda. 2012. Pneumocystis jirovecii pneumonia in non-HIV-infected patients in the era of novel immunosuppressive therapies. J. Infect. Chemother.

Thomas, S., R.Vivancos, C. Corless, G. Wood, N.J. Beeching and M.B. Beadsworth. 2011. Increasing frequency of Pneumocystis jirovecii pneumonia in renal transplant recipients in the United Kingdom: clonal variability, clusters, and geographic location. Clin. Infect. Dis. 53: 307–8.

Tomblyn, M., T. Chiller, H. Einsele, R. Gress, K. Sepkowitz, J. Storek, J.R. Wingard, J.A. Young and M.J. Boeckh. 2009. Guidelines for preventing infectious complications among hematopoietic cell transplantation recipients: a global perspective. Biol. Blood Marrow Transplant 15: 1143–238.

Toruner, M., E.V. Loftus, Jr., W.S. Harmsen, A.R. Zinsmeister, R. Orenstein, W.J. Sandborn, J.F. Colombel and L.J. Egan. 2008. Risk factors for opportunistic infections in patients with inflammatory bowel disease. Gastroenterology. 134: 929–36.

Trnka, Y.M. and J.T. Lamont. 1981. Association of Clostridium difficile toxin with symptomatic relapse of chronic inflammatory bowel disease. Gastroenterology. 80: 693–6.

Ueno, Y., S. Tanaka, M. Shimamoto, Y. Miyanaka, T. Hiyama, M. Ito, Y. Kitadai, M. Yoshihara, M. Sumii and K. Chayama. 2005. Infliximab therapy for Crohn's disease in a patient with chronic hepatitis B. Dig. Dis. Sci. 50: 163–6.

Velayos, F.S. and W.J. Sandborn. 2004. Pneumocystis carinii pneumonia during maintenance anti-tumor necrosis factor-alpha therapy with infliximab for Crohn's disease. Inflamm. Bowel Dis. 10: 657–60.

Vida Perez, L., F. Gomez Camacho, V. Garcia Sanchez, E.M. Iglesias Flores, L. Castillo Molina, A. Cerezo Ruiz, L. Casais Juanena and J.F. De Dios Vega. 2009. [Adequate rate of response to hepatitis B virus vaccination in patients with inflammatory bowel disease]. Med. Clin. (Barc). 132: 331–5.

Yazaki, H., N. Goto, K. Uchida, T. Kobayashi, H. Gatanaga and S. Oka. 2009. Outbreak of Pneumocystis jiroveci pneumonia in renal transplant recipients: P. jiroveci is contagious to the susceptible host. Transplantation. 88: 380–5.

Zeitz, J., B. Mullhaupt, H. Fruehauf, G. Rogler and S.R. Vavricka. 2009. Hepatic failure due to hepatitis B reactivation in a patient with ulcerative colitis treated with prednisone. Hepatology. 50: 653–4.

9

Opportunistic Pathogens in Inflammatory Bowel Disease, The Case of Adherent-Invasive *E. coli*

Arlette Darfeuille-Michaud

Introduction

Crohn's disease (CD) and ulcerative colitis (UC) are two major forms of idiopathic inflammatory bowel disease (IBD) affecting 1.4 million individuals in the United States and 2.2 million in Europe (Economou and Pappas 2008). IBD, especially CD, are now widely accepted as complex multifactor diseases that occur in individuals with genetic predisposition in whom environmental and microbial triggers cause a deleterious chronic immune response (Stappenbeck et al. 2011; Kaser et al. 2010; Xavier and Podolsky 2007; Chassaing and Darfeuille-Michaud 2011). Over the last decade, genetic studies have emphasized the role of host susceptibility in IBD onset with the identification of about one hundred risk loci. It is striking that most susceptibility genes identified belong to a restricted number of pathways involved in intestinal barrier function, autophagy, microbial

Clermont Université, Université d'Auvergne, Clermont-Ferrand, France; Inserm U1071, Clermont-Ferrand, France; INRA, USC2018, Clermont-Ferrand, France; and Centre Hospitalier Universitaire, Clermont-Ferrand, France.
E-mail: arlette.darfeuille-michaud@u-clermont1.fr

defense, innate immune regulation, lymphocyte differentiation, reactive oxygen species (ROS) generation or ER stress, for a review (Khor et al. 2011). It is widely admitted that risk variants alter the regulation or efficiency of these key molecular pathways following environmental stresses, in an acute or chronic manner, leading to chronic gut disorders. Interestingly, these pathways are mostly engaged in host response against microbes and intracellular bacteria handling, indicating a possible involvement of pathogens in CD etiology.

Microbial Trigger in CD Etiology

Since the first description of CD and UC, it was thought that commensal microbiota or infectious agents are involved in the pathogenesis and clinical course of IBD. Indeed, the two major sites of lesions in CD, the terminal ileum and colon, overlap with gut regions of greater microorganism density. In addition, reducing the level of mucosal associated bacteria, either by fecal stream diversion (Rutgeerts et al. 1991) or broad-spectrum antibiotics (such as metronidazole or ciprofloxacin) induces a clinical improvement in a subset of active CD patients (Selby et al. 2007) and for a review (Sartor 2004). Metagenomic analyses have revealed a breakdown in the balance between "protective" and "harmful" intestinal bacteria in the gut of CD patients, with a decrease in the abundance of the Firmicutes (*Clostridium* and *Bacillus* species Gram-positive bacteria), and an increase in Proteobacteria (Gram-negative bacteria, including *Escherichia* spp.) (Sokol et al. 2008; Zoetendal et al. 2008). Aside dysbiosis, some specific microorganisms have been suspected to act as an infectious trigger in CD etiology, including *Mycobacterium avium* subsp. *paratuberculosis* (MAP), (Sartor 2005) yeasts (Seibold et al. 2004) and adherent-invasive *Escherichia coli* (AIEC) (Darfeuille-Michaud et al. 2004). Interestingly, the increase in *Escherichia coli*, found by metagenomic analyses, has also been reported by several independent groups using classical culture techniques or molecular biology based-approaches (Baumgart et al. 2007; Darfeuille-Michaud et al. 1998; Eaves-Pyles et al. 2008; Martinez-Medina et al. 2009; Sasaki et al. 2007). This abnormal colonization by *E. coli* strains is especially observed on the ileal mucosa of CD patients. The exact origin of the dysbiosis, and the associated *E. coli* abnormal colonization, in CD patient remains to be elucidated. Various explanations have been suggested, including the hygiene hypothesis linked to frequent use of antibiotics or microbicidal substances, a breakdown of microbiota homeostasis following acute gastroenteritis by pathogens such as *Salmonella* spp. or *Campylobacter* spp. (Gradel et al.

2009) or the impact of dietary compounds (refined sugar for example) used in the food of developed countries that could favor the growth of certain bacterial species (Brown 2008). Another attractive hypothesis is that the genetic polymorphisms observed in CD, in association with other environmental factors, could lead in some individuals to an inappropriate immune response, creating niches for abnormal, long-lasting colonization by detrimental bacteria, such as *E. coli*.

Characterization of Adherent-invasive *E. coli* (AIEC)

Several independent studies have reported increased numbers of mucosa-associated *E. coli* with invasive properties or the presence of intramucosal *E. coli* in IBD patients (Darfeuille-Michaud et al. 2004; Baumgart et al. 2007; Darfeuille-Michaud et al. 1998; Eaves-Pyles et al. 2008; Martinez-Medina et al. 2009; Sasaki et al. 2007). This could be related to the very high levels of anti-*E. coli* outer membrane protein antibodies observed in patients with ileal involvement of Crohn's disease (Mow et al. 2004). The CD-associated *Escherichia coli*, compared to commensal *E. coli*, have acquired specific virulence factors that increase their ability to adapt to new niches and allow them to cause disease. They are highly adherent to intestinal epithelial cells and are also invasive. On the basis of their pathogenic traits, a new pathogenic group of *E. coli* was designated AIEC, for adherent-invasive *Escherichia coli* (Boudeau et al. 1999). The interaction between AIEC and intestinal epithelial cultured cells induces inflammatory responses such as upregulated expression of IL-8 and CCL20, leading to transmigration of polymorphonuclear leukocytes and dendritic cells as shown in co-culture models (Sasaki et al. 2007). AIEC can also disrupt the integrity of the polarized epithelial cell monolayer, allowing bacteria to breach the intestinal barrier and to penetrate into the gut mucosa (Boudeau et al. 1999). In addition, AIEC are able to survive and to replicate extensively within a large vacuole in macrophages sharing features with phagolysosomes (Glasser et al. 2001; Bringer et al. 2006). To survive and replicate in the harsh environment encountered inside these compartments, including acid pH and proteolytic activity of cathepsin D, bacteria have developed adaptation mechanisms in which acidity constitutes a key signal for activating the expression of virulence genes (Bringer et al. 2005). Macrophages infected with AIEC release large amounts of TNF-α (Glasser et al. 2001) and in an *in vitro* model of human granuloma, AIEC-infected macrophages aggregate, fuse to form multinucleated giant cells, and subsequently recruit lymphocytes (Meconi et al. 2007).

Host factors involved in the abnormal predominance of AIEC associated with gut mucosa of patients with Crohn's disease

AIEC were isolated from ileal specimens of 36.4% of CD patients verses 6% of controls (Darfeuille-Michaud et al. 2004). The higher prevalence of AIEC bacteria in CD patients arise from 1) an abnormal expression of host factors such as CEACAM6 and the ER stress response protein Gp96 involved in AIEC gut colonization, 2) an increase ability to cross the intestinal barrier and 3) a loss of control of intracellular AIEC replication related to autophagy deficiencies.

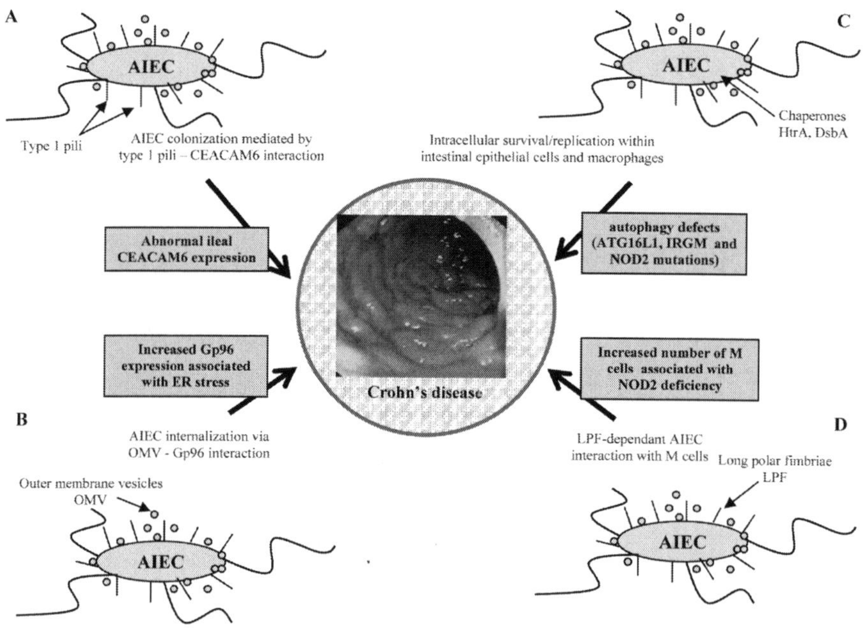

Figure 1. Host-AIEC Interactions with Ileal Mucosa in Crohn's Disease. A: Adherent-invasive *E. coli* colonize the gut by adhering to the ileal mucosa via type 1 pili variants recognizing mannose residues of carcinoembryonic antigen-related cell adhesion molecule 6 (CEACAM6), which is abnormally expressed on the apical surface of ileal epithelial cells in patients with ileal CD. **B**: The endoplasmic reticulum stress response protein Gp96, overexpressed on the apical surface of ileal epithelial cells in patients with CD, acts as a host cell receptor for AIEC outer membrane vesicles (OMVs), promoting AIEC invasion. **C:** AIEC are able to survive within intestinal epithelial cells and macrophages and there is an uncontrolled replication of intracellular AIEC bacteria in host cells with autophagy defects (ATG16L1, IRGM, NOD2 mutations). **D:** AIEC express Long Polar Fimbriae (LPF) allowing the bacteria to efficiently interact with Peyer's patches and to translocate across M cells whose number is increased in NOD2 deficiency context.

Color image of this figure appears in the color plate section at the end of the book.

1. Abnormal expression of CECAM6

Among host factors that facilitate AIEC colonization is the abnormal ileal expression of carcinoembryonic antigen-related cell adhesion molecule 6 (CEACAM6), which has been observed in patients with ileal CD (Barnich et al. 2007). Related to abnormal expression of CEACAM6, AIEC adhere to the brush border of ileal enterocytes isolated from CD patients but not from individuals without IBD. In addition, most AIEC strains associated with CD ileal mucosa express type 1 pili variants that increase the interaction between AIEC and ileal epithelial cells. Increased expression of CEACAM6 result from interferon (IFN)-γ or tumor necrosis factor (TNF)-α stimulation and also from infection with AIEC bacteria, indicating that AIEC could promote their own colonization in CD patients. In addition AIEC bacteria are able to persist in the gut of transgenic CEABAC10 mice that express human CEACAMs and induce severe colitis (Carvalho et al. 2009).

2. Increased expression of the endoplasmic reticulum (ER) stress response protein Gp96

Another pathway that has also recently emerged in IBD pathophysiology is the unfolded protein response (UPR), which is induced by endoplasmic reticulum (ER) stress (Kaser and Blumberg 2010) ER stress occurs upon the accumulation of misfolded or unfolded proteins in the ER, which initiates the UPR (Schroder and Kaufman 2005). An increased expression of the ER stress response protein Gp96 on the apical surface of ileal epithelial cells in patients with CD (Rolhion et al. 2010). This glycoprotein is able to bind to *E. coli* outer membrane protein A present on the surface of AIEC outer membrane vesicles (OMVs), a possible way for the AIEC bacteria to deliver bacterial effectors into host cells and to promote bacterial invasion. Thus, AIEC bacteria could take advantage of the ER stress occurring in patients with IBD.

3. Increased number of M cells in the follicle-associated epithelium lying over Peyer's patches

Increased number of membranous/microfold cells (M cells) have been observed in *NOD2* knock-out mice (Chassaing et al. 2011; Barreau, et al. 2007). AIEC express long polar fimbriae (LPF) that allow the bacteria to increase their interaction with mouse and human Peyer's patches and their translocation across M cell monolayers (Barreau et al. 2007). Thus, in the presence of NOD2 mutations as observed in CD, higher number of AIEC bacteria could interact with Peyer's patches. This observation is particularly relevant since recurrent ileal CD originates with small erosions in the follicle-associated epithelium (FAE) that lies over Peyer's patches (Morson 1972).

4. Mutations in autophagy-related genes

Several combinations of genetic predisposing factors to CD have been identified. The most well-replicated IBD genetic association is the *NOD2* (nucleotide-binding oligomerization domain 2) gene occurring in CD (Hugot et al. 2001; Ogura et al. 2001). Functional consequences of NOD2 mutations could be related to defects in Paneth cells function and therefore the secretion of antimicrobial peptides (Wehkamp et al. 2004) but also to synergic effects of NOD2/Toll-like receptor (TLR) signalling and autophagy deficiency (Travassos et al. 2010; Cooney et al. 2010). Indeed, the intracellular sensors NOD1 and NOD2, by recruiting autophagy proteins to the plasma membrane at the bacterial entry site, are critical for the autophagic response, autophagy being an efficient innate host defense mechanism against invading micro-organisms. In addition, genome-wide association (GWA) studies indicated a highly significant and replicated association between CD and variants in two separate autophagy genes, *ATG16L1* (Autophagy-related like 1) and *IRGM* (Immunity-related GTPase family M) (Hampe et al. 2007; Parkes et al. 2007; Rioux et al. 2007; Wellcome 2007). A coding polymorphism in the autophagy gene *ATG16L1* was identified as a risk factor in CD with ATG16L1 polymorphism T300A predisposing to ileal disease and polymorphisms of the IRGM locus might predispose to fistulising behaviour in CD (Glas et al. 2013)[7]. When intracellular ATG16L1 and IRGM expression was modified in epithelial cells the number of intracellular AIEC bacteria was greatly increased, indicating that functional autophagy has a key role in controlling intracellular AIEC replication (Lapaquette et al. 2010; Brest et al. 2011). In addition, impaired ATG16L1, IRGM or NOD2 expression induced increased intramacrophagic AIEC and increased secretion of IL-6 and TNF-a in response to AIEC infection (Lapaquette P et al. 2012). In contrast, forced induction of autophagy decreased the numbers of intramacrophagic AIEC and pro-inflammatory cytokine release, even in a NOD2-deficient context, indicating that stimulating autophagy in CD patients could be a powerful therapeutic strategy to concomitantly restrain AIEC replication and slow down the inflammatory response.

Key words: IBD, Crohn's disease, Adherent-invasive *E. coli*, Host susceptibility, CEACAM, ER stress, Autophagy

References Cited

Barnich, N., F.A. Carvalho, A.L. Glasser et al. 2007. CEACAM6 acts as a receptor for adherent-invasive *E. coli*, supporting ileal mucosa colonization in Crohn disease. J. Clin. Invest. 117: 1566–74.

Barreau, F., U. Meinzer, F. Chareyre et al. 2007. CARD15/NOD2 is required for Peyer's patches homeostasis in mice. PLoS ONE. 2: e523.

Baumgart, M., B. Dogan, M. Rishniw et al. 2007. Culture independent analysis of ileal mucosa reveals a selective increase in invasive *Escherichia coli* of novel phylogeny relative to depletion of Clostridiales in Crohn's disease involving the ileum. Isme J. 1: 403–18.

Boudeau, J., A.L. Glasser, E. Masseret et al. 1999. Invasive ability of an *Escherichia coli* strain isolated from the ileal mucosa of a patient with Crohn's disease. Infect. Immun. 67: 4499–509.

Brest, P., P. Lapaquette, M. Souidi et al. 2011. A synonymous variant in IRGM alters a binding site for miR-196 and causes deregulation of IRGM-dependent xenophagy in Crohn's disease. Nat. Genet. 43: 242–5.

Bringer, M.A., A.L. Glasser, C.H. Tung et al. 2006. The Crohn's disease-associated adherent-invasive *Escherichia coli* strain LF82 replicates in mature phagolysosomes within J774 macrophages. Cell Microbiol. 8: 471–84.

Bringer, M.A., N. Barnich, A.L. Glasser et al. 2005. HtrA stress protein is involved in intramacrophagic replication of adherent and invasive *Escherichia coli* strain LF82 isolated from a patient with Crohn's disease. Infect. Immun. 73: 712–21.

Brown, S.A., K.L. Palmer and M. Whiteley. 2008. Revisiting the host as a growth medium. Nat. Rev. Microbiol. 6: 657–66.

Carvalho, F.A., N. Barnich, A. Sivignon et al. 2009. Crohn's disease adherent-invasive *Escherichia coli* colonize and induce strong gut inflammation in transgenic mice expressing human CEACAM. J. Exp. Med. 206: 2179–89.

Chassaing, B. and A. Darfeuille-Michaud. 2011. The commensal microbiota and enteropathogens in the pathogenesis of inflammatory bowel diseases. Gastroenterology. 140: 1720–28.

Chassaing, B., N. Rolhion, A. De Vallée et al. 2011. Crohn's Disease-Associated Adherent-Invasive *Escherichia coli* Target mouse and human Peyer's Patches via Long Polar Fimbriae. J. Clin. Invest. 121: 966–75.

Cooney, R., J. Baker, O. Brain et al. 2010. NOD2 stimulation induces autophagy in dendritic cells influencing bacterial handling and antigen presentation. Nat. Med. 16: 90–7.

Darfeuille-Michaud, A., C. Neut, N. Barnich et al. 1998. Presence of adherent *Escherichia coli* strains in ileal mucosa of patients with Crohn's disease. Gastroenterology. 115: 1405–13.

Darfeuille-Michaud, A., J. Boudeau, P. Bulois et al. 2004. High prevalence of adherent-invasive *Escherichia coli* associated with ileal mucosa in Crohn's disease. Gastroenterology. 127: 412–21.

Denizot, J., A. Sivignon, F. Barreau et al. 2012. Adherent-invasive *Escherichia coli* induce claudin-2 expression and barrier defect in CEABAC10 mice and Crohn's disease patients. Inflamm. Bowel Dis. 18: 294–304.

Eaves-Pyles, T., C.A. Allen, J. Taormina et al. 2008. *Escherichia coli* isolated from a Crohn's disease patient adheres, invades, and induces inflammatory responses in polarized intestinal epithelial cells. Int. J. Med. Microbiol. 298: 397–409.

Economou, M. and G. Pappas. 2008. New global map of Crohn's disease: Genetic, environmental, and socioeconomic correlations. Inflamm. Bowel Dis. 14: 709–20.

Glas, J., J. Seiderer, S. Bues, J. Stallhofer, C. Fries, T. Olszak, E. Tsekeri, M. Wetzke, F. Beigel, C. Steib, M. Friedrich, B. Göke, J. Diegelmann, D. Czamara and S. Brand. 2013. IRGM variants and susceptibility to inflammatory bowel disease in the German population. PLoS One. 8(1): e54338. doi: 10.1371/journal.pone.0054338. Epub 2013 Jan 24.

Glasser, A.L., J. Boudeau, N. Barnich et al. 2001. Adherent invasive *Escherichia coli* strains from patients with Crohn's disease survive and replicate within macrophages without inducing host cell death. Infect. Immun. 69: 5529–37.

Gradel, K.O., H.L. Nielsen, H.C. Schonheyder et al. 2009. Increased short- and long-term risk of inflammatory bowel disease after *Salmonella* or *Campylobacter* gastroenteritis. Gastroenterology. 137: 495–501.

Hampe, J., A. Franke, P. Rosenstiel et al. 2007. A genome-wide association scan of nonsynonymous SNPs identifies a susceptibility variant for Crohn disease in ATG16L1. Nat. Genet. 39: 207–11.

Hugot, J.P., M. Chamaillard, H. Zouali et al. 2001. Association of NOD2 leucine-rich repeat variants with susceptibility to Crohn's disease. Nature. 411: 599–603.

Kaser, A. and R.S. Blumberg. 2010. Endoplasmic reticulum stress and intestinal inflammation. Mucosal Immunol. 3: 11–6.

Kaser, A., S. Zeissig and R.S. Blumberg. 2010. Inflammatory bowel disease. Annu. Rev. Immunol. 28: 573–621.

Khor, B., A. Gardet and R.J. Xavier. 2011. Genetics and pathogenesis of inflammatory bowel disease. Nature. 474: 307–17.

Lapaquette, P., A.L. Glasser, A. Huett et al. 2010. Crohn's disease-associated adherent-invasive *E. coli* are selectively favoured by impaired autophagy to replicate intracellularly. Cell Microbiol. 12: 99–113.

Lapaquette, P., M.A. Bringer and A. Darfeuille-Michaud. 2012. Defects in autophagy favour adherent-invasive *Escherichia coli* persistence within macrophages leading to increased pro-inflammatory response. Cell Microbiol. 14: 791–807.

Martinez-Medina, M., X. Aldeguer, M. Lopez-Siles et al. 2009. Molecular diversity of *Escherichia coli* in the human gut: new ecological evidence supporting the role of adherent-invasive E. coli (AIEC) in Crohn's disease. Inflamm. Bowel Dis. 15: 872–82.

Meconi, S., A. Vercellone, F. Levillain et al. 2007. Adherent-invasive *Escherichia coli* isolated from Crohn's disease patients induce granulomas *in vitro*. Cell Microbiol. 5: 1252–61.

Morson, B.C. 1972. The early histological lesion of Crohn's disease. Proc. R Soc. Med. 65: 71–2.

Mow, W.S., E.A. Vasiliauskas, Y.C. Lin et al. 2004. Association of antibody responses to microbial antigens and complications of small bowel Crohn's disease. Gastroenterology. 126: 414–24.

Ogura, Y., D.K. Bonen, N. Inohara et al. 2001. A frameshift mutation in NOD2 associated with susceptibility to Crohn's disease. Nature. 411: 603–6.

Parkes, M., J.C. Barrett, N.J. Prescott et al. 2007. Sequence variants in the autophagy gene IRGM and multiple other replicating loci contribute to Crohn's disease susceptibility. Nat. Genet. 39: 830–2.

Rioux, J.D., R.J. Xavier, K.D. Taylor et al. 2007. Genome-wide association study identifies new susceptibility loci for Crohn disease and implicates autophagy in disease pathogenesis. Nat. Genet. 39: 596–604.

Rolhion, N., N. Barnich, M.A. Bringer et al. 2010. Abnormally expressed ER stress response chaperone Gp96 in CD favours adherent-invasive *Escherichia coli* invasion. Gut. 59: 1355–62.

Rutgeerts, P., K. Goboes, M. Peeters et al. 1991. Effect of faecal stream diversion on recurrence of Crohn's disease in the neoterminal ileum. Lancet. 338: 771–4.

Sartor, R.B. 2004. Therapeutic manipulation of the enteric microflora in inflammatory bowel diseases: antibiotics, probiotics, and prebiotics. Gastroenterology. 126: 1620–33.

Sartor, R.B. 2005. Does *Mycobacterium avium* subspecies paratuberculosis cause Crohn's disease? Gut. 54: 896–8.

Sasaki, M., S.V. Sitaraman, B.A. Babbin et al. 2007. Invasive *Escherichia coli* are a feature of Crohn's disease. Lab Invest. 87: 1042–54.

Schroder, M. and R.J. Kaufman. 2005. The mammalian unfolded protein response. Annu. Rev. Biochem. 74: 739–89.

Seibold, F., A. Konrad, B. Flogerzi et al. 2004. Genetic variants of the mannan-binding lectin are associated with immune reactivity to mannans in Crohn's disease. Gastroenterology. 127: 1076–84.

Selby, W., P. Pavli, B. Crotty et al. 2007. Two-year combination antibiotic therapy with clarithromycin, rifabutin, and clofazimine for Crohn's disease. Gastroenterology. 132: 2313–9.

Shanahan, F. and C.N. Bernstein. 2009. The evolving epidemiology of inflammatory bowel disease. Curr. Opin. Gastroenterol. 25: 301–5.

Sokol, H., C. Lay, P. Seksik et al. 2008. Analysis of bacterial bowel communities of IBD patients: what has it revealed? Inflamm. Bowel Dis. 14: 858–67.

Stappenbeck, T.S., J.D. Rioux, A. Mizoguchi et al. 2011. Crohn disease: a current perspective on genetics, autophagy and immunity. Autophagy. 7: 355–74.

Travassos, L.H., L.A. Carneiro, M. Ramjeet et al. 2010. Nod1 and Nod2 direct autophagy by recruiting ATG16L1 to the plasma membrane at the site of bacterial entry. Nat. Immunol. 11: 55–62.

Wehkamp, J., J. Harder, M. Weichenthal et al. 2004. NOD2 (CARD15) mutations in Crohn's disease are associated with diminished mucosal alpha-defensin expression. Gut. 53: 1658–64.

Wellcome. 2007. Genome-wide association study of 14,000 cases of seven common diseases and 3,000 shared controls. Nature. 447: 661–78.

Xavier, R.J. and D.K. Podolsky. 2007. Unravelling the pathogenesis of inflammatory bowel disease. Nature. 448: 427–34.

Zoetendal, E.G., M. Rajilic-Stojanovic and W.M. de Vos. 2008. High-throughput diversity and functionality analysis of the gastrointestinal tract microbiota. Gut. 57: 1605–15.

10

Correction of Microbiota Disturbances or Antagonism Against Specific Pathogens in IBD

Paul K. Flanagan and *Jonathan M. Rhodes**

Introduction

The causative role of *H. pylori* in "peptic" ulcer causation was viewed with scepticism by most gastroenterologists until it could be shown that eradication of the bacterium led to cure of the condition. We now have a similar situation in inflammatory bowel disease—we have evidence of intriguing changes in the faecal and mucosal microbiota in both Crohn's disease and ulcerative colitis—particularly the former—but until we can show that treatments which target these changes improve the patient's condition, many will think that the altered microbiota just reflects changes that are secondary to inflammation. In this chapter we review the results of attempts to treat inflammatory bowel disease by targeting the microbiota as well as discussing appropriate strategies that might be followed.

Department of Gastroenterology, Institute of Translational Medicine, University of Liverpool, Nuffield Building, Crown Street, Liverpool L69 3GE, UK.
E-mail: paulflanagan@nhs.net
* Corresponding author: J.M.Rhodes@liverpool.ac.uk

We Need to Know What Bacteria to Target and Where They Are

If bacteria have a pathogenic role in inflammatory bowel disease, the most obvious therapeutic approach is to try antibiotics. If this approach is to be successful though, we need to know what bacteria we are targeting, what are their antibiotic sensitivities, and where they are—are they in the colonic lumen, on the mucosal surface or deep within the tissues. Are they even, as is likely to be the case for Adherent and Invasive *E. coli* or *Mycobacterium paratuberculosis*, hidden within vesicles inside macrophages? Are the bacteria present as a result of a "one-off" infection in which case a single course of antibiotics might be curative or are they bacteria that are normally resident in the colon and likely to be a recurring problem? If the latter, then antibiotic resistance will rapidly become a problem if a single antibiotic is used long term.

Another major problem is that it is looking increasingly likely on the basis of genetic studies that what we currently call Crohn's disease is more than one condition and it is entirely feasible that one microbial factor might be causative in one group of Crohn's patients and a completely different agent could be causative in another population. We therefore need not only to stratify clinical trials carefully according to disease phenotype but also to investigate better ways of using biomarkers, such as anti-bacterial antibodies perhaps, to identify subpopulations of patients with inflammatory bowel disease that may have a specific microbial aetiology. Meanwhile, there is now clear evidence that isolated colonic Crohn's disease is not associated with the same genetic or serological markers as other forms of Crohn's disease with small intestinal involvement and should be treated as a separate condition in therapeutic trials.

Approaches that Target Specific Bacterial Pathogens

Mycobacterium paratuberculosis

The possible association between Crohn's disease and *Mycobacterium paratuberculosis*, the causative agent of Johne's disease, a granulomatous intestinal disease of cattle, has been debated since its first mention by Dalziel (Dalziel 1913), predating even Burrill Crohn's classic paper. In its favour are the many, but varied, reports of its culture or identification of its DNA in Crohn's tissue samples (Chiodini et al. 2012; Ryan et al. 2002). Against it is the good response of Crohn's disease to treatment with anti-TNF agents that are known to be dangerously harmful in the presence of infection with *Mycobacterium tuberculosis*. Our own group has noted that *M. paratuberculosis* possesses the mannan epitope of the Anti-*Saccharomycesboulardii* antibody

(ASCA) that is commonly found in the sera of patients with Crohn's disease and has shown that *M. paratuberculosis* releases a glycoconjugate, that contains the mannanASCAepitope and which suppresses the ability of macrophages to kill phagocytosed *E. coli* (Mpofu et al. 2007). This raises the possibility that *M. paratuberculosis* might have a role in Crohn's disease pathogenesis by having an indirect pathogenic effect. The lack of direct pathogenicity of *M. paratuberculosis* in humans might then explain the lack of exacerbation with anti-TNF treatment.

Past trials of anti-tuberculous treatment have proved unsuccessful at inducing remission in Crohn's disease (Borgaonkar et al. 2000; Feller et al. 2010) but these used combinations of therapy that would be suboptimal against *M. paratuberculosis*. Following uncontrolled reports of prolonged remission in Crohn's disease inresponse to therapy targeting *M. paratuberculosis* (Borody et al. 2002), a controlled trial of clarithromycin, rifabutin and clofazimine antibioticsmaintained for three years was conducted in Australia (Selby et al. 2007). This showed no sustained benefit over three years in comparison with placebo, i.e., no lasting cure was achieved, however there was some short term benefit during the first year. Adherents to the *M. paratuberculosis* hypothesis have criticized the trial on the grounds firstly that no attempt was made to identify the (minority?) patients with Crohn's disease who had *M. paratuberculosis* infection (Kuenstner 2007) and, secondly, on the uncertain basis that the antibiotic regimen used was sub-optimal (Lipton and Barash 2007).

Adherent and Invasive *E. coli*

Many researchers have now independently shown a strong statistical association between Crohn's disease and an increase in mucosa-associated *E. coli* (Flanagan et al. 2011). These bacteria have the ability to adhere to and invade intestinal epithelial cell lines in culture and to replicate inside macrophages—properties that have been used to define them as Adherent and Invasive *E. coli* (Glasser et al. 2001). *In vitro* the presence of these *E. coli* within macrophages has been shown to induce granuloma formation (Meconi et al. 2007) and in Crohn's tissue *E. coli* have been identified within granulomas using both immunohistochemical (Liu et al. 1995) and DNA-based techniques such as laser-capture microscopy followed by PCR (Ryan et al. 2004).

Antibiotics targeting these *E. coli* will therefore have to be able to penetrate macrophages—gentamicin and penicillins, for example do not have this property and will be ineffective. In the laboratory, clarithromycin is effective at killing *E. coli* within macrophages and we speculated that it might have been the inclusion of clarithromycin in the regimens used to target *M. paratuberculosis* that was responsible for some of the possible short-term benefit seen. Following earlier promising results in an uncontrolled study, we conducted a placebo-controlled trial of clarithromycin in active Crohn's disease. This was negative in respect of its primary endpoint (remission at three months) but some benefit was seen at one month that was not sustained (Leiper et al. 2008).

Ciprofloxacin is also very effective at killing *E. coli* within macrophages in the laboratory and has been extensively used in Crohn's disease although clinical trials have generally been underpowered and inconclusive (Table 1). Our anecdotal experience with it is that it, like clarithromycin, can be quite effective for up to a month but that the response is usually not maintained. Long term use of the antibiotic (e.g., more than three months) should be discouraged not only because of lack of evidence for efficacy but also because of the risk of tendon rupture.

Multi-drug resistance has, not surprisingly, been demonstrated for Crohn's disease *E. coli* isolates (Dogan et al. 2012). It seems probable therefore that effective therapy to target intra-macrophage *E. coli* will need to use combinations of antibiotics. We have in our laboratory assessed various combinations and several of these show greater efficacy than ciprofloxacin alone, particularly when tested at low concentrations corresponding to 10% of Cmax (the peak concentration likely to be present with conventional doing *in vivo*). Possible combinations include ciprofloxacin plus either trimethoprim or doxycycline (Subramanian et al. 2008).

We have recently become very interested in the possibility of using the anti-malarial hydroxychloroquine as an adjunct to antibiotic therapy in Crohn's disease. There are two conditions: Whipple's disease and Q fever, where the pathogenesis is unequivocally related to growth of the pathogenic agent (*Tropherymawhipplei* and *Coxiellaburnetii* respectively) within macrophage vesicles. The combination of hydroxychloroquine and an antibiotic, usually doxycycline, has greatly improved the chance of prolonged cure in both these conditions (Raoult et al. 1999; Rolain et al. 2007). The mechanism by which the hydroxychloroquine works is unclear. Possible mechanisms include a rise in the pH of the macrophage vesicles—since the relevant bacteria require an acid environment to proliferate, and a reduction in availability of iron within the vesicles. We are about to start an open-label trial of combination hydroxychloroquine, ciprofloxacin, and doxycycline therapy in active Crohn's disease.

Non-targeted Empiric Use of Antibiotics

Many of the features of Crohn's disease—the formation of abscesses and fistulae and the frequency of night sweats, are compatible with bacterial infection so many clinicians have used antibiotics empirically. In recent years the most commonly used antibiotics have probably been ciprofloxacin and

Table 1. Clinical Trials of Antibiotics in the Treatment of Inflammatory Bowel Disease.

Indication	N=	Endpoint	Blinding	Treatment duration	Treatment	Control	Reached endpoint (treatment)	Reached endpoint (control)	Significance	Author
Active Crohn's	22	Point score	Double	2 months	Metronidazole 250mg qds	Placebo	No difference	No difference	NS	(Blichfeldt et al. 1978)
Active Crohn's	105	CDAI <150	Double	16 weeks	Metronidazole 20mg/kg OR Metronidazole 10mg/kg	Placebo	36% 27%	25%	NS	(Sutherland et al. 1991)
Active Crohn's	40	CDAI <150	Nil	6 weeks	Ciprofloxacin 500mg bd	Pentasa 2g bd	56%	55%	NS	(Colombel et al. 1999)
Active Crohn's	47	Mean CDAI	Nil	6 months	Ciprofloxacin 500mg bd	Placebo	Mean CDAI 111	Mean CDAI 205	$P<0.001$	(Arnold et al. 2002)
Active Crohn's	72	HBI≤3	Nil	10 weeks	Ciprofloxacin 500mg bd + Metronidazole 250mg tds	Uncontrolled	68%	n/a	n/a	(Greenbloom et al. 1998)
Active Crohn's	134	CDAI <150	Double	8 weeks	Metronidazole 500mg bd + Ciprofloxacin 500mg bd + Budesonide 9mg od	Budesonide 9mg od	33%	38%	NS	(Steinhart et al. 2002)
Active Crohn's	7	Improvement	Nil	4 weeks	Metronidazole 250mg bd + Ciprofloxacin 200mg tds	Uncontrolled	n/a	n/a	n/a	(Ishikawa et al. 2003)

Indication	N=	Endpoint	Blinding	Treatment duration	Treatment	Control	Reached endpoint (treatment)	Reached endpoint (control)	Significance	Author
Active Crohn's	72	Clinical Improvement	Nil	2 weeks	Metronidazole 400mg bd OR Cotrimoxazole 480mg bd OR Metronidazole + Cotrimox	Placebo	67% 17% 71%	35%	NS	(Ambrose et al. 1985)
Active Crohn's	25	HBI≤3	Nil	4-12 weeks	Clarithromycin 250mg bd	Uncontrolled	4 week 48% 12 week 44%	n/a	n/a	(Leiper et al. 2000)
Active Crohn's	14	CDAI <150	Nil	4-24 weeks	Clarithromycin 200mg bd	Uncontrolled	4 weeks 36% 24 week 29%	n/a	n/a	(Inoue et al. 2007)
Active Crohn's	41	CDAI <150	Nil	3months	Clarithromycin 1g od	Placebo	26%	27%	NS	(Leiper et al. 2008)
Active Crohn's	83	CDAI <150	Double	12 weeks	Rifaxamin 800mg bd OR Rifaxamin 800mg od	Placebo	52% 32%	33%	NS	(Prantera et al. 2006)
Active Crohn's	402	CDAI <150	Double	12 weeks	Rifaximin ER 400mg bd OR Rifaximin ER 800mg bd OR Rifaximin ER 1200mg bd	Placebo	54% 62% 47%	43%	800mg dose only, $p<0.005$	(Prantera et al. 2012)

Table 1. contd....

Table 1. contd.

Indication	N=	Endpoint	Blinding	Treatment duration	Treatment	Control	Reached endpoint (treatment)	Reached endpoint (control)	Significance	Author
MAP in Crohn's	213	Relapse	Double	2 years or until relapse	Clarithromycin 750 mg/day + Rifabutin 450 mg/day + Clofazimine 50 mg/day	Placebo	12 months 39% 24 months 26% 36 months 59%	52% 43% 50%	NS NS NS	(Selby et al. 2007)
MAP in Crohn's	46	HBI <5	Nil	Mean of 18 months	Rifabutin 300mg/150mg + Clarithromycin 250mg bd +/- Ciprofloxacin, Ofloxacin or Clofazamine	Uncontrolled	93.5%	n/a	n/a	(Gui et al. 1997)
MAP in Crohn's	40	Treatment failure	Double	9 months	Rifampicin 600mg once + Ethambutol 15mg/kg/day + Dapsone 100mg 6 d/week + Clofazamine 100mg alt days	Placebo	13% relapse	61% relapse	P<0.05	(Prantera et al. 1994)
MAP in Crohn's	31	Persistent active disease	Double	3 months	Ethambutol 15mg/kg/day + Clarithromycin 500mg bd	Placebo	33% active	44% active	NS	(Goodgame et al. 2001)
MAP in Crohn's	130	Composite effects	Double	2 years	Rifamipicin 450mg/day Isoniazid 300mg/day Ethambutol 15mg/kg/day	Placebo	No difference	No difference	NS	(Swift et al. 1994)

MAP in Crohn's	130	Relapse	Double	5 years follow up	Rifamipicin 450mg/day Isoniazid 300mg/day Ethambutol 15mg/kg/day	Placebo	39%	43%	NS	(Thomas et al. 1998)
MAP in Crohn's	27	Mean CDAI	Double	1 year	Ethambutol 15mg/kg/day + Rifampicin 10mg/kg/day	Placebo	183	156	NS	(Shaffer et al. 1984)
MAP in Crohn's	36	Fall in CDAI	Nil	4-17 months	Rifabutin 150mg bd + Clarithromycin 250mg bd	Uncontrolled	58.3%	n/a	n/a	(Shafran et al. 2002)
MAP in Crohn's	15	Remission	Nil	3 months	Clarithromycin 500mg bd	Uncontrolled	71%	n/a	n/a	(Graham et al. 1995)
MAP in Crohn's	49	Remission	Nil	12 months	Clofazamine 100mg bd + Corticosteroids	Placebo	64%	50%	NS	(Afdhal et al. 1991)
Perianal Crohn's	52	Response	Nil	8 weeks	Ciprofloxacin 500mg bd + Metronidazole 500mg tds + Azathioprine	Uncontrolled	35%	n/a	n/a	(Dejaco et al. 2003)
Perianal Crohn's	24	50% reduced fistulae	Double	12 weeks	Ciprofloxacin 500mg bd + Infliximab	Placebo + Infliximab	73%	39%	NS	(West et al. 2004)
Perianal Crohn's	25	Remission	Nil	10 weeks	Ciprofloxacin 500mg bd OR Metronidazole 500mg bd	Placebo	30% 0%	12.5%	NS	(Thia et al. 2009)

Table 1. contd....

Table 1. contd.

Indication	N=	Endpoint	Blinding	Treatment duration	Treatment	Control	Reached endpoint (treatment)	Reached endpoint (control)	Significance	Author
Perianal Crohn's	74	Fall in PCDAI	Double	4 weeks	10% Metronidazole ointment	Placebo	Equal	Equal	NS	(Maeda et al. 2010)
Post-op Crohn's	60	Recurrence	Double	3 months	Metronidazole 20mg/ kg/day	Placebo	52%	75%	NS	(Rutgeerts et al. 1995)
Post-op Crohn's	81	Recurrence	Double	1 year	Ornidazole 1g/day	Placebo	7.9%	37.5%	P<0.005	(Rutgeerts et al. 2005)
Post-op Crohn's	81	Recurrence	Nil	3 months	Metronidazole 250mg tds + Azathioprine	Metronidazole 250mg tds	55%	78%	P<0.05	(D'Haens et al. 2008)
Chronic active UC	83	Treatment failure	Double	6 months	Ciprofloxacin 500mg bd + Prednisolone	Placebo + Prednisolone	21% failure	46% failure	P<0.05	(Turunen et al. 1998)
Chronic active UC	20	Remission	Nil	2 weeks	Amoxycillin 500mg tds Metronidazole 250mg tds Tetracycline 500mg tds	Untreated	85%	65%	p<0.05	(Ohkusa et al. 2005)
Chronic active UC	48	Steroid withdrawal	Nil	2 weeks	Amoxycillin 500mg tds Metronidazole 250mg tds Tetracycline 500mg tds	Uncontrolled	70.8%	n/a	n/a	(Terao et al. 2011)
Chronic active UC	25	Fall in Lichtiger's	Nil	2 weeks	Amoxycillin 500mg tds Metronidazole 250mg tds Tetracycline 500mg tds	Uncontrolled	73.3%	n/a	n/a	(Uehara et al. 2010)

Chronic active UC	210	Response	Double	2 weeks	Amoxycillin 500mg tds Metronidazole 250mg tds Tetracycline 500mg tds	Placebo	44.8%	22.8%	$p<0.005$	(Ohkusa et al. 2010)
UC flare	33	Response	Double	7 days	Vancomycin 500mg qds + Prednisolone 40mg od	Placebo + Prednisolone	89%	46%	NS	(Dickinson et al. 1985)
Acute severe UC	39	Response	Double	5 days	iv Metronidazole 500mg tds + Methylprednisolone	Placebo + Methylpred	74%	70%	NS	(Chapman et al. 1986)
Acute severe UC	84	Remission	Double	7 days	Tobramycin 120mg tds + Corticosteroids	Placebo + corticosteroids	74%	43%	$p<0.01$	(Burke et al. 1990)
Acute severe UC	39	Response	Nil	10 days	Metronidazole 500mg tds + Tobramycin 4mg/kg tds + iv Hydrocortisone 100mg qds	Placebo + iv Hydrocortisone	63.1%	65%	NS	(Mantzaris et al. 1994)
Acute severe UC	70	Remission	Single	2 weeks	Ciprofloxacin 250mg bd + Prednisolone	Placebo + Prednisolone	70.5%	72%	NS	(Mantzaris et al. 1997)
Acute severe UC	28	Improvement	Single	10 days	Rifaxamin 400mg bd + Methylprednisolone	Placebo + Methylpred	64.3%	41.7%	NS	(Gionchetti et al. 1999)
Acute severe UC	55	Improvement	Double	10 days	iv Ciprofloxacin 400mg bd + iv Hydrocortisone 100mg qds	Placebo + iv Hydrocortisone	79.3%	77%	NS	(Mantzaris et al. 2001)

metronidazole—either separately or in combination (Table 1). The largest placebo-controlled trial (134 patients) failed to show any benefit overall when the combination of ciprofloxacin and metronidazole was added to budesonide but there was substantial benefit in isolated colonic Crohn's disease when this group was analysed separately (Steinhart et al. 2002). Metronidazole used as a single agent has also proved effective in patients with Crohn's disease who have colonic involvement—although treatment had to be maintained for up to 16 weeks (Sutherland et al. 1991). A small controlled trial has suggested (non-significant) benefit with ciprofloxacin but not metronidazole in the treatment of peri-anal Crohn's disease (Thia et al. 2009) although topical metronidazole may reduce pain and discharge without achieving fistula closure (Maeda et al. 2010). An impressive and intriguing trial showed that immediate post-operative treatment with metronidazole maintained for three months reduced the rate of relapse in patients with Crohn's disease undergoing resective surgery from 25% in the first 12 months, with placebo, to 4% (Rutgeerts et al. 1995). Given the specificity of metronidazole for anaerobic bacteria and the sparcity of evidence for a pathogenic role of any specific anaerobes in Crohn's pathogenesis this is hard to explain but see comments later regarding possible anti-inflammatory effects of metronidazole.

In severe ulcerative colitis intravenous tetracycline was part of the "Truelove regimen" (Truelove and Jewell 1974). Its use fell out of favour because of its toxicity and it was never put formally to trial but metronidazole was subsequently shown not to improve outcome in patients treated for severe colitis with intravenous steroids (Chapman et al. 1986). It is still our practice though to include metronidazole in the treatment of all newly diagnosed cases of severe colitis because of the possibility of confusion with amoebic colitis.

Other antibiotics, including ciprofloxacin, oral tobramycin, which is not absorbed, and vancomycinhave been used to induce remission in active ulcerative colitis. Overall results are unimpressive (Khan et al. 2011) although a response was seen with tobramycin in acute treatment (Burke et al. 1990) that was not sustained with maintenance (Lobo et al. 1993). A small controlled trial of the non-absorbed antibiotic rifaximin was not significant although some benefit was seen for secondary endpoints (Gionchetti et al. 1999) with a larger trial using a modified release preparation finding benefit only at a dose of 800 mg (Prantera et al. 2012).

It needs to be remembered that traveller's diarrhoea, most commonly due to enterotoxigenic *E. coli*, can induce relapse of inflammatory bowel disease (Isgar et al. 1983) and that a single stool culture has less than 50% sensitivity for detecting a diarrhoea-causing pathogen (Matsumoto et al. 1994). Pathogens shown to induce relapse of ulcerative colitis include *Salmonella* spp. and Campylobacter. *Clostridium difficile* is more commonly present in the faeces of inflammatory bowel disease patients than controls and probably increases severity and duration of attacks although it seems not to initiate relapse (Goodhand et al. 2011).

Antibiotics May Have Anti-inflammatory Effects and Immunosuppressives May Have Antibiotic Effects!

Metronidazole has been shown to have quite potent inhibitory activity against leucocyte recruitment (Arndt et al. 1994) and this might contribute to a therapeutic effect in inflammatory bowel disease—although it would be expected that this might be of more value in ulcerative colitis where neutrophil chemotaxis seems at least normal rather than in Crohn's disease where it is already suppressed (Marks et al. 2006).

Azathioprine has antibiotic effects. A Swedish study showed that tissue taken at the time of Crohn's surgery in patients receiving azathioprine had antibiotic activity against anaerobes (Danielsson and Persson 1977) and it has also been shown to be active against mycobacteria (Shin and Collins 2008). Laboratory studies by our group have confirmed an antibiotic effect for azathioprine and shown that it, and its metabolite 6-thioguanine, have some efficacy against *E. coli* replicating within macrophages (Knight et al. 2011). Intriguingly, corticosteroids, which are notorious for increasing the risk of sepsis, had a similar effect. It seems likely that their harmful effects in sepsis relate to their inhibition of leukocyte chemotaxis (Rhodes et al. 1981) rather than to any action on leukocyte killing of phagocytosed bacteria.

Approaches that Aim to Prevent Bacterial-Epithelial Interaction

Soluble plant fibres that have a "contrabiotic" effect—blocking bacteria-epithelial interaction

In Crohn's disease there is evidence that bacteria penetrate the mucosa and can be found within the tissue of the intestine and mesenteric lymph nodes (Flanagan et al. 2011). In ulcerative colitis there is no evidence of bacterial invasion but strong circumstantial evidence that bacteria may have a pathogenic role by interacting with epithelial cells. In either case any agent that could prevent interaction between bacteria and the intestinal epithelial cells might be therapeutic.

Having shown that a high proportion of the *E. coli* isolated from the colonic mucosa of patients with inflammatory bowel disease (and colon cancer) possess haemagglutinins that correlate with their ability to adhere to and invade intestinal epithelial cells, we investigated a range of simple and complex carbohydrates for their ability to block adherence (Martin et al. 2004). We found that the monosaccharides and oligosaccharides tested were ineffective but that more complex glycans such as bovine submaxillarymucin and soluble plant fibres were strongly inhibitory. Several plant fibres were inhibitory, notably plantain and broccoli, whereas others such as apple or leek had variable or little effect. We have subsequently shown that these soluble plant fibres can also block the epithelial adherence of well established pathogens such as Salmonella, Shigella, Enterotoxigenic *E. coli* and *Clostridium difficile* (Roberts et al. 2012) and have termed this potentially beneficial effect "contrabiotic" to distinguish it from the prebiotic effects of plant components that are mediated by encouraging the growth of probiotic bacteria.

Although the Crohn's *E. coli* isolates have been shown to invade intestinal epithelial cell lines in culture the epithelial cells used in these assays are very abnormal—usually cancerous, and there is no clear evidence that the Adherent and Invasive *E. coli* invade through normal human gut epithelial cells *in vivo*—even in Crohn's disease. For other *bona fide* pathogens such as Salmonella and Mycobacteria there is good evidence that initial invasion *in vivo* has to be through the specialized microfold (M) cells that account for about 5% of the cells in the "dome" epithelium overlying Peyer'spatches in the distal small intestine and lymphoid follicles in the colon. There is also evidence that these are the sites of the initial lesions in Crohn's disease (Fujimura et al. 1996). We have therefore evaluated the ability of soluble plant fibres to inhibit bacterial translocation across cultured simulated M cell monolayers and across human Peyer's patches cultured *ex vivo* in Ussing chambers. This has confirmed the ability of plant fibres such as plantain and broccoli to inhibit bacterial translocation including that of the adherent and invasive *E. coli* (Roberts et al. 2010). A clinical trial has been running to assess the efficacy of dietary supplementation with soluble plantain fibre in maintaining remission in Crohn's disease but this is sadly currently halted for commercial reasons.

Enhancing the mucosal barrier

In inflammatory bowel diseases, both Crohn's and ulcerative colitis, there is increasing evidence to suggest that impairment of the mucosal barrier is a factor in pathogenesis. In ulcerative colitis an animal model commonly used to mimic the disease is oral administration to rodents of dextran sulphate. This weakens the mucus barrier and results in direct apposition between bacteria and the epithelium, whereas in health the bacteria are kept away from the colonic epithelium by a continuous mucus barrier (Johansson et al. 2010). Treatments that aim to restore this barrier might be expected to have efficacy. An intriguing treatment based on a delayed release oral formulation of the natural cell wall component,phosphatidyl choline (lecithin), is showing considerable promise in early clinical trials in ulcerative colitis (Stremmel et al. 2010).

Ulcerative colitis is characterized, when active, by mucus depletion. This is probably secondary to the disease process but it nevertheless must weaken the mucosal defence against bacterial attack or interaction. Whereas in Crohn's disease there is considerable evidence of tissue invasion by bacteria, probably as discussed by initial ingress through M cells overlying Peyer's patches or lymphoid follicles, in ulcerative colitis there is very little evidence for bacterial invasion. Instead, evidence from gene association studies suggests that subtle defects in the mucosal barrier may be very important. There is a potential to see dramatic changes in the interactions between non-pathogenic bacteria and the epithelium if the barrier is disturbed. Bacterial flagellin, for example, has no significant effect when infused rectally into healthy mice but once the mucosal barrier has been damaged by oral administration of dextran sulphate, subsequent rectal administration of flagellin causes a very severe, usually fatal colitis. The likely mechanism is access by flagellin to its epithelial receptor TLR5 that is expressed on the baso-lateral but not on the apical (luminal) aspect of intestinal epithelial cells (Rhee et al. 2005) although this has been disputed (Ivison et al. 2010). Access by luminal bacteria to the epithelial cell, including its baso-lateral membrane, will in the healthy colon be blocked not only by the integrity of the epithelium, including its tight junctions, but also by the overlying mucus layer, which in the healthy colon, but not the small intestine, forms a continuous protective layer that is largely impervious to bacteria (Johansson et al. 2011).

The mucus depletion within the tissue of the active colitic colon is driven by the inflammatory process—probably mainly by interaction between neutrophils and goblet cells with neutrophil elastase acting as a potent secretagogue and mucus synthesis failing to keep up with the result that the goblets become depleted of mucus (Leiper et al. 2001). Once secreted, the mucus is subjected to intense attack by bacterial enzymes. Colonic bacteria feed on the secreted mucus using it asan energy source and some bacteria, notably *Bacteroides thetaiotamicron*, secrete an array of enzymes that are able to degrade the mucus (Tsai et al. 1991). It is its sulphation that most clearly distinguishes colonic mucus from that in the smallintestine. Ester-sulphate is typically present attached to terminal carbohydrates in the mucus oligosaccharides and has to be removed before bacterial glycosidases can degrade the remainder of the side chains. We have previously shown that relapse of ulcerative colitis is associated with increased faecal concentrationsof mucus-degrading sulphatases and proteases (Dwarakanath et al. 1995; Tsai et al. 1995; Tsai et al. 1992). These represent therapeutic targets that deserve further study. They can be inhibited by bismuth salts and this may explain the therapeutic effects of bismuth enemas (Ryder et al. 1990) and perhaps, since bismuth and arsenic are closely related elements, also that of the long used arsenical suppositories that are still sometimes helpful in resistant proctitis.

It seems feasible that defects in the mucosal barrier might also be inducible by ingested emulsifiers. Dietary emulsifiers are detergents that are commonly present in processed foods. We have shown that they can enhance bacterial translocation both *in vitro* in cell monolayers and also in *ex vivo* cultured explants of human terminal ileum (Roberts et al. 2010). Bacterial translocation in these models occurs through non-M cell epithelial cells and is transcellular, although in the presence of emulsifiers some translocation is seen through non-M cell epithelial cells. Emulsifiers seem plausible environmental factors that might promote Crohn's disease so we currently advise our patients to avoid processed food but further *in vivo* studies are clearly needed.

Approaches that Aim to Restore the Healthy Microbiota

Specific organisms—*F. prauznitzii*

A powerful association has been shown between the presence of a probiotic bacterium, *Faecalibacterium prauznitzii*, in the ileum of patients with Crohn's disease and lack of subsequent relapse after respective surgery (Sokol et al. 2008). Indeed, Crohn's patients are less likely than healthy controls to have this bacterium present in their faeces (Sokol et al. 2009). The bacterium secretes a substance, not yet identified in the literature, which has a potent anti-inflammatory effect in animal models of inflammatory bowel disease. It is to be hoped that a therapy based on this might be developed before too long.

Faecal transplantation

The most reproducible finding in respect of the faecal microbiota in inflammatory bowel disease has been a reduction in diversity (Manichanh et al. 2006; Andoh et al. 2007). In animal models of inflammatory bowel disease, whether based on immune mechanisms or on the introduction of specific pathogens, similar reduction in diversity occurs implying that this is likely to be a secondary phenomenon related to the underlying inflammation (Lupp et al. 2007). Similarly, in human inflammatory bowel diseases, dysbiosis is shown to be more marked at sites of inflammation (Walker et al. 2011). It is unclear whether this reduction in diversity is itself harmful but it is a not unreasonable assumption that it might be. Any intervention that could restore the normal diversity quickly might therefore be therapeutic. Faecal transplantation ie infusion of faeces from a healthy subject, is increasingly being explored. Faecal suspensions have been introduced either per-anum by enema or colonoscope and also orally, e.g., by infusion at upper GI endoscopy. Reports are so far largely small and anecdotal (Landy et al. 2011) but include isolated reports of dramatic success in ulcerative colitis (Bennet and Brinkman 1989; Borody et al. 2003).

Future Strategies

It could be argued that we need a better understanding of the changes in gut microbiota before we will be able to identify a novel therapeutic approach that targets the microbiota. Alternatively, it can also be argued that we will never know which of the microbiota changes seen in inflammatory bowel disease are important until we know the results of therapeutic trials that target them. Possible strategies might target any of the layers of interaction between the gut and the microbiota, as outlined in Table 2. If any of these approaches prove successful we could have exciting new lines of treatment for inflammatory bowel disease that would likely be relatively cheap and safe. We would also then have gained a much clearer understanding of the pathogenesis of this complex condition.

Table 2. Strategies for Targeting Microbiota-related Therapies in Inflammatory Bowel Disease.

Potential Therapeutic Targets	Therapeutic Approaches
Luminal microbiota	Probiotics Faecal transplantation Non-absorbed antibiotics
Mucus (continuous in colon, discontinuous in ileum)	Inhibition of mucus-degrading enzymes Synthetic defensins
Epithelium	Stregnthening of mucosal barrier—phosphatidyl choline therapy, avoidance of ingestion of dietary emulsifiers Contrabiotics—soluble plant fibres that block bacteria-epithelial interaction
Local intestinal and lymphoid tissue	Antibiotics—particularly those that work within macrophages—possibly with hydroxychloroquine as adjunct
Immune system	Improved innate immunity—possibly with vitamin D supplementation Future—possible vaccination against specific pathogens?

References Cited

Afdhal, N.H., A. Long, J. Lennon, J. Crowe and D.P. O'donoghue. 1991. Controlled trial of antimycobacterial therapy in crohn's disease. Clofazimine versus placebo. Dig. Dis. Sci. 36: 449–53.

Ambrose, N.S., R.N. Allan, M.R. Keighley, D.W. Burdon, D. Youngs, P. Barnes and J.E. Lennard-Jones. 1985. Antibiotic therapy for treatment in relapse of intestinal crohn's disease. A prospective randomized study. Dis. Colon Rectum. 28: 81–5.

Andoh, A., S. Sakata, Y. Koizumi, K. Mitsuyama, Y. Fujiyama and Y. Benno. 2007. Terminal restriction fragment length polymorphism analysis of the diversity of fecal microbiota in patients with ulcerative colitis. Inflamm. Bowel Dis. 13: 955–62.

Arndt, H., K.D. Palitzsch, M.B. Grisham and D.N. Granger. 1994. Metronidazole inhibits leukocyte-endothelial cell adhesion in rat mesenteric venules. Gastroenterology. 106: 1271–6.

Arnold, G.L., M.R. Beaves, V.O. Pryjdun and W.J. Mook. 2002. Preliminary study of ciprofloxacin in active crohn's disease. Inflamm. Bowel Dis. 8: 10–5.

Bennet, J.D. and M. Brinkman. 1989. Treatment of ulcerative colitis by implantation of normal colonic flora. Lancet. 1: 164.

Blichfeldt, P., J.P. Blomhoff, E. Myhre and E. Gjone. 1978. Metronidazole in crohn's disease. A double blind cross-over clinical trial. Scand. J. Gastroenterol. 13: 123–7.

Borgaonkar, M.R., D.G. Macintosh and J.M. Fardy. 2000. A meta-analysis of antimycobacterial therapy for crohn's disease. Am. J. Gastroenterol. 95: 725–9.

Borody, T.J., S. Leis, E.F. Warren and R. Surace. 2002. Treatment of severe crohn's disease using antimycobacterial triple therapy—approaching a cure? Dig. Liver Dis. 34: 29–38.

Borody, T.J., E.F. Warren, S. Leis, R. Surace and O. Ashman. 2003. Treatment of ulcerative colitis using fecal bacteriotherapy. J. Clin. Gastroenterol. 37: 42–7.

Burke, D.A., A.T. Axon, S.A. Clayden, M.F. Dixon, D. Johnston and R.W. Lacey. 1990. The efficacy of tobramycin in the treatment of ulcerative colitis. Aliment Pharmacol. Ther. 4: 123–9.

Chapman, R.W., W.S. Selby and D.P. Jewell. 1986. Controlled trial of intravenous metronidazole as an adjunct to corticosteroids in severe ulcerative colitis. Gut. 27: 1210–2.

Chiodini, R.J., W.M. Chamberlin, J. Sarosiek and R.W. Mccallum. 2012. Crohn's disease and the mycobacterioses: A quarter century later. Causation or simple association? Crit. Rev. Microbiol. 38: 52–93.

Colombel, J.F., M. Lemann, M. Cassagnou, Y. Bouhnik, B. Duclos, J.L. Dupas, B. Notteghem and J.Y. Mary. 1999. A controlled trial comparing ciprofloxacin with mesalazine for the treatment of active crohn's disease. Groupe d'etudes therapeutiques des affections inflammatoires digestives (getaid). Am. J. Gastroenterol. 94: 674–8.

D'haens, G.R., S. Vermeire, G. Van Assche, M. Noman, I. Aerden, G. Van Olmen and P. Rutgeerts. 2008. Therapy of metronidazole with azathioprine to prevent postoperative recurrence of crohn's disease: A controlled randomized trial. Gastroenterology. 135: 1123–9.

Dalziel, T.K. 1913. Chronic interstitial enteritis. British Medical Journal. 2: 1068–70.

Danielsson, D. and S. Persson. 1977. Antibacterial activity of azathioprine (imuran) against anaerobic intestinal bacteria: Implications in crohn's disease. Current chemotherapy proceedings of the 10th international congress of chemotherapy.

Dejaco, C., M. Harrer, T. Waldhoer, W. Miehsler, H. Vogelsang and W. Reinisch. 2003. Antibiotics and azathioprine for the treatment of perianal fistulas in crohn's disease. Aliment Pharmacol. Ther. 18: 1113–20.

Dickinson, R.J., H.J. O'connor, I. Pinder, I. Hamilton, D. Johnston and A.T. Axon. 1985. Double blind controlled trial of oral vancomycin as adjunctive treatment in acute exacerbations of idiopathic colitis. Gut. 26: 1380–4.

Dogan, B., E. Scherl, B. Bosworth, R. Yantiss, C. Altier, P.L. Mcdonough, Z.D. Jiang, H.L. Dupont, P. Garneau, J. Harel, M. Rishniw and K.W. Simpson. 2012. Multidrug resistance is common in *Escherichia coli* associated with ileal crohn's disease. Inflamm. Bowel Dis.

Dwarakanath, A.D., B.J. Campbell, H.H. Tsai, D. Sunderland, C.A. Hart and J.M. Rhodes. 1995. Faecal mucinase activity assessed in inflammatory bowel disease using 14c threonine labelled mucin substrate. Gut. 37: 58–62.

Feller, M., K. Huwiler, A. Schoepfer, A. Shang, H. Furrer and M. Egger. 2010. Long-term antibiotic treatment for crohn's disease: Systematic review and meta-analysis of placebo-controlled trials. Clin. Infect. Dis. 50: 473–80.

Flanagan, P., B.J. Campbell and J.M. Rhodes. 2011. Bacteria in the pathogenesis of inflammatory bowel disease. Biochem. Soc. Trans. 39: 1067–72.

Fujimura, Y., R. Kamoi and M. Iida. 1996. Pathogenesis of aphthoid ulcers in crohn's disease: Correlative findings by magnifying colonoscopy, electron microscopy, and immunohistochemistry. Gut. 38: 724–32.

Gionchetti, P., F. Rizzello, A. Ferrieri, A. Venturi, C. Brignola, M. Ferretti, S. Peruzzo, M. Miglioli and M. Campieri. 1999. Rifaximin in patients with moderate or severe ulcerative colitis refractory to steroid-treatment: A double-blind, placebo-controlled trial. Dig. Dis. Sci. 44: 1220–1.

Glasser, A.L., J. Boudeau, N. Barnich, M.H. Perruchot, J.F. Colombel and A. Darfeuille-Michaud. 2001. Adherent invasive *Escherichia coli* strains from patients with crohn's disease survive and replicate within macrophages without inducing host cell death. Infect. Immun. 69: 5529–37.

Goodgame, R.W., K. Kimball, S. Akram, E. Ike, C.N. Ou, F. Sutton and D. Graham. 2001. Randomized controlled trial of clarithromycin and ethambutol in the treatment of crohn's disease. Aliment Pharmacol. Ther. 15: 1861–6.

Goodhand, J.R., W. Alazawi and D.S. Rampton. 2011. Systematic review: Clostridium difficile and inflammatory bowel disease. Aliment Pharmacol. Ther. 33: 428–41.

Graham, D.Y., M.T. Ai-Assi and M. Robinson. 1995. Prolonged remission in crohn's disease following therapy for mycobacterium paratuberculosis infection. Gastroenterology. 108: A826.

Greenbloom, S.L., A.H. Steinhart and G.R. Greenberg. 1998. Combination ciprofloxacin and metronidazole for active crohn's disease. Can. J. Gastroenterol. 12: 53–6.

Gui, G.P., P.R. Thomas, M.L. Tizard, J. Lake, J.D. Sanderson and J. Hermon-Taylor. 1997. Two-year-outcomes analysis of crohn's disease treated with rifabutin and macrolide antibiotics. J. Antimicrob. Chemother. 39: 393–400.

Inoue, S., H. Nakase, M. Matsuura, S. Ueno, N. Uza, H. Kitamura, S. Mikami, H. Tamaki, K. Kasahara and T. Chiba. 2007. Open label trial of clarithromycin therapy in japanese patients with crohn's disease. J. Gastroenterol. Hepatol. 22: 984–988.

Isgar, B., M. Harman and P.J. Whorwell. 1983. Factors preceding relapse of ulcerative colitis. Digestion. 26: 236–8.

Ishikawa, T., S. Okamura, H. Oshimoto, R. Kobayashi and M. Mori. 2003. Metronidazole plus ciprofloxacin therapy for active crohn's disease. Intern. Med. 42: 318–21.

Ivison, S.M., M.E. Himmel, G. Hardenberg, P.A. Wark, A. Kifayet, M.K. Levings and T.S. Steiner. 2010. Tlr5 is not required for flagellin-mediated exacerbation of dss colitis. Inflamm. Bowel Dis. 16: 401–9.

Johansson, M.E., D. Ambort, T. Pelaseyed, A. Schutte, J.K. Gustafsson, A. Ermund, D.B. Subramani, J.M. Holmen-Larsson, K.A. Thomsson, J.H. Bergstrom, S. Van Der Post, A. Rodriguez-Pineiro, H. Sjovall, M. Backstrom and G.C. Hansson. 2011. Composition and functional role of the mucus layers in the intestine. Cell Mol. Life Sci. 68: 3635–41.

Johansson, M.E., J.K. Gustafsson, K.E. Sjoberg, J. Petersson, L. Holm, H. Sjovall and G.C. Hansson. 2010. Bacteria penetrate the inner mucus layer before inflammation in the dextran sulfate colitis model. PLoS One. 5: e12238.

Khan, K.J., T.A. Ullman, A.C. Ford, M.T. Abreu, A. Abadir, J.K. Marshall, N.J. Talley and P. Moayyedi. 2011. Antibiotic therapy in inflammatory bowel disease: A systematic review and meta-analysis. Am. J. Gastroenterol. 106: 661–73.

Knight, P., B.J. Campbell and J.M. Rhodes. 2011. Azathioprine and 6-thioguanine but not 6-mercapopurine inhibit intra-macrophage replication of crohn's disease *Escherichia coli*. Gut. 60: A134.

Kuenstner, J.T. 2007. The australian antibiotic trial in crohn's disease: Alternative conclusions from the same study. Gastroenterology. 133: 1742–3; author reply 1745–6.

Landy, J., H.O. Al-Hassi, S.D. Mclaughlin, A.W. Walker, P.J. Ciclitira, R.J. Nicholls, S.K. Clark and A.L. Hart. 2011. Review article: Faecal transplantation therapy for gastrointestinal disease. Aliment Pharmacol. Ther. 34: 409–15.

Leiper, K., B.J. Campbell, M.D. Jenkinson, J. Milton, L.G. Yu, J. Democratis and J. Rhodes. 2001. Interaction between bacterial peptides, neutrophils and goblet cells: A possible mechanism for neutrophil recruitment and goblet cell depletion in colitis. Clin. Sci. (Lond.). 101: 395–402.

Leiper, K., K. Martin, A. Ellis, A.J. Watson, A.I. Morris and J.M. Rhodes. 2008. Clinical trial: Randomized study of clarithromycin versus placebo in active crohn's disease. Aliment Pharmacol. Ther. 27: 1233–9.

Leiper, K., A.I. Morris and J.M. Rhodes. 2000. Open label trial of oral clarithromycin in active crohn's disease. Aliment Pharmacol. Ther. 14: 801–6.

Lipton, J.E. and D.P. Barash. 2007. Flawed australian cd study does not end map controversy. Gastroenterology. 133: 1742; author reply 1745–6.

Liu, Y., H.J. Van Kruiningen, A.B. West, R.W. Cartun, A. Cortot and J.F. Colombel. 1995. Immunocytochemical evidence of listeria, escherichia coli, and streptococcus antigens in crohn's disease. Gastroenterology. 108: 1396–404.

Lobo, A.J., D.A. Burke, G.M. Sobala and A.T. Axon. 1993. Oral tobramycin in ulcerative colitis: Effect on maintenance of remission. Aliment Pharmacol. Ther. 7: 155–8.

Lupp, C., M.L. Robertson, M.E. Wickham, I. Sekirov, O.L. Champion, E.C. Gaynor and B.B. Finlay. 2007. Host-mediated inflammation disrupts the intestinal microbiota and promotes the overgrowth of enterobacteriaceae. Cell Host Microbe. 2: 204.

Maeda, Y., S.C. Ng, P. Durdey, C. Burt, J. Torkington, P.K. Rao, J. Mayberry, T. Moshkovska, C.D. Stone, E. Carapeti and C.J. Vaizey. 2010. Randomized clinical trial of metronidazole ointment versus placebo in perianal crohn's disease. Br. J. Surg. 97: 1340–7.

Manichanh, C., L. Rigottier-Gois, E. Bonnaud, K. Gloux, E. Pelletier, L. Frangeul, R. Nalin, C. Jarrin, P. Chardon, P. Marteau, J. Roca and J. Dore. 2006. Reduced diversity of faecal microbiota in crohn's disease revealed by a metagenomic approach. Gut. 55: 205–11.

Mantzaris, G.J., E. Archavlis, P. Christoforidis, D. Kourtessas, P. Amberiadis, N. Florakis, K. Petraki, C. Spiliadi and G. Triantafyllou. 1997. A prospective randomized controlled trial of oral ciprofloxacin in acute ulcerative colitis. Am. J. Gastroenterol. 92: 454–6.

Mantzaris, G.J., A. Hatzis, P. Kontogiannis and G. Triadaphyllou. 1994. Intravenous tobramycin and metronidazole as an adjunct to corticosteroids in acute, severe ulcerative colitis. Am. J. Gastroenterol. 89: 43–6.

Mantzaris, G.J., K. Petraki, E. Archavlis, P. Amberiadis, D. Kourtessas, A. Christidou and G. Triantafyllou. 2001. A prospective randomized controlled trial of intravenous ciprofloxacin as an adjunct to corticosteroids in acute, severe ulcerative colitis. Scand. J. Gastroenterol. 36: 971–4.

Marks, D.J., M.W. Harbord, R. Macallister, F.Z. Rahman, J. Young, B. Al-Lazikani, W. Lees, M. Novelli, S. Bloom and A.W. Segal. 2006. Defective acute inflammation in crohn's disease: A clinical investigation. Lancet. 367: 668–78.

Martin, H.M., B.J. Campbell, C.A. Hart, C. Mpofu, M. Nayar, R. Singh, H. Englyst, H.F. Williams and J.M. Rhodes. 2004. Enhanced *Escherichia coli* adherence and invasion in crohn's disease and colon cancer. Gastroenterology. 127: 80–93.

Matsumoto, T., M. Iida, Y. Kimura and M. Fujishima. 1994. Culture of colonoscopically obtained biopsy specimens in acute infectious colitis. Gastrointest. Endosc. 40: 184–7.

Meconi, S., A. Vercellone, F. Levillain, B. Payre, T. Al Saati, F. Capilla, P. Desreumaux, A. Darfeuille-Michaud and F. Altare. 2007. Adherent-invasive escherichia coli isolated from crohn's disease patients induce granulomas *in vitro*. Cell Microbiol. 9: 1252–61.

Mpofu, C.M., B.J. Campbell, S. Subramanian, S. Marshall-Clarke, C.A. Hart, A. Cross, C.L. Roberts, A. Mcgoldrick, S.W. Edwards and J.M. Rhodes. 2007. Microbial mannan inhibits bacterial killing by macrophages: A possible pathogenic mechanism for crohn's disease. Gastroenterology. 133: 1487–98.

Ohkusa, T., K. Kato, S. Terao, T. Chiba, K. Mabe, K. Murakami, Y. Mizokami, T. Sugiyama, A. Yanaka, Y. Takeuchi, S. Yamato, T. Yokoyama, I. Okayasu, S. Watanabe, H. Tajiri and N. Sato. 2010. Newly developed antibiotic combination therapy for ulcerative colitis: A double-blind placebo-controlled multicenter trial. Am. J. Gastroenterol. 105: 1820–9.

Ohkusa, T., T. Nomura, T. Terai, H. Miwa, O. Kobayashi, M. Hojo, Y. Takei, T. Ogihara, S. Hirai, I. Okayasu and N. Sato. 2005. Effectiveness of antibiotic combination therapy in patients with active ulcerative colitis: A randomized, controlled pilot trial with long-term follow-up. Scand. J. Gastroenterol. 40: 1334–42.

Prantera, C., A. Kohn, R. Mangiarotti, A. Andreoli and C. Luzi. 1994. Antimycobacterial therapy in crohn's disease: Results of a controlled, double-blind trial with a multiple antibiotic regimen. Am. J. Gastroenterol. 89: 513–8.

Prantera, C., H. Lochs, M. Campieri, M.L. Scribano, G.C. Sturniolo, F. Castiglione and M. Cottone. 2006. Antibiotic treatment of crohn's disease: Results of a multicentre, double blind, randomized, placebo-controlled trial with rifaximin. Aliment Pharmacol. Ther. 23: 1117–25.

Prantera, C., H. Lochs, M. Grimaldi, S. Danese, M.L. Scribano and P. Gionchetti. 2012. Rifaximin-extended intestinal release induces remission in patients with moderately active crohn's disease. Gastroenterology. 142: 473–481 e4.

Raoult, D., P. Houpikian, H. Tissot Dupont, J.M. Riss, J. Arditi-Djiane and P. Brouqui. 1999. Treatment of q fever endocarditis: Comparison of 2 regimens containing doxycycline and ofloxacin or hydroxychloroquine. Arch. Intern. Med. 159: 167–73.

Rhee, S.H., E. Im, M. Riegler, E. Kokkotou, M. O'brien and C. Pothoulakis. 2005. Pathophysiological role of toll-like receptor 5 engagement by bacterial flagellin in colonic inflammation. Proc. Natl. Acad. Sci. USA 102: 13610–5.

Rhodes, J.M., T.C. Bartholomew and D.P. Jewell. 1981. Inhibition of leucocyte motility by drugs used in ulcerative colitis. Gut. 22: 642–7.

Roberts, C.L., A.V. Keita, S.H. Duncan, N. O'kennedy, J.D. Soderholm, J.M. Rhodes and B.J. Campbell. 2010. Translocation of crohn's disease escherichia coli across m-cells: Contrasting effects of soluble plant fibres and emulsifiers. Gut. 59: 1331–9.

Roberts, C.L., A.V. Keita, B.N. Parsons, M. Prorok-Hamon, P. Knight, C. Winstanley, O.K. N, J.D. Soderholm, J.M. Rhodes and B.J. Campbell. 2012. Soluble plantain fibre blocks adhesion and m-cell translocation of intestinal pathogens. J. Nutr. Biochem.

Rolain, J.M., P. Colson and D. Raoult. 2007. Recycling of chloroquine and its hydroxyl analogue to face bacterial, fungal and viral infections in the 21st century. Int. J. Antimicrob. Agents 30: 297–308.

Rutgeerts, P., M. Hiele, K. Geboes, M. Peeters, F. Penninckx, R. Aerts and R. Kerremans. 1995. Controlled trial of metronidazole treatment for prevention of crohn's recurrence after ileal resection. Gastroenterology. 108: 1617–21.

Rutgeerts, P., G. Van Assche, S. Vermeire, G. D'haens, F. Baert, M. Noman, I. Aerden, G. De Hertogh, K. Geboes, M. Hiele, A. D'hoore and F. Penninckx. 2005. Ornidazole for prophylaxis of postoperative crohn's disease recurrence: A randomized, double-blind, placebo-controlled trial. Gastroenterology. 128: 856–61.

Ryan, P., M.W. Bennett, S. Aarons, G. Lee, J.K. Collins, G.C. O'sullivan, J. O'connell and F. Shanahan. 2002. Pcr detection of mycobacterium paratuberculosis in crohn's disease granulomas isolated by laser capture microdissection. Gut. 51: 665–70.

Ryan, P., R.G. Kelly, G. Lee, J.K. Collins, G.C. O'sullivan, J. O'connell and F. Shanahan. 2004. Bacterial DNA within granulomas of patients with crohn's disease—detection by laser capture microdissection and pcr. Am. J. Gastroenterol. 99: 1539–43.

Ryder, S.D., R.J. Walker, H. Jones and J.M. Rhodes. 1990. Rectal bismuth subsalicylate as therapy for ulcerative colitis. Aliment Pharmacol. Ther. 4: 333–8.

Selby, W., P. Pavli, B. Crotty, T. Florin, G. Radford-Smith, P. Gibson, B. Mitchell, W. Connell, R. Read, M. Merrett, H. Ee and D. Hetzel. 2007. Two-year combination antibiotic therapy with clarithromycin, rifabutin, and clofazimine for crohn's disease. Gastroenterology. 132: 2313–9.

Shaffer, J.L., S. Hughes, B.D. Linaker, R.D. Baker and L.A. Turnberg. 1984. Controlled trial of rifampicin and ethambutol in crohn's disease. Gut. 25: 203–5.

Shafran, I., L. Kugler, F.A. El-Zaatari, S.A. Naser and J. Sandoval. 2002. Open clinical trial of rifabutin and clarithromycin therapy in crohn's disease. Dig. Liver Di., 34: 22–8.

Shin, S.J. and M.T. Collins. 2008. Thiopurine drugs azathioprine and 6-mercaptopurine inhibit mycobacterium paratuberculosis growth *in vitro*. Antimicrob. Agents Chemother. 52: 418–26.

Sokol, H., B. Pigneur, L. Watterlot, O. Lakhdari, L.G. Bermudez-Humaran, J.J. Gratadoux, S. Blugeon, C. Bridonneau, J.P. Furet, G. Corthier, C. Grangette, N. Vasquez, P. Pochart, G. Trugnan, G. Thomas, H.M. Blottiere, J. Dore, P. Marteau, P. Seksik and P. Langella. 2008. Faecalibacterium prausnitzii is an anti-inflammatory commensal bacterium identified by gut microbiota analysis of crohn disease patients. Proc. Natl. Acad. Sci. USA 105: 16731–6.

Sokol, H., P. Seksik, J.P. Furet, O. Firmesse, I. Nion-Larmurier, L. Beaugerie, J. Cosnes, G. Corthier, P. Marteau and J. Dore. 2009. Low counts of faecalibacterium prausnitzii in colitis microbiota. Inflamm. Bowel Dis. 15: 1183–9.

Steinhart, A.H., B.G. Feagan, C.J. Wong, M. Vandervoort, S. Mikolainis, K. Croitoru, E. Seidman, D.J. Leddin, A. Bitton, E. Drouin, A. Cohen and G.R. Greenberg. 2002. Combined budesonide and antibiotic therapy for active crohn's disease: A randomized controlled trial. Gastroenterology. 123: 33–40.

Stremmel, W., A. Hanemann, A. Braun, S. Stoffels, M. Karner, S. Fazeli and R. Ehehalt. 2010. Delayed release phosphatidylcholine as new therapeutic drug for ulcerative colitis—a review of three clinical trials. Expert Opin. Investig. Drugs 19: 1623–30.

Subramanian, S., C.L. Roberts, C.A. Hart, H.M. Martin, S.W. Edwards, J.M. Rhodes and B.J. Campbell. 2008. Replication of colonic crohn's disease mucosal *Escherichia coli* isolates within macrophages and their susceptibility to antibiotics. Antimicrob. Agents Chemother. 52: 427–34.

Sutherland, L., J. Singleton, J. Sessions, S. Hanauer, E. Krawitt, G. Rankin, R. Summers, H. Mekhjian, N. Greenberger, M. Kelly et al. 1991. Double blind, placebo controlled trial of metronidazole in crohn's disease. Gut. 32: 1071–5.

Swift, G.L., E.D. Srivastava, R. Stone, R.D. Pullan, R.G. Newcombe, J. Rhodes, S. Wilkinson, P. Rhodes, G. Roberts, B.W. Lawrie et al. 1994. Controlled trial of anti-tuberculous chemotherapy for two years in crohn's disease. Gut. 35: 363–8.

Terao, S., K. Yamashiro, I. Tamura, T. Hirano, T. Ohkusa and K. Kato. 2011. Antibiotic combination therapy for steroid withdrawal in steroid-dependent ulcerative colitis. Digestion. 83: 198–203.

Thia, K.T., U. Mahadevan, B.G. Feagan, C. Wong, A. Cockeram, A. Bitton, C.N. Bernstein and W.J. Sandborn. 2009. Ciprofloxacin or metronidazole for the treatment of perianal fistulas in patients with crohn's disease: A randomized, double-blind, placebo-controlled pilot study. Inflamm. Bowel Dis. 15: 17–24.

Thomas, G.A., G.L. Swift, J.T. Green, R.G. Newcombe, C. Braniff-Mathews, J. Rhodes, S. Wilkinson, G. Strohmeyer and G. Kreuzpainter. 1998. Controlled trial of antituberculous chemotherapy in crohn's disease: A five year follow up study. Gut. 42: 497–500.

Truelove, S.C. and D.P. Jewell. 1974. Intensive intravenous regimen for severe attacks of ulcerative colitis. Lancet. 1: 1067–70.

Tsai, H.H., A.D. Dwarakanath, C.A. Hart, J.D. Milton and J.M. Rhodes. 1995. Increased faecal mucin sulphatase activity in ulcerative colitis: A potential target for treatment. Gut. 36: 570–6.

Tsai, H.H., C.A. Hart and J.M. Rhodes. 1991. Production of mucin degrading sulphatase and glycosidases by bacteroides thetaiotaomicron. Letters in Applied Microbiology. 13: 97–101.

Tsai, H.H., D. Sunderland, G.R. Gibson, C.A. Hart and J.M. Rhodes. 1992. A novel mucin sulphatase from human faeces: Its identification, purification and characterization. Clin. Sci. (Lond.). 82: 447–54.

Turunen, U.M., M.A. Farkkila, K. Hakala, K. Seppala, A. Sivonen, M. Ogren, M. Vuoristo, V.V. Valtonen and T.A. Miettinen. 1998. Long-term treatment of ulcerative colitis with ciprofloxacin: A prospective, double-blind, placebo-controlled study. Gastroenterology. 115: 1072–8.

Uehara, T., K. Kato, T. Ohkusa, M. Sugitani, Y. Ishii, N. Nemoto and M. Moriyama. 2010. Efficacy of antibiotic combination therapy in patients with active ulcerative colitis, including refractory or steroid-dependent cases. J. Gastroenterol. Hepatol. 25 Suppl 1: S62–6.

Walker, A.W., J.D. Sanderson, C. Churcher, G.C. Parkes, B.N. Hudspith, N. Rayment, J. Brostoff, J. Parkhill, G. Dougan and L. Petrovska. 2011. High-throughput clone library analysis of the mucosa-associated microbiota reveals dysbiosis and differences between inflamed and non-inflamed regions of the intestine in inflammatory bowel disease. BMC Microbiol. 11: 7.

West, R.L., C.J. Van Der Woude, B.E. Hansen, R.J. Felt-Bersma, A.J. Van Tilburg, J.A. Drapers and E.J. Kuipers. 2004. Clinical and endosonographic effect of ciprofloxacin on the treatment of perianal fistulae in crohn's disease with infliximab: A double-blind placebo-controlled study. Aliment Pharmacol. Ther. 20: 1329–36.

11

Probiotics in Inflammatory Bowel Disease: Modulation of the Pathological Immune or Inflammatory Activation

Their Potential Benefits in the Different Phases of the Disease

Siew C. Ng[1,*] and *Ailsa L. Hart*[2]

Introduction

Clinical trials of probiotics in inflammatory bowel disease (IBD) provide the proof of concept that modulating the gut microbiota can impact on intestinal inflammation to variable degrees. There are few clinical trials that demonstrate major clinical benefits of probiotics in the setting of active intestinal inflammation. Benefits are however demonstrated in the setting of maintaining remission, once active inflammation has settled with other treatment modalities, or preventing onset of inflammation in the case of patients with ileo-anal pouches. New mechanisms by which probiotics exert

[1] Institute of Digestive Disease, Department of Medicine and Therapeutics, Prince of Wales Hospital, NganShing Street, Chinese University of Hong Kong, Hong Kong.
[2] Department of Gastroenterology, St. Mark's Hospital, Northwick Park, Watford Road, Harrow Middlesex, HA1 3UJ, London, United Kingdom.
* Corresponding author

their beneficial effects have been identified and differences exist between different probiotic bacterial species and strains. Mechanisms contributing to altered immune function and downregulation of inflammation induced by probiotic bacteria include modulation of the microbiota itself, improved barrier function with consequent reduction in immune exposure to microbiota and direct effects of bacteria on different epithelial and immune cell types.

The challenge is to marry the complex underlying disease pathogenesis with the likely multifactorial mechanisms of action of the probiotic on the host. Questions remain of which organism(s) to use and should organisms be derived from commensal bacterial concentrations (e.g., *F. prausnitzii*) or fermented foods; what dose should be used; for how long should probiotics be used; what about use of concomitant treatment, for example with antibiotics as were used in the pouchitis trials to open up an ecological niche; what about use in particular clinical or indeed genetic subsets; and what about synergistic versus antagonistic activities of strains?

In this chapter, clinical trial data are reviewed for ulcerative colitis (UC), pouchitis and Crohn's disease (CD) and the trials are divided according to whether they assess induction of remission, maintenance of remission or prevention of onset of inflammation. Mechanistic effects of probiotics are discussed with an emphasis on those organisms that have been used to treat human IBD in controlled clinical trials.

Rationale for the Use of Probiotic Bacteria in IBD

There is strong evidence for the involvement of microbes in the development of IBD. Experiments in murine models and clinical observations in IBD suggest that luminal contents provide a constant antigenic stimulus for intestinal inflammation (Sartor 1997; Sellon et al. 1998). The rationale for using probiotics in IBD stems from studies showing dysbiosis in UC, CD and pouchitis detected using conventional anaerobic culture and culture-independent techniques (Swidsinski et al. 2002; Manichanh et al. 2006; Ley et al. 2008; Joossens et al. 2011). The composition and number of commensal bacteria are altered in IBD with increased bacteroides, adherent or invasive *Escherichia coli*, and enterococci, and reduced bifidobacterium and lactobacillus species (Neut et al. 2002; Swidsinski et al. 2002). Recently, lower counts of *Faecalibacterium prausnitzii* have been reported in patients with CD (Sokol et al. 2009). *F. prausnitzii* exhibits anti-inflammatory effects on cellular and TNBS colitis models and these data suggest that counterbalancing dysbiosis using *F. prausnitzii* as a probiotic may be a promising strategy in CD treatment (Sokol et al. 2008).

Sonnenburg et al. showed that manipulating the microbiota with probiotics can influence the host. When germ free mice were colonized with a prominent component of the adult human gut microbiota (*Bifidobacteria thetaiotaomicron*) and a commonly used probiotic (*Bifidobacterium longum*), the probiotic bacteria repressed *B. thetaiotaomicron* expression of antibacterial proteins to promote its own survival, and influenced the composition, structure, and function of its microbial community (Sonnenburg et al. 2006). In an animal model of experimental colitis, a single microbial molecule from *Bacterial Fragillis* can protect its host from inflammatory disease caused by *Helicobacter hepaticus*, suggesting that natural anti-inflammatory molecules from the bacteria microbiota can actively promote human health, and may potentially be therapies for human inflammatory disorders (Mazmanian et al. 2008). Metabonomic studies have demonstrated that probiotics can modulate the gut microbiome and the metabolism of short chain fatty acids, amino acids, bile acids and plasma lipoproteins (Martin et al. 2008).

Ulcerative Colitis

Induction of remission

The probiotic combination VSL#3™ demonstrates some benefit in inducing remission in patients with mild to moderate ulcerative colitis (UC). VSL#3 contains live lyophilized *Bifidobacterium breve, Bifidobacterium longum, Bifidobacterium infantis, Lactobacillus acidophilus, Lactobacillus plantarum, Lactobacillus paracasei, Lactobacillus bulgaricus* and *Streptococcus thermophilus.* In a multicenter, randomized, double blind, placebo-controlled trial, Sood et al. studied the probiotic combination VSL#3™ in adults with mild-to-moderate UC (Sood et al. 2009). Patients were given 3.6×10^{12} CFU VSL#3™ (N = 77) or placebo (n = 70) twice daily for 12 weeks. A primary end point of 50% decrease in the Ulcerative Colitis Disease Activity Index (UCDAI) at 6 weeks was achieved in a greater number of those receiving probiotic (32.5%) than the group given placebo (10%) ($p = 0.001$). A secondary end point of remission by 12 weeks was achieved in 33 subjects given probiotic (42.9%) compared with 11 subjects given placebo (15.7%) ($p = 0.001$). In another similar study, Tursi et al. assessed effects of the probiotic combination VSL#3™ in adults with mild-to-moderate UC (Tursi et al. 2010). A total of 144 consecutive patients were treated with 3.6×10^{12} CFU VSL#3™ daily (N = 71) or placebo (n = 73) for 8 weeks. The decrease in UCDAI scores of 50% or more was higher in the VSL#3™ group than in the placebo group (63.1 vs. 40.8; per protocol (PP) P=0.010; intention to treat (ITT) P=0.031).

In a trial in children with moderate-to-severe ulcerative colitis, VSL#3™ or placebo was administered along with standard IBD therapy (corticosteroids and mesalazine) (Miele et al. 2009) . The corticosteroids were tapered after a month if subjects were in remission. In this study, remission was achieved in 13 of 14 participants (92.8%) treated with VSL#3™ and

IBD therapy and in 4 of 15 patients (36.4%) treated with placebo and IBD therapy ($p < 0.001$). However, it should be noted that the response rate to corticosteroids and mesalazine in the placebo treated group is lower than that published.

In addition, a bifidobacteria-fermented milk supplementation has been assessed in small randomised trials in treating active UC, both showing some benefit (Kato et al. 2004; Ishikawa et al. 2003). In the trial by Kato et al. twenty patients with mild to moderate UC randomly received 100 ml/day of bifidobacteria-fermented milk or placebo for 12 weeks with conventional treatment. Improvements were significant in both groups, but the clinical activity index was significantly lower in the bifidobacteria-fermented milk than in the placebo group after treatment (Kato et al. 2004). The post-treatment endoscopic activity index and histological score were significantly reduced in the bifidobacteria-fermented milk, but not the placebo group. However, the trials are limited by the small numbers of patients participating.

Maintenance of remission

Efficacy of a non-pathogenic *E. coli* strain, *E. coli* Nissle, has been assessed in maintaining remission of UC. Nissle had observed a decrease in these organisms in patients with non-infectious bowel disorders and he also observed that administration of the non-pathogenic *E. coli* strain, referred to as Nissle 1917 serotype 06:K5:H1, resulted in improvement in patients' symptoms. Kruis et al. performed a double-blind, double-dummy trial in 103 patients with inactive ulcerative colitis, in which patients were randomised to receive either 1,500 mg/day mesalazine of 200mg/day of *E. coli* Nissle strain 1917, containing 5×10^{10} viable bacteria for 12 weeks (Kruis et al. 1997). The relapse rate was 16% in the *E. coli* Nissle group and 11% in the mesalazine group. A larger, longer term study randomised 327 patients with inactive colitis to either 1,500 mg/day mesalazine or 200mg/day of *E. coli* Nissle strain 1917 for one year (Kruis et al. 2004). Relapse rates were 45% in the *E. coli* Nissle group and 36% in the mesalazine group, with the difference in relapse rates suggesting equivalence in maintaining remission.

Rembacken et al. performed a single-centre, randomised, placebo-controlled trial involving 120 patients with UC (Rembacken et al. 1999). Initially all patients had active disease and received medical therapy to induce remission (oral prednisolone or topical steroids). In addition patients received gentamicin 80mg three times a day and either mesalazine 2,400 mg/day or *E. coli* Nissle strain 1917, containing 10×10^{10} viable bacteria. In the maintenance part of the study, patients who were in remission at 3 months were randomised to receive either 1,200 mg/day mesalazine or *E. coli* Nissle strain 1917, containing 5×10^{10} viable bacteria and were observed for 1 year. In the maintenance part of the study, relapse was frequent in both groups: 67% in the *E. coli* Nissle group and 73% in the mesalazine group. These relapse rates are similar to those reported in placebo-treated patients in most published maintenance studies, suggesting that both treatments in this study were ineffective in maintaining remission.

In a systematic review, there is an overall risk ratio of 1.08 (95% CI 0.86, 1.37) for preventing relapses in inactive UC with *Escherichia coli* Nissle 1917 versus standard treatment (n=3) (Jonkers et al. 2012).

Pouchitis

Pouchitis is the most common complication after ileal pouch-anal anastomosis for UC. Approximately 50% of all patients with a pouch will experience inflammation in their pouch at some point (Navaneethan and Shen 2010). Bacteria appear to play a pivotal role in pouchitis, in particular there is a decreased diversity of bacteria (Landy et al. 2012). Most patients respond to antibiotics, but 10–15% of patients experience refractory symptoms. Prolonged courses of antibiotics may be beneficial in patients with refractory pouchitis, but are associated with side effects and resistance. An alternative way to modify the bacterial milieu involves the use of probiotics.

Induction of remission

There is limited evidence for a role of probiotics as monotherapy for mild/moderate pouchitis at the present time.

Maintenance of remission

In pouchitis, once healing has been achieved with antibiotics, remission needs to be maintained and probiotics have been assessed in this setting. Forty patients with chronic pouchitis, defined by clinical, endoscopic and histologic criteria were treated with antibiotics to achieve remission (Gionchetti et al. 2000) and then were randomised in a double-blind, placebo-controlled trial to receive the probiotic combination VSL#3™ (3g twice a day) or placebo for 9 months. In the group of patients receiving VSL#3™, three (15%) patients relapsed compared with 20 (100%) patients in the placebo group ($p<0.001$). Within 4 months of stopping the active therapy, 100% of the patients who had responded to the active therapy relapsed. All of the patients treated with VSL#3™ had an increase in the concentration of lactobacilli, bifidobacteria and Streptococcus thermophilus in their stool samples that returned to basal levels within one month of stopping VSL#3™.

A two centre study confirmed these results (Mimura et al. 2004). Thirty six patients with recurrent or chronic active pouchitis had remission induced with ciprofloxacin and metronidazole, and they were then randomised to VSL#3™ probiotics or placebo for 1 year. Although the same total dose of VSL#3™ was used (6g per day), it was administered in a more convenient dosing regimen of once daily rather than twice daily. Of the 36 randomised patients, 20 received VSL#3™ and 16 received placebo. In the VSL#s group, two (10%) patients relapsed compared with 15 (94%) in the placebo group ($p<0.0001$). Quality of life measured using the Inflammatory Bowel Disease Questionnaire was improved in the VSL#3™ treated group. In contrast, the benefits suggested by these trials are not borne out in an open label trial by Shen et al. in which 31 patients were prescribed a 2-week treatment of a single antibiotic (ciprofloxacin) followed by VSL#3™ (Shen et al. 2005). Probiotic therapy was stopped by 9 of 31 (29%) seven weeks into therapy and 25 of 31 (81%) by 8 months in view of failure to prevent pouchitis (n = 23) or side effects (n = 2). 6 of 31 (19%) did not develop clinical evidence of pouchitis by the end of the 8-month trial period, although some endoscopic activity was noted in patients. Overall, given the number needed to treat with oral probiotic therapy to prevent one additional relapse of 2, it has been suggested that for those patients with prompt recurrence of pouchitis following antibiotic usage or having multiple recurrences of pouchitis despite antibiotics, either VSL#3™ or chronic use of antibiotics is indicated. However, probiotics are not suggested for acute treatment of pouchitis

Post-operative prophylaxis

The probiotic combination VSL#3™ has also been used in the context of prevention of onset of pouchitis. A large proportion of episodes of pouchitis occur in the first year after a pouch is formed. Forty patients with a newly formed ileo-anal pouch were randomised to receive VSL#3™ (3g per day) or placebo for one year (Gionchetti et al. 2003). In the VSL#3™ group, two (10%) of twenty patients experienced an episode of pouchitis, confirmed by clinical, endoscopic and radiologic assessment, compared with 8 (40%) of the twenty patients treated with placebo ($p<0.01$).

Crohn's Disease

Induction of remission

There are limited studies which have assessed the role of probiotics in induction of remission in CD. Twenty patients with CD with moderately active inflammation were randomised to *S. boulardii* (750 mg = 3×10^{11}

organisms) or placebo for 7 weeks in a double-blind study (Plein and Hotz 1993). Only the group receiving *S. boulardii* had a significant reduction in disease activity, as assessed by Crohn's Disease Activity Index. One placebo-controlled trial used *L. rhamnosus* GG and enrolled a total of 11 participants. Four of 5 patients in the probiotic group achieved remission compared to 5 of 6 in the placebo group (OR 0.80; 95% CI 0.04 to 17.20). However, overall there is insufficient evidence to make any conclusions about the effectiveness of probiotics for treatment of active CD (Jonkers et al. 2012).

Maintenance of Remission

Malchow et al. conducted a double-blind study in which 28 patients with Crohn's disease were randomised to receive *E. coli* Nissle strain 1917, containing 109 rising to 5 x 1010 viable bacteria or placebo for 1 year (Malchow et al. 1997). The relapse rate was 30% in the *E. coli* group compared with 7-% on the placebo group. In the largest maintenance trial to date, Bousvaros et al. reported no difference in the proportion of those developing relapse on *L. rhamnosus* strain GG 2 × 1010 CFU/day (31%; 12 of 39) or placebo (17%; 6 of 36, p = 0.18) (Bousvaros et al. 2005). There appeared to be a trend towards higher relapse rate in subjects treated with probiotics than those on placebo.

Post-operative prophylaxis

Probiotics have been assessed in the context of prevention of recurrence of CD following surgical resection in 5 randomised trials. Three of the trials involved single probiotic strains being administered and 2 trials included combination probiotics. Prantera et al. used *L. rhamnosus* GG in 45 patients who had undergone resections for CD. Patients were allocated to receive 12 billion colony forming units of *L. rhamnosus* GG placebo for one year. There was clinical recurrence in three (16.6%) patients who received Lactobacillus and in two (10.5%) who received placebo. There was a trend towards more patients in the probiotic-treated group having endoscopic evidence of disease compared with placebo-treated patients (Prantera et al. 2002). Two trials have assessed *L. johnsonii* LA1. In the trial by Marteau et al. 98 patients were enrolled and at 6 months endoscopic recurrence was seen in 21 of 43 (43%) in the *L. johnsonii* LA1 group and 30 of 47 (64%) of placebo group and (p = 0.15) (Marteau et al. 2006). In the trial by van Gossum et al., 70 patients were enrolled. Overall, *L. johnsonii* LA1 failed to prevent early endoscopic recurrence at 12 weeks after ileo-caecal resection. Similarly, a study using synbiotics failed to reveal any differences between treatment and placebo groups for endoscopic recurrence (Chermesh et al. 2007).

A meta-analysis of effects of probiotics as a class in the setting of prevention of post-operative recurrence of CD suggested that their effect was no different than placebo (Doherty et al. 2010). The relative risk of clinical recurrence with any probiotic relative to placebo (n = 213) was 1.41 (95% CI 0.59 to 3.36), any endoscopic recurrence (n = 333) was 0.98 (95% CI 0.74 to 1.29) and severe endoscopic recurrence (n = 213) was 0.96 (95% CI 0.58 to 1.59). Thus far, there is no evidence to support the use of probiotics in CD.

Mechanisms of Action of Probiotics

Probiotic bacteria have multiple and diverse influences on the host. Different organisms can influence the luminal environment, epithelial and mucosal barrier function, and the mucosal immune system. They can exert their effects on different cell types including epithelial cells, dendritic cells, monocytes or macrophages, B cells, T cells with regulatory properties and NK cells (Zhang et al. 2007; Neurath 2007). In animal models including dextran sodium sulphate-and hapten-induced colitis, HLA-B27 transgenic rats, and IL-10 and IL-2-deficient mice (Rachmilewitz et al. 2002; Dieleman et al. 2003; Madsen et al. 1999; Waidmann et al. 2003), favourable action of probiotics on intestinal inflammation has been demonstrated. These animal models of colitis have been useful in providing mechanistic and therapeutic proof of concept. However, not all commensal or probiotic bacteria have similar actions on gut immune function. For example, in the IL-10 knockout mouse model of colitis, *lactobacillus* and *bifidobacterium* species are effective in reducing established intestinal inflammation (Madsen et al. 1999; McCarthy et al. 2003), whereas in HLA-B27 transgenic mice, Lactobacillus GG effectively prevents relapse of colitis after antibiotic treatment (Dieleman et al. 2003).

Modification of the Intestinal Microbiota

Probiotic bacteria can modify the intestinal microbiota by producing anti-bacterial substances and defensins, inhibiting bacterial adherence and translocation, or directly reducing luminal pH.

Probiotic bacteria have been reported to produce antimicrobial compounds called bacteriocins. The inhibitory activity of these bacteriocins varies; some inhibit other lactobacilli or taxonomically-related Gram positive bacteria, whereas some are active against a much wider range of Gram positive and Gram negative bacteria as well as yeasts (Nemcova 1997). For example, the probiotic *L. salivarius* subsp. *salivarius* UCC118 produces a peptide which inhibits a broad range of pathogens including Bacillus,

Staphylococcus, Enterococcus, Listeria and Salmonella species (Flynn et al. 2002) *Lactobacillus* species release bacteriocins which reduce adherence of *Helicobacter pylori* infection on epithelial cells (Gotteland et al. 2006), and Lacticin 3147, a broad-spectrum bacteriocin produced by *Lactococcuslactis* subsp., inhibits a range of genetically distinct *Clostridium difficile* isolates from healthy subjects and patients with IBD (Rea et al. 2007).

Pre-incubation of intestinal epithelial cell line isolated from patients with CD using the probiotic strain *E. coli* Nissle 1917 resulted in diminished adhesion and invasion by adherent invasive *E. coli*. These findings demonstrate that probiotics prevent epithelial injury induced by attaching-effacing bacteria (Malchow 1997; Rembacken et al. 1999). The pretreatment of intestinal (T84) cells with lactic acid-producing bacteria reduced the ability of pathogenic *E. coli* to inject virulence factors into the cells or to breach the intracellular tight junctions (Sherman et al. 2005).

In patients with UC the ingestion of the probiotic preparation VSL#3 resulted in a decrease in the luminal pH (Venturi et al. 1999). In a fatal mouse Shiga toxin-producing *Escherichia coli* O157:H7 infection model, the probiotic *B. breve* produced a high concentration of acetic acid, consequently lowering the luminal pH. This pH reduction was associated with improved animal survival (Asahara et al. 2004).

Defensins are anti-microbial peptides involved in innate defence mechanisms. The probiotic *E. coli Nissle* strain induced expression of human beta-defensin 2 (hBD-2) in Caco-2 intestinal epithelial cells (Wehkamp et al. 2004); this type of effect may contribute to an improved mucosal barrier and provide a means of limiting access of enteric pathogens. Induction of human beta-defensin 2 by *E coli Nissle 1917* is dependent on the NF-kB and AP-1-pathways, mediated through bacterial flagellin (Schlee et al. 2007).

Enhancement of Barrier Function

Disruption of the epithelial barrier function has been described in IBD, both in active and inactive disease, and in healthy relatives of patients with IBD (Schmitz et al. 1999; Hilsden et al. 1996; Peeters et al. 1997; Wyatt et al. 1993; May et al. 1993; McCole 2012). Recent data indicate that probiotics can improve integrity of the intestinal epithelial barrier under certain pathological circumstances (Ukena et al. 2007; Meddings 2008; Hering et al. 2012; Girardin and Seidman 2011). Enhancement of barrier function by probiotic bacteria has been observed both in *in vitro* models and *in vivo* in the whole animal. The probiotic mixture VSL#3 normalised barrier integrity, as assessed by transepithelial potential differences and mannitol fluxes in excised tissue from mice (Madsen et al. 2001). Furthermore, in an *in vitro* culture using T84 epithelial cells, VSL#3 decreased monolayer permeability and conductance, indicating that the increase in resistance was specific to one or more of the bacteria in VSL#3. In IL-10 deficient mice with chronic colitis lactobacillus improved barrier function *in vivo* (Madsen et al. 1999). In a methotrexate-induced model of colitis, *L. plantarum* and *L. reuteri* enhanced barrier function (Mao et al. 1996). However, enhancement of barrier function was not observed in all models of colitis studies (Kennedy et al. 2000). In addition barrier protection is not exhibited by all probiotic or commensals and strain-specific molecules involved remain to be clarified (Miyauchi et al. 2012).

The mechanisms by which probiotics bacteria enhance gut mucosal barrier function are unclear, but may relate to alterations in mucus or chloride secretion or changes in tight junction protein expression by epithelial cells. Some probiotic bacteria modify MUC gene expression and mucus secretion. *L. plantarum* 299v increased MUC2 and MUC3 mRNA expression when incubated with the epithelial cell line HT-29 (Mack et al. 1999). VSL#3 and *E. coli* Nissle strain increased MUC2, MUC3 and MUC5AC gene and protein expression (Otte and Podolsky 2004). Some probiotic bacteria limit chloride and water secretion. For example, *S. Thermophilus*and *L. acidophilus* reversed the increase in enteroinvasive *E. coli*-induced chloride secretion by an epithelial cell line (Resta-Lenert and Barrett 2003).

Tight junction proteins are dynamic structures subject to structural changes. In epithelial cells, the tight junction protein, zonula occludens-1 (ZO-1) redistributes when exposed to pathogenic bacteria such as *S. Dublin* (Otte and Podolsky 2004). The co-culture of epithelial cells with VSL#3 probiotic bacteria in addition to *S. dublin* prevented the redistribution of ZO-1 and stabilised the barrier function, suggesting that probiotic bacteria may be important in preservation of the cytoskeleton structure. *E. coli Nissle 1917* cancounteract the disruptive effects of *enteropathic* *E. coli* on T-84 epithelial cells monolayers. This effect is achieved by altering protein kinase C signalling and increasing the redistribution and expression of ZO-2 which acts as a crucial factor in maintaining epithelial tight junction function (Zyrek et al. 2007).

Immunomodulation

Epithelial cells have been shown to sense commensal or probiotic bacteria and pathogenic bacteria in different ways. Pathogenic bacteria induce pro-inflammatory responses in intestinal epithelial cells by activating the transcription factor NFκB. In contrast, non-pathogenic species attenuate pro-inflammatory responses by inhibiting degradation of counter-regulatory factor IκB. This method of blocking proinflammatory responses has been shown by non-pathogenic *Salmonella pullorum* which attenuates IL-8 secretion elicited by pathogenic *Salmonella typhimurium* (Neish et al. 2000). Another method of avoiding pro-inflammatory responses to commensal bacteria has been demonstrated for *Bacteroidesthetaiotaomicron* which induced an anti-inflammatory response in epithelial cells by shuttling transcription factor NK-kappa B out of the nucleus by a pathway involving the nuclear hormone receptor PPARγ, resulting in attenuation of NFκB-mediated inflammatory gene expression (Kelly et al. 2004).

Lammers and co-workers and Otte and Podolsky showed that probiotic bacteria in the VSL#3 combination did not induce IL-8 secretion by epithelial cells compared with intestinal pathogens such as enteropathogenic *E. coli, Salmonella dublin, Shigelladysenteriae*and *Listeria monocytogenes,* all of which can induce secretion of IL-8. VSL #3 probiotic bacteria produced soluble factors which inhibit the NFκB pathway and induce expression of cytoprotective heat shock proteins(Petrof et al. 2004). Furthermore, DNA derived from the probiotic mixture VSL#3 delayed NFκB activation, stabilised levels of IκB, and inhibited proteasome function (Jijon et al. 2004). VSL#3 probiotic bacteria decreased IL-8 secretion seen with the pathogenic bacteria alone indicating that probiotic bacteria can override the effects of pathogenic bacteria. The probiotic bacteria, *E. coli Nissle 1917,* induced IL-8 secretion by intestinal epithelial cell lines in a dose dependent manner suggesting that the ability to prevent secretion of IL-8 from epithelial cells is not a feature of all probiotic bacteria (Lammers et al. 2002; Otte and Podolsky 2004).

One other effect of probiotic bacteria on epithelial cells is the ability of commensal organisms to act through pattern recognition molecules or Toll-like receptors (TLR), such as TLR-2 and TLR-4, possibly on epithelial cells. Such interactions induce the production of protective cytokines that enhance epithelial cell regeneration and inhibit epithelial cell apoptosis (Rakoff-Nahoum et al. 2004). *L. casei* prevents the development of acute DSS-induced colitis in TLR-4 mutant mice by inhibiting myeloperoxidase activity and IL-12p40, and increasing TGF beta and IL-10 mRNA. These effects suggest that the mechanism of action of *L. casei* depends largely on TLR-4 status (Chung et al. 2008).

Probiotic bacteria may also enable epithelial recovery or prevent apoptosis, as suggested by a study in which cytokine-induced apoptosis was prevented in intestinal epithelial cells in the presence of *Lactobacillus rhamnosus* GG (Yan and Polk 2002). Culture of probiotic bacteria with either mouse or human colon cells activated anti-apoptotic Akt/protein kinase B and inhibited activation of the pro-apoptotic p38/mitogen-activated protein kinase. An inhibition of apoptosis may enhance survival of intestinal cells and promote proliferation during recovery from epithelial injury. In addition, a non-pathogenic commensal bacterium *S. pullorum,* influenced epithelial cell proliferation by "injecting" factors into the gut epithelium that blocked β-catenin degradation, a substance which has been implicated in epithelial growth control (Neish et al. 2000).

Effects of Probiotic Bacteria on Dendritic Cells

Dendritic cells (DC) are antigen presenting cells that are important in the early stage of bacterial recognition and in shaping downstream T-cell responses. The effects of different probiotic bacteria on DC have been studied in different experimental systems including whole blood DC, freshly isolated lamina propria DC, monocyte-derived DC and bone-marrow derived DC and in different species (human and mouse). DC in the gut have specialised functions, contributing to oral tolerance induction by generating regulatory T-cells and IgA-producing B cells through production of cytokines such as IL-10 and TGF-beta (Akbari et al. 2001; Williamson et al. 2002). Intestinal DC interact with bacteria via M-cells and by passing their dendrites between epithelial tight junctions into the gut lumen (Stagg et al. 2003; Rescigno et al. 2001). Rol et al. showed in a mouse model that the entry of commensal bacteria in Peyer's Patches via M cells was mediated by secretory IgA (SIgA). Selected SIgA-mediated targeting of bacteria was restricted to tolerogenic DC subset located in the subepithelial dome of Peyer's Patches, which may represent a mechanism whereby the host mucosal immune system controls the continuous dialogue between the host and commensal bacteria (Rol et al. 2012).

In humans the probiotic combination VSL#3 was a potent inducer of IL-10 by both blood and lamina propria DC *in vitro* (Hart et al. 2004). Patients with acute UC when treated with oral VSL#3 have increased IL-10 and reduced IL-12p40 production by colonic DC; these effects were not seen in placebo-treated patients (Ng et al. 2010a). In patients with pouchitis treated with the probiotic combination VSL#3 there were enhanced mucosa levels of IL-10 and decreased levels of TNF-α, IL-1, nitric oxide synthase and matrix metalloproteinase (Ulisse et al. 2001), suggesting that the *in vitro* findings correlate with those *in vivo*, although the cellular source of the IL-10 *in vivo* was not clear. In agreement with the observation that VSL#3 was a potent inducer of IL-10 by DC, Drakes et al. showed that murine bone marrow-derived DC incubated with VSL#3 increased IL-10 detected by ELISA (Drakes et al. 2004).

Individual strains within VSL#3 displayed distinct immunomodulatory effects on DC; the most marked anti-inflammatory effects were produced by bifidobacteria strains (*B. longum, B. infantis* and *B. breve*) which upregulated IL-10 production by DC. Using DC derived from human cord blood monocytes, Young et al. demonstrated that *B. longum, B. bifidum* and *B. pseudocatenulatum*, but not *B. infantis*, induced high levels of IL-10 (Young et al. 2004). Rigby et al. showed that murine freshly isolated lamina propria DC incubated with *B. longum* secreted a greater proportion of IL-10 thanIL-12 (Rigby et al. 2002). In a different experiment, purified human monocytes and monocyte-derived DC were stimulated with Ultra Violet-inactivated gram-positive (*Lactobacillus plantarum* and Bifidobacteriumadolescentis) and gram-negative (*Escherichia coli* and Veillonellaparvula) bacterial strains. *B. adolescentis* induce low amounts of IL-12, TNF-alpha, IL-6 and IL-8; *L. reuteri* and *L. casei*, but not *L. plantarum*, primed monocyte-derived DC to drive the development of T cells with regulatory properties (Karlsson et al. 2004; Braat et al. 2004b). These regulatory T cells produced increased levels of IL-10 and were able to inhibit the proliferation of bystander T cells in an IL-10-dependent fashion. Strikingly, both *L. reuteri* and *L. casei*, but not *L. plantarum*, bind the C-type lectin DC-specific intercellular adhesion molecule 3-grabbing non-integrin (DC-SIGN). Blocking antibodies to DC-SIGN inhibited the induction of the "T reg" cells by these probiotic bacteria, implying that ligation of DC-SIGN can actively prime DC to induce "T reg" cells. Thus the targeting of DC-SIGN by certain probiotic bacteria might explain their beneficial effect in the treatment of CD (Smits et al. 2005).

In contrast to the anti-inflammatory activity of some strains of *bifidobacteria* and *lactobacilli*, other *lactobacilli* strains have proinflammatory activity. When cultured with human monocyte-derived DC, *L. reuteri, L. gasseri*and *L. johnsonii*, induced activation and maturation of DC, enhanced IL-12 production and induced allogeneic T cell priming (Mohamadzadeh et al. 2005). However, Christensen et al. demonstrated that *L. reuteri* induced little or no IL-12 production by DC in an experimental system using murine bone-marrow derived DC and inhibited proinflammatory cytokine production (IL-12, IL-6 and TNF-alpha) by *L. casei* which was a potent inducer of IL-12. These varying effects of probiotic bacteria highlight differences which arise when different experimental systems in different animals are used. There appear to be different responses of different bacterial strains even within a genus. In human monocyte-derived DC, *L. rhamnosus* and *L. plantarum* induced no or low levels of IL-12 (Veckman et al. 2004; Karlsson et al. 2004; Braat et al. 2004b; Braat et al. 2004a) in contrast to the high levels of IL-12 noted with *L. gasseri* and *L. johnsonii* (Mohamadzadeh et al. 2005).

Braat et al. demonstrated that *L. rhamnosus* may "educate" DC to stimulate proliferation of peripheral CD4+T cells and reduce CD3/CD28 stimulated cytokine production *in vitro*. *L. rhamnosus* resulted in a reduction of IL-4 from peripheral CD4+ T cells of normal individuals, and a decrease in IFN-γ and IL-2 production by CD4+ T cells from patients with CD. This observation suggests that a comparable mechanism may exist in humans. It supports an anti-inflammatory mechanism of action for probiotics that, apart from influencing both Th1 and Th2 immune responses, also has an indirect effect via antigen presenting cells in the gut (Braat et al. 2004b).

Effects of Probiotic Bacteria on other Immune Cells

Blood monocytes and tissue macrophages are effective secondary presenters of antigens to memory T cells. *L. plantarum* increased IL-10 synthesis and secretion in macrophages derived from the inflamed colon (Pathmakanthan et al. 2004). In contrast, *L. rhamnosus* GG promoted the production of IFN-gamma, IL-12 and IL-18 (Miettinen et al. 1998), and induced NFκB and STAT DNA-binding in primary human macrophages. He et al. showed that *B. bifidum*, *B. breve* and *B. infantis* stimulated more IL-10 and less IL-12 and TNFα from a murine macrophage-like cell line than *B. adolescentis* (He et al. 2002), again suggesting strain differences within a genus. DNA derived from the probiotic mixture VSL#3 activated NFκB and induced low levels of IL-6 and IL-12 by bone-marrow derived macrophages compared with immunostimulatory oligonucleotides (Rachmilewitz et al. 2004).

Probiotic bacteria may affect lymphocytes directly or secondarily via changes in stimulation induced by alterations in antigen presenting DC or macrophages. These effects have been demonstrated for different types of lymphocytes. They may also exert beneficial effects and modulate the immune response to potentially harmful antigens via B lymphocytes and antibody production. For example, *L. rhamnosus* GG administered to children with acute gastroenteritis enhanced a non-specific humoral immune response, reflected by an increase in IgG, IgA and IgM secretion from circulating lymphocytes (Kaila et al. 1992). *B. bifidum* enhanced the antibody response to ovalbumin, and yoghurts containing *L. acidophilus*, *L. bulgaricus*, *S. thermophilus*, *B. bifidum* and *B. infantis* stimulated the IgA response to cholera toxin in mice (Tejada-Simon et al. 1999).

The symbiotic *L. casei* ssp. *Casei* with dextran (prebiotic) significantly elevated the NK cell activities in spleen mononuclear cells from BALB/c mice, and oral administration of this symbiotic to healthy volunteers induced NK cell activities, and increased production of IL-12 in human peripheral blood mononuclear cells (Ogawa et al. 2006). Takeda et al. showed that *L. casei* Shirota can enhance NK cell activity *in vivo* and *in vitro* in humans; this effect may be dependent on IL-12 (Takeda et al. 2006).

Probiotic bacteria also induce regulatory T cells in the context of intestinal inflammation. In TNBS colitis, the probiotic combination VSL#3 ameliorated intestinal inflammation, the beneficial effect of VSL#3 was attributable to IL-10-dependent-regulatory CD4+ T cells bearing surface TGFβ (Di Giacinto et al. 2005). These cells appear to be similar to CD25+ regulatory T cells that inhibit cell-transfer colitis by a TGFβ-dependent mechanism (Oida et al. 2003; Nakamura et al. 2004).

A novel biological property of probiotics recently reported is the capacity to affect immune cell redistribution by improving the competence of lymphatic endothelial cells to trap T lymphocytes. *S. boulardii* has a unique action on inflammation by a specific alteration of the migratory behavior of T cells causing accumulation of these cells in mesenteric lymph nodes. In *Sacharomyces boulardii*-fed mice, IFN-γ production by CD4+ T cells was reduced in the colon but increased in the mesenteric lymph nodes (Dalmasso et al. 2006).

Mechanisms of Probiotics in Clinical Diseases

A number of studies in IBD have shown that probiotics can induce regulatory cytokines, including IL-10 and TGF-ß, and suppress pro-inflammatory cytokines, such as TNF, in the mucosa of patients with CD and pouchitis (Borruel et al. 2002; Pathmakanthan et al. 2004). VSL#3 induces IL-10 and downregulates IL-12p40 production by lamina propria DC in patients with UC; similar cytokine changes were seen in patients who were treated with corticosteroids (Ng et al. 2010a).

In a study aimed at assessing the anti-inflammatory effects of probiotics, healthy controls and IBD patients consumed *L. rhamnosus GR-1* and *L. reuteri RC-14* supplemented yogurt for 30 days. The proportion of putative regulatory CD4(+) CD25(high) T cells in peripheral blood increased significantly in IBD patients, after treatment with probiotics but not in controls. In addition serum IL-12 concentrations and the proportion of IL-2(+) and CD69(+) T cells from stimulated cells decreased in IBD patients. The increase in CD4(+) CD25(high) T cells correlated with the decrease in the percentage of TNF-alpha- or IL-12-producing monocytes and DC. Thus probiotic yogurt intake was associated with significant anti-inflammatory effects that paralleled the expansion of the peripheral pool of putative "T reg" cells in IBD patients with few effects in controls (Lorea et al. 2007).

Adhesive *E. coli* have been implicated in the pathogenesis of UC. Studies using a 16S rRNA technique have shown reductions in bifidobacteria (Mylonaki et al. 2005; Macfarlane et al. 2005) and lactobacilli in patients with UC (Bullock et al. 2004). There was an increased bifidobacterial colonization of the rectal mucosa with a short term synbiotic treatment. In a separate experimental system, colonic biopsies from UC were co-cultured for 24 hours with *B. longum*. The concentrations of TNF and IL-8 in supernatants of inflamed UC tissue co-cultured with probiotics were lower than those cultured alone. The number of lamina propria mononuclear cells with nuclear factor-kappa B (NK-κB) p65 positive in co-cultured tissues was also reduced (Bai et al. 2006). Release of TNF-α by inflamed CD mucosa was significantly reduced by co-culture *L. casei* or *L. bulgaris*, but not *L. crispatus* and *E. coli*. No change in TNF-α production was seen in experiments with non-inflamed Crohn's mucosa and control mucosa. Certain probiotic bacteria may therefore interact with immunocompetent cells at the mucosal interface and thus modulate local production of pro-inflammatory cytokines by inflamed tissue (Borruel et al. 2002).

Probiotic Effects—Live Bacteria or Bacterial DNA?

Accumulating data suggest that bacterial DNA sequences may provide the same effects as live bacteria (Jijon et al. 2004; Rachmilewitz et al. 2002). Bacterial DNA contains non-methylated CpG motifs which bind to TLR-9. In the presence of both TLR-9 and MyD88, non-viable bacteria may have the ability to signal and elicit beneficial effects. In an experiment that used methylated and non-methylated genomic DNA isolated from the probiotic preparation VSL#3, DNAse-treated probiotics, and *E. coli* genomic DNA, the authors demonstrated that genomic DNA (but not methylated DNA, calf thymus DNA or DNAse-treated probiotics) ameliorated the severity of colitis in DSS-induced, TNBS-induced and spontaneous colitis in IL-10 knock-out mice. In the same study, intragastric and subcutaneous administration of gamma-irradiated nonviable bacteria and live bacteria had similar beneficial effects (Rachmilewitz et al. 2004). Lammers et al. demonstrated that Bifidobacterium genomic DNA induced the secretion of IL-10 by peripheral blood mononuclear cells from healthy donors, demonstrating the immunomodulatory effects of bacteria DNA (Lammers et al. 2003). By using a VSL#3 conditioned medium, Petrof et al. showed that early proteasome inhibition may account for NFkB inhibition and heat shock protein induction in cell line from the intestinal epithelium of mice. Based on the beneficial effects of VSL#3 conditioned medium, these data further challenge the concept that live bacteria is required to exert beneficial effects of probiotics (Petrof et al. 2004).

Questions that Remain Unanswered

Studies in animal models of colitis and experimental studies with probiotics are characterized by a high level of heterogeneity due to the wide variety of animal models used. Furthermore, in most currently available animal models, disease mostly occurs as a result of experimental manipulation. Such models only present limited specific pathogenic features of the complex, multi-faceted nature of IBD in humans, and may not accurately reflect human disease. Nonetheless, it is increasingly apparent that different probiotic bacteria act through multiple and contemporaneous pathways rather than by a single common mechanism. Ibnou-Zekri et al. highlighted that the activity of probiotic strains *in vitro* may not parallel similar *in vivo* behaviour. In their study, two strains of Lactobacillus that exhibited very similar *in vitro* growth, survival and adherence properties, showed distinct differences in colonization patterns and resultant host immune responses at both the mucosal and systemic levels (Ibnou-Zekri et al. 2003). In contrast, we have demonstrated consistent *in vitro* and *in vivo* disease effects, with VSL#3 inducing IL-10 and downregulating IL-12p40 production by colonic DC, both *in vitro* and *in vivo*. Even such consistent *in vitro* and *in vivo* mechanistic data, however, only correlate with the clinical outcome in some patients (Ng et al. 2010b).

To determine further the role of probiotics in the IBD treatment armamentarium, large well designed multicentre controlled clinical trials are needed. Not all *lactobacillus* and *bifidobacterium* species are equally beneficial; each may have individual mechanisms of action that is dependent on host characteristics. Different bacteria may have dominant effects in different genetic backgrounds and in diseases which vary in their pathogenesis (Ng et al. 2011). Thus optimal use of various probiotics may depend on identifying patient subsets by genoypte, phenotype, stool microbiologic, serologic or T cells immune response criteria. The concept of synbiotics (simultaneous use of prebiotics and probiotics) may reduce the dose, frequency, or duration of each treatment. Recent work has focused on the bacteria-host interaction showing that the host, a component of its microbiota, and probiotic bacteria can adapt their substrate utilization in response to one another. Our current challenge is to marry complex underlying disease pathogenesis of IBD with multifactorial mechanisms of action of the probiotic on the host. Questions remain of which organism(s) to use and whether organisms should be derived from live preparation, commensal bacterial concentrations (e.g., *F. prausnitzii*) or fermented foods; the optimal dose and duration for treatment, the benefit of concomitant treatment, for example with antibiotics as were used in the pouchitis trials and the possibility of targeted therapy in particular clinical or genetic subsets.

Conclusion

The intestinal bacteria flora contributes significantly to the pathogenesis of IBD, along with genetic susceptibility and mucosal immune dysregulation. Probiotics are likely to become an integral component of the therapeutic armamentarium of IBD in combination with traditional anti-inflammatory and immunosuppressive agents. Positive controlled clinical trials have demonstrated that the probiotic mixture, VSL#3, is beneficial in maintaining remission in pouchitis and preventing the development of pouchitis in patients following ileo-anal pouch formation, and may also be effective in the treatment of acute mild to moderately active UC. *Sacharomyces bourladii* appears useful in maintaining remission in CD but Lactobacillus GG and *Lactobacillus johnsonii* LA1are not beneficial in preventing post-operative recurrence in CD. Understanding probiotic action may permit modulation of the immune system, both locally and systemically. Knowledge of probiotics on the host immune system has entered a new and fascinating phase of research and progression in this field is likely to offer novel and useful means to modulate host immunity for protection from, or treatment of, a wide variety of human disorders, including IBD.

Summary

Ulcerative Colitis

- VSL#3 is effective in the treatment of acute, mild to moderately active ulcerative colitis in adults and children.
- *E. coli Nissle 1917* isusefulin the maintenance of remissionof ulcerativecolitis.

Pouchitis

- VSL#3 is effective in the prevention ofpouchitis following ileo-anal pouch formation, and
- in the maintenance of remission ofpouchitis.
- There is limited evidence for the use of probiotics in the treatment of acute pouchitis.

Crohn's disease

- Probiotics are not effective in the treatment or prevention of recurrence inCrohn's disease. It is likely that the pathogenesis of ulcerative colitis and Crohn's disease, particularly the role of microbes-host interaction, is different between the two disease entities.

Mechanisms of probiotics

- Mechanisms by probiotic bacteria in inflammatory bowel disease include the modulation of the microbiota itself, improved barrier function and direct effects of bacteria on different epithelial and immune cell types, including dendritic cells and T cells.

Reference Cited

Akbari, O., R.H. DeKruyff and D.T. Umetsu. 2001. Pulmonary dendritic cells producing IL-10 mediate tolerance induced by respiratory exposure to antigen. Nat. Immunol. 2: 725–731.

Asahara, T., K. Shimizu, K. Nomoto, T. Hamabata, A. Ozawa and Y. Takeda. 2004. Probiotic bifidobacteria protect mice from lethal infection with Shiga toxin-producing *Escherichia coli* O157:H7. Infect. Immun. 72: 2240–2247.

Bai, A.P., Q. Ouyang, X.R. Xiao and S.F. Li. 2006. Probiotics modulate inflammatory cytokine secretion from inflamed mucosa in active ulcerative colitis. Int. J. Clin. Pract. 60: 284–288.

Borruel, N., M. Carol, F. Casellas, M. Antolin, F. de Lara, E. Espin, J. Naval, F. Guarner and J.R. Malagelada. 2002. Increased mucosal tumour necrosis factor alpha production in Crohn's disease can be downregulated *ex vivo* by probiotic bacteria. Gut. 51: 659–664.

Bousvaros, A., S. Guandalini, R.N. Baldassano, C. Botelho, J. Evans, G.D. Ferry, B. Goldin, L. Hartigan, S. Kugathasan, J. Levy, K.F. Murray, M. Oliva-Hemker, J.R. Rosh, V. Tolia, A. Zholudev, J.A. Vanderhoof and P.L. Hibberd. 2005. A randomized, double-blind trial of Lactobacillus GG versus placebo in addition to standard maintenance therapy for children with Crohn's disease. Inflamm. Bowel Dis. 11: 833–839.

Braat, H., E.C. de Jong, J.M. van den Brande, M.L. Kapsenberg, M.P. Peppelenbosch, E.A. van Tol and S.J. Van Deventer. 2004a. Dichotomy between *Lactobacillus rhamnosus* and Klebsiella pneumoniae on dendritic cell phenotype and function. J. Mol. Med. 82: 197–205.

Braat, H., B.J. van den, E. van Tol, D. Hommes, M. Peppelenbosch and S. van Deventer. 2004b. *Lactobacillus rhamnosus* induces peripheral hyporesponsiveness in stimulated CD4+ T cells via modulation of dendritic cell function. Am. J. Clin. Nutr. 80: 1618–1625.

Bullock, N.R., J.C. Booth and G.R. Gibson. 2004. Comparative composition of bacteria in the human intestinal microflora during remission and active ulcerative colitis. Curr. Issues Intest. Microbiol. 5: 59–64.

Chermesh, I., A. Tamir, R. Reshef, Y. Chowers, A. Suissa, D. Katz, M. Gelber, Z. Halpern, S. Bengmark and R. Eliakim. 2007. Failure of Synbiotic 2000 to prevent postoperative recurrence of Crohn's disease. Dig. Dis. Sci. 52: 385–389.

Christensen, H.R., H. Frøkiaer and J.J. Pestka. 2002. Lactobacilli differentially modulate expression of cytokines and maturation surface markers in murine dendritic cells. J. Immunol. 168(1): 171–8.

Chung, Y.W., J.H. Choi, T.Y. Oh, C.S. Eun and D.S. Han. 2008. Lactobacillus casei prevents the development of dextran sulphate sodium-induced colitis in Toll-like receptor 4 mutant mice. Clin. Exp. Immunol. 151: 182–189.

Dalmasso, G., F. Cottrez, V. Imbert, P. Lagadec, J.F. Peyron, P. Rampal, D. Czerucka, H. Groux, A. Foussat and V. Brun. 2006. *Saccharomyces boulardii* inhibits inflammatory bowel disease by trapping T cells in mesenteric lymph nodes. Gastroenterology. 131: 1812–1825.

Di Giacinto, C., M. Marinaro, M. Sanchez, W. Strober and M. Boirivant. 2005. Probiotics ameliorate recurrent Th1-mediated murine colitis by inducing IL-10 and IL-10-dependent TGF-beta-bearing regulatory cells. J. Immunol. 174: 3237–3246.

Dieleman, L.A., M.S. Goerres, A. Arends, D. Sprengers, C. Torrice, F. Hoentjen, W.B. Grenther and R.B. Sartor. 2003. *Lactobacillus* GG prevents recurrence of colitis in HLA-B27 transgenic rats after antibiotic treatment. Gut. 52: 370–376.

Doherty, G.A., G.C. Bennett, A.S. Cheifetz and A.C. Moss. 2010. Meta-analysis: targeting the intestinal microbiota in prophylaxis for post-operative Crohn's disease. Aliment. Pharmacol. Ther. 31: 802–809.

Drakes, M., T. Blanchard and S. Czinn. 2004. Bacterial probiotic modulation of dendritic cells. Infect. Immun. 72: 3299–3309.

Flynn, S., D. van Sinderen, G.M. Thornton, H. Holo, I.F. Nes and J.K. Collins. 2002. Characterization of the genetic locus responsible for the production of ABP-118, a novel bacteriocin produced by the probiotic bacterium *Lactobacillus salivarius* subsp. salivarius UCC118. Microbiology. 148: 973–984.

Gionchetti, P., F. Rizzello, U. Helwig, A. Venturi, K.M. Lammers, P. Brigidi, B. Vitali, G. Poggioli, M. Miglioli and M. Campieri. 2003. Prophylaxis of pouchitis onset with probiotic therapy: a double-blind, placebo-controlled trial. Gastroenterology. 124: 1202–1209.

Gionchetti, P., F. Rizzello, A. Venturi, P. Brigidi, D. Matteuzzi, G. Bazzocchi, G. Poggioli, M. Miglioli and M. Campieri. 2000. Oral bacteriotherapy as maintenance treatment in patients with chronic pouchitis: a double-blind, placebo-controlled trial. Gastroenterology. 119: 305–309.

Girardin, M. and E.G. Seidman. 2011. Indications for the use of probiotics in gastrointestinal diseases. Dig. Dis. 29: 574–587.

Gotteland, M., O. Brunser and S. Cruchet. 2006. Systematic review: are probiotics useful in controlling gastric colonization by Helicobacter pylori? Aliment. Pharmacol. Ther. 23: 1077–1086.

Hart, A.L., K. Lammers, P. Brigidi, B. Vitali, F. Rizzello, P. Gionchetti, M. Campieri, M.A. Kamm, S.C. Knight and A.J. Stagg. 2004. Modulation of human dendritic cell phenotype and function by probiotic bacteria. Gut. 53: 1602–1609.

He, F., H. Morita, A.C. Ouwehand, A. Hosoda, M. Hiramatsu, J. Kurisaki, F. Isolauri, Y. Benno and S. Salminen. 2002. Stimulation of the secretion of pro-inflammatory cytokines by Bifidobacterium strains. Microbiol. Immunol. 46: 781–785.

Hering, N.A., M. Fromm and J.D. Schulzke. 2012. Determinants of colonic barrier function in inflammatory bowel disease and potential therapeutics. J. Physiol. 590: 1035–1044.

Hilsden, R.J., J.B. Meddings and L.R. Sutherland. 1996. Intestinal permeability changes in response to acetylsalicylic acid in relatives of patients with Crohn's disease. Gastroenterology. 110: 1395–1403.

Ibnou-Zekri, N., S. Blum, E.J. Schiffrin and W.T. von der. 2003. Divergent patterns of colonization and immune response elicited from two intestinal Lactobacillus strains that display similar properties *in vitro*. Infect. Immun. 71: 428–436.

Ishikawa, H., I. Akedo, Y. Umesaki, R. Tanaka, A. Imaoka and T. Otani. 2003. Randomized controlled trial of the effect of bifidobacteria-fermented milk on ulcerative colitis. J. Am. Coll. Nutr. 22: 56–63.

Jijon, H., J. Backer, H. Diaz, H. Yeung, D. Thiel, C. McKaigney, C. De Simone and K. Madsen. 2004. DNA from probiotic bacteria modulates murine and human epithelial and immune function. Gastroenterology. 126: 1358–1373.

Jonkers, D., J. Penders, A. Masclee and M. Pierik. 2012. Probiotics in the management of inflammatory bowel disease: a systematic review of intervention studies in adult patients. Drugs. 72L 803–823.

Joossens, M., G. Huys, M. Cnockaert, P. De, V, K. Verbeke, P. Rutgeerts, P. Vandamme and S. Vermeire. 2011. Dysbiosis of the faecal microbiota in patients with Crohn's disease and their unaffected relatives. Gut. 60: 631–637.

Kaila, M., E. Isolauri, E. Soppi, E. Virtanen, S. Laine and H. Arvilommi. 1992. Enhancement of the circulating antibody secreting cell response in human diarrhea by a human Lactobacillus strain. Pediatr. Res. 32: 141–144.

Karlsson, H., P. Larsson, A.E. Wold and A. Rudin. 2004. Pattern of cytokine responses to gram-positive and gram-negative commensal bacteria is profoundly changed when monocytes differentiate into dendritic cells. Infect. Immun. 72: 2671–2678.

Kato, K., S. Mizuno, Y. Umesaki, Y. Ishii, M. Sugitani, A. Imaoka, M. Otsuka, O. Hasunuma, R. Kurihara, A. Iwasaki and Y. Arakawa. 2004. Randomized placebo-controlled trial assessing the effect of bifidobacteria-fermented milk on active ulcerative colitis. Aliment. Pharmacol. Ther. 20: 1133–1141.

Kelly, D., J.I. Campbell, T.P. King, G. Grant, E.A. Jansson, A.G. Coutts, S. Pettersson and S. Conway. 2004. Commensal anaerobic gut bacteria attenuate inflammation by regulating nuclear-cytoplasmic shuttling of PPAR-gamma and RelA. Nat. Immunol. 5: 104–112.

Kennedy, R.J., M. Hoper, K. Deodhar, S.J. Kirk, K.R. Gardiner. 2000. Probiotic therapy fails to improve gut permeability in a hapten model of colitis. Scand. J. Gastroenterol. 35: 1266–1271.

Kruis, W., P. Fric, J. Pokrotnieks, M. Lukas, B. Fixa, M. Kascak, M.A. Kamm, J. Weismueller, C. Beglinger, M. Stolte, C. Wolff and J. Schulze. 2004. Maintaining remission of ulcerative colitis with the probiotic *Escherichia coli* Nissle 1917 is as effective as with standard mesalazine. Gut. 53: 1617–1623.

Kruis, W., E. Schutz, P. Fric, B. Fixa, G. Judmaier and M. Stolte. 1997. Double-blind comparison of an oral *Escherichia coli* preparation and mesalazine in maintaining remission of ulcerative colitis. Aliment. Pharmacol. Ther. 11: 853–858.

Lammers, K.M., P. Brigidi, B. Vitali, P. Gionchetti, F. Rizzello, E. Caramelli, D. Matteuzzi and M. Campieri. 2003. Immunomodulatory effects of probiotic bacteria DNA: IL-1 and IL-10 response in human peripheral blood mononuclear cells. FEMS Immunol. Med. Microbiol. 38: 165–172.

Lammers, K.M., U. Helwig, E. Swennen, F. Rizzello, A. Venturi, E. Caramelli, M.A. Kamm, P. Brigidi, P. Gionchetti and M. Campieri. 2002. Effect of probiotic strains on interleukin 8 production by HT29/19A cells. Am. J. Gastroenterol. 97: 1182–1186.

Landy, J., H.O. Al-Hassi, S.D. McLaughlin, S.C. Knight, P.J. Ciclitira, R.J. Nicholls, S.K. Clark and A.L. Hart. 2012. Etiology of pouchitis. Inflamm. Bowel Dis. 18: 1146–1155.

Ley, R.E., M. Hamady, C. Lozupone, P.J. Turnbaugh, R.R. Ramey, J.S. Bircher, M.L. Schlegel, T.A. Tucker, M.D. Schrenzel, R. Knight and J.I. Gordon. 2008. Evolution of mammals and their gut microbes. Science. 320: 1647–1651.

Lorea, B.M., P.V. Kirjavainen, S. Hekmat and G. Reid. 2007. Anti-inflammatory effects of probiotic yogurt in inflammatory bowel disease patients. Clin. Exp. Immunol. 149: 470–479.

Macfarlane, S., E. Furrie, A. Kennedy, J.H. Cummings and G.T. Macfarlane. 2005. Mucosal bacteria in ulcerative colitis. Br. J. Nutr. 93 Suppl 1: S67–S72.

Mack, D.R., S. Michail, S. Wei, L. McDougall and M.A. Hollingsworth. 1999. Probiotics inhibit enteropathogenic *E. coli* adherence *in vitro* by inducing intestinal mucin gene expression. Am. J. Physiol. 276: G941–G950.

Madsen, K., A. Cornish, P. Soper, C. McKaigney, H. Jijon, C. Yachimec, J. Doyle, L. Jewell and C. De Simone. 2001. Probiotic bacteria enhance murine and human intestinal epithelial barrier function. Gastroenterology. 121: 580–591.

Madsen, K.L., J.S. Doyle, L.D. Jewell, M.M. Tavernini and R.N. Fedorak. 1999. *Lactobacillus* species prevents colitis in interleukin 10 gene-deficient mice. Gastroenterology. 116: 1107–1114.

Malchow, H.A. 1997. Crohn's disease and *Escherichia coli*. A new approach in therapy to maintain remission of colonic Crohn's disease? J. Clin. Gastroenterol. 25: 653–658.

Manichanh, C., L. Rigottier-Gois, E. Bonnaud, K. Gloux, E. Pelletier, L. Frangeul, R. Nalin, C. Jarrin, P. Chardon, P. Marteau, J. Roca and J. Dore. 2006. Reduced diversity of faecal microbiota in Crohn's disease revealed by a metagenomic approach. Gut. 55: 205–211.

Mao, Y., S. Nobaek, B. Kasravi, D. Adawi, U. Stenram, G. Molin and B. Jeppsson. 1996. The effects of *Lactobacillus* strains and oat fiber on methotrexate-induced enterocolitis in rats. Gastroenterology. 111: 334–344.

Marteau, P., M. Lemann, P. Seksik, D. Laharie, J.F. Colombel, Y. Bouhnik, G. Cadiot, J.C. Soule, A. Bourreille, E. Metman, E. Lerebours, F. Carbonnel, J.L. Dupas, M. Veyrac, B. Coffin, J. Moreau, V. Abitbol, S. Blum-Sperisen and J.Y. Mary. 2006. Ineffectiveness of *Lactobacillus johnsonii* LA1 for prophylaxis of postoperative recurrence in Crohn's disease: a randomised, double blind, placebo controlled GETAID trial. Gut. 55: 842–847.

Martin, F.P., Y. Wang, N. Sprenger, I.K. Yap, T. Lundstedt, P. Lek, S. Rezzi, Z. Ramadan, P. van Bladeren, L.B. Fay, S. Kochhar, J.C. Lindon, E. Holmes and J.K. Nicholson. 2008. Probiotic modulation of symbiotic gut microbial-host metabolic interactions in a humanized microbiome mouse model. Mol. Syst. Biol. 4: 157.

May, G.R., L.R. Sutherland and J.B. Meddings. 1993. Is small intestinal permeability really increased in relatives of patients with Crohn's disease? Gastroenterology. 104: 1627–1632.

Mazmanian, S.K., J.L. Round and D.L. Kasper. 2008. A microbial symbiosis factor prevents intestinal inflammatory disease. Nature. 453: 620–625.

McCarthy, J., L. O'Mahony, L. O'Callaghan, B. Sheil, E.E. Vaughan, N. Fitzsimons, J. Fitzgibbon, G.C. O'Sullivan, B. Kiely, J.K. Collins and F. Shanahan. 2003. Double blind, placebo controlled trial of two probiotic strains in interleukin 10 knockout mice and mechanistic link with cytokine balance. Gut. 52: 975–980.

McCole, D.F. 2012. Regulation of epithelial barrier function by the inflammatory bowel disease candidate gene, PTPN2. Ann. N.Y. Acad. Sci. 1257: 108–114.

Meddings, J. 2008. The significance of the gut barrier in disease. Gut. 57: 438–440.

Miele, E., F. Pascarella, E. Giannetti, L. Quaglietta, R.N. Baldassano and A. Staiano. 2009. Effect of a probiotic preparation (VSL#3) on induction and maintenance of remission in children with ulcerative colitis. Am. J. Gastroenterol. 104: 437–443.

Miettinen, M., S. Matikainen, J. Vuopio-Varkila, J. Pirhonen, K. Varkila, M. Kurimoto and I. Julkunen. 1998. Lactobacilli and streptococci induce interleukin-12 (IL-12), IL-18, and gamma interferon production in human peripheral blood mononuclear cells. Infect. Immun. 66: 6058–6062.

Mimura, T., F. Rizzello, U. Helwig, G. Poggioli, S. Schreiber, I.C. Talbot, R.J. Nicholls, P. Gionchetti, M. Campieri and M.A. Kamm. 2004. Once daily high dose probiotic therapy (VSL#3) for maintaining remission in recurrent or refractory pouchitis. Gut. 53: 108–114.

Miyauchi, E., J. O'Callaghan, L.F. Butto, G. Hurley, S. Melgar, S. Tanabe, F. Shanahan, K. Nally and P.W. O'Toole. 2012. Mechanism of protection of transepithelial barrier function by *Lactobacillus salivarius*: strain-dependence and attenuation by bacteriocin production. Am. J Physiol. Gastrointest. Liver Physiol.

Mohamadzadeh, M., S. Olson, W.V. Kalina, G. Ruthel, G.L. Demmin, K.L. Warfield, S. Bavari and T.R. Klaenhammer. 2005. Lactobacilli activate human dendritic cells that skew T cells toward T helper 1 polarization. Proc. Natl. Acad. Sci. USA 102: 2880–2885.

Mylonaki, M., N.B. Rayment, D.S. Rampton, B.N. Hudspith and J. Brostoff. 2005. Molecular characterization of rectal mucosa-associated bacterial flora in inflammatory bowel disease. Inflamm. Bowel.Dis. 11: 481–487.

Nakamura, K., A. Kitani, I. Fuss, A. Pedersen, N. Harada, H. Nawata and W. Strober. 2004. TGF-beta 1 plays an important role in the mechanism of CD4+CD25+ regulatory T cell activity in both humans and mice. J. Immunol. 172: 834–842.

Navaneethan, U. and B. Shen. 2010. Diagnosis and management of pouchitis and ileoanal pouch dysfunction. Curr. Gastroenterol Rep. 12: 485–494.

Neish, A.S., A.T. Gewirtz, H. Zeng, A.N. Young, M.E. Hobert, V. Karmali, A.S. Rao and J.L. Madara. 2000. Prokaryotic regulation of epithelial responses by inhibition of IkappaB-alpha ubiquitination. Science. 289: 1560–1563.

Nemcova, R. 1997. [Criteria for selection of lactobacilli for probiotic use]. Vet. Med. (Praha). 42: 19–27.

Neurath, M.F. 2007. IL-23: a master regulator in Crohn disease. Nat. Med. 13: 26–28.

Neut, C., P. Bulois, P. Desreumaux, J.M. Membre, E. Lederman, L. Gambiez, A. Cortot, P. Quandalle, H. van Kruiningen and J.F. Colombel. 2002. Changes in the bacterial flora of the neoterminal ileum after ileocolonic resection for Crohn's disease. Am. J. Gastroenterol. 97: 939–946.

Ng, S.C., J.L. Benjamin, N.E. McCarthy, C.R. Hedin, A. Koutsoumpas, S. Plamondon, C.L. Price, A.L. Hart, M.A. Kamm, A. Forbes, S.C. Knight, J.O. Lindsay, K. Whelan and A.J. Stagg . 2011. Relationship between human intestinal dendritic cells, gut microbiota, and disease activity in Crohn's disease. Inflamm. Bowel Dis. 17: 2027–2037.

Ng, S.C., S. Plamondon, M.A. Kamm, A.L. Hart, H.O. Al-Hassi, T. Guenther, A.J. Stagg and S.C. Knight. 2010a. Immunosuppressive effects via human intestinal dendritic cells of probiotic bacteria and steroids in the treatment of acute ulcerative colitis. Inflamm. Bowel Dis. 16: 1286–1298.

Ng, S.C., S. Plamondon, M.A. Kamm, A.L. Hart, H.O. Al-Hassi, T. Guenther, A.J. Stagg and S.C. Knight. 2010b. Immunosuppressive effects via human intestinal dendritic cells of probiotic bacteria and steroids in the treatment of acute ulcerative colitis. Inflamm. Bowel Dis.

Ogawa, T., Y. Asai, R. Tamai, Y. Makimura, H. Sakamoto, S. Hashikawa and K. Yasuda. 2006. Natural killer cell activities of synbiotic *Lactobacillus casei* ssp. casei in conjunction with dextran. Clin. Exp. Immunol. 143: 103–109.

Oida, T., X. Zhang, M. Goto, S. Hachimura, M. Totsuka, S. Kaminogawa and H.L. Weiner. 2003. CD4+. J. Immunol. 170: 2516–2522.

Otte, J.M. and D.K. Podolsky. 2004. Functional modulation of enterocytes by gram-positive and gram-negative microorganisms. Am. J. Physiol. Gastrointest. Liver Physiol. 286: G613–G626.

Pathmakanthan, S., C.K. Li, J. Cowie and C.J. Hawkey. 2004. *Lactobacillus plantarum* 299: beneficial *in vitro* immunomodulation in cells extracted from inflamed human colon. J. Gastroenterol. Hepatol. 19: 166–173.

Peeters, M., B. Geypens, D. Claus, H. Nevens, Y. Ghoos, G. Verbeke, F. Baert, S. Vermeire, R. Vlietinck and P. Rutgeerts. 1997. Clustering of increased small intestinal permeability in families with Crohn's disease. Gastroenterology. 113: 802–807.

Petrof, E.O., K. Kojima, M.J. Ropeleski, M.W. Musch, Y. Tao, C. De Simone and E.B. Chang. 2004. Probiotics inhibit nuclear factor-kappaB and induce heat shock proteins in colonic epithelial cells through proteasome inhibition. Gastroenterology. 127: 1474–1487.

Plein, K. and J. Hotz. 1993. Therapeutic effects of *Saccharomyces boulardii* on mild residual symptoms in a stable phase of Crohn's disease with special respect to chronic diarrhea--a pilot study. Z. Gastroenterol. 31: 129–134.

Prantera, C., M.L. Scribano, G. Falasco, A. Andreoli and C. Luzi. 2002. Ineffectiveness of probiotics in preventing recurrence after curative resection for Crohn's disease: a randomised controlled trial with Lactobacillus GG. Gut. 51: 405–409.

Rachmilewitz, D., F. Karmeli, K. Takabayashi, T. Hayashi, L. Leider-Trejo, J. Lee, L.M. Leoni and E. Raz. 2002. Immunostimulatory DNA ameliorates experimental and spontaneous murine colitis. Gastroenterology. 122: 1428–1441.

Rachmilewitz, D., K. Katakura, F. Karmeli, T. Hayashi, C. Reinus, B. Rudensky, S. Akira, K. Takeda, J. Lee, K. Takabayashi and E. Raz. 2004. Toll-like receptor 9 signaling mediates the anti-inflammatory effects of probiotics in murine experimental colitis. Gastroenterology. 126: 520–528.

Rakoff-Nahoum, S., J. Paglino, F. Eslami-Varzaneh, S. Edberg and R. Medzhitov. 2004. Recognition of commensal microflora by toll-like receptors is required for intestinal homeostasis. Cell. 118: 229–241.

Rea, M.C., E. Clayton, P.M. O'Connor, F. Shanahan, B. Kiely, R.P. Ross and C. Hill. 2007. Antimicrobial activity of lacticin 3,147 against clinical *Clostridium difficile* strains. J. Med. Microbiol. 56: 940–946.

Rembacken, B.J., A.M. Snelling, P.M. Hawkey, D.M. Chalmers and A.T. Axon. 1999. Non-pathogenic *Escherichia coli* versus mesalazine for the treatment of ulcerative colitis: a randomised trial. Lancet. 354: 635–639.

Rescigno, M., M. Urbano, B. Valzasina, M. Francolini, G. Rotta, R. Bonasio, F. Granucci, J.P. Kraehenbuhl and P. Ricciardi-Castagnoli. 2001. Dendritic cells express tight junction proteins and penetrate gut epithelial monolayers to sample bacteria. Nat. Immunol. 2: 361–367.

Resta-Lenert, S. and K.E. Barrett. 2003. Live probiotics protect intestinal epithelial cells from the effects of infection with enteroinvasive *Escherichia coli* (EIEC). Gut. 52: 988–997.

Rigby, R., M.A. Kamm, S.C. Knight, A.L. Hart and A.J. Stagg. 2002. Pathogenic bacteria stimulate colonic dendritic cells to produce pro-inflammatory IL-12 while the response to probiotic bacteria is to produce anti-inflammatory IL-10. Gut. 50: A70.

Rol, N., L. Favre, J. Benyacoub and B. Corthesy. 2012. The Role of Secretory Immunoglobulin A in the Natural Sensing of Commensal Bacteria by Mouse Peyer's Patch Dendritic Cells. J. Biol. Chem. 287: 40074–40082.

Sartor, R.B. 1997. The influence of normal microbial flora on the development of chronic mucosal inflammation. Res. Immunol. 148: 567–576.

Schlee, M.,J. Wehkamp, A. Altenhoefer, T.A. Oelschlaeger, E.F. Stange and K. Fellermann. 2007. Induction of human beta-defensin 2 by the probiotic *Escherichia coli* Nissle 1917 is mediated through flagellin. Infect. Immun. 75: 2399–2407.

Schmitz, H., C. Barmeyer, M. Fromm, N. Runkel, H.D. Foss, C.J. Bentzel, E.O. Riecken and J.D. Schulzke. 1999. Altered tight junction structure contributes to the impaired epithelial barrier function in ulcerative colitis. Gastroenterology. 116: 301–309.

Sellon, R.K., S. Tonkonogy, M. Schultz, L.A. Dieleman, W. Grenther, E. Balish, D.M. Rennick and R.B. Sartor. 1998. Resident enteric bacteria are necessary for development of spontaneous colitis and immune system activation in interleukin-10-deficient mice. Infect. Immun. 66: 5224–5231.

Shen, B., A. Brzezinski, V.W. Fazio, F.H. Remzi, J.P. Achkar, A.E. Bennett, K. Sherman and B.A. Lashner. 2005. Maintenance therapy with a probiotic in antibiotic-dependent pouchitis: experience in clinical practice. Aliment. Pharmacol. Ther. 22: 721–728.

Sherman, P.M., K.C. Johnson-Henry, H.P. Yeung, P.S. Ngo, J. Goulet and T.A. Tompkins 2005. Probiotics reduce enterohemorrhagic *Escherichia coli* O157:H7- and enteropathogenic E. coli O127:H6-induced changes in polarized T84 epithelial cell monolayers by reducing bacterial adhesion and cytoskeletal rearrangements. Infect. Immun. 73: 5183–5188.

Smits, H.H., A. Engering, K.D. van der, E.C. de Jong, K. Schipper, T.M. van Capel, B.A. Zaat, M. Yazdanbakhsh, E.A. Wierenga, Y. van Kooyk and M.L. Kapsenberg. 2005. Selective probiotic bacteria induce IL-10-producing regulatory T cells *in vitro* by modulating dendritic cell function through dendritic cell-specific intercellular adhesion molecule 3-grabbing nonintegrin. J. Allergy Clin. Immunol. 115: 1260–1267.

Sokol, H., B. Pigneur, L. Watterlot, O. Lakhdari, L.G. Bermudez-Humaran, J.J. Gratadoux, S. Blugeon, C. Bridonneau, J.P. Furet, G. Corthier, C. Grangette, N. Vasquez, P. Pochart, G. Trugnan, G. Thomas, H.M. Blottiere, J. Dore, P. Marteau, P. Seksik and P. Langella. 2008. *Faecalibacterium prausnitzii* is an anti-inflammatory commensal bacterium identified by gut microbiota analysis of Crohn disease patients. Proc. Natl. Acad. Sci. USA 105: 16731–16736.

Sokol, H., P. Seksik, J.P. Furet, O. Firmesse, I. Nion-Larmurier, L. Beaugerie, J. Cosnes, G. Corthier, P. Marteau and J. Dore. 2009. Low counts of Faecalibacterium prausnitzii in colitis microbiota. Inflamm. Bowel.Dis. 15: 1183–1189.

Sonnenburg, J.L., C.T. Chen and J.I. Gordon. 2006. Genomic and metabolic studies of the impact of probiotics on a model gut symbiont and host. PLoS. Biol. 4: e413.

Sood, A., V. Midha, G.K. Makharia, V. Ahuja, D. Singal, P. Goswami and R.K. Tandon. 2009. The probiotic preparation, VSL#3 induces remission in patients with mild-to-moderately active ulcerative colitis. Clin. Gastroenterol. Hepatol. 7: 1202–9, 1209.

Stagg, A.J., A.L. Hart, S.C Knight and M.A. Kamm. 2003. The dendritic cell: its role in intestinal inflammation and relationship with gut bacteria. Gut. 52: 1522–1529.

Swidsinski, A., A. Ladhoff, A. Pernthaler, S. Swidsinski, V. Loening-Baucke, M. Ortner, J. Weber, U. Hoffmann, S. Schreiber, M. Dietel and H. Lochs. 2002. Mucosal flora in inflammatory bowel disease. Gastroenterology. 122: 44–54.

Takeda, K., T. Suzuki, S.I. Shimada, K. Shida, M. Nanno and K. Okumura. 2006. Interleukin-12 is involved in the enhancement of human natural killer cell activity by *Lactobacillus casei* Shirota. Clin. Exp. Immunol. 146: 109–115.

Tejada-Simon, M.V., J.H. Lee, Z. Ustunol and J.J. Pestka. 1999. Ingestion of yogurt containing Lactobacillus acidophilus and Bifidobacterium to potentiate immunoglobulin A responses to cholera toxin in mice. J. Dairy Sci. 82: 649–660.

Tursi, A., G. Brandimarte, A. Papa, A. Giglio, W. Elisei, G.M. Giorgetti, G. Forti, S. Morini, C. Hassan, M.A. Pistoia, M.E. Modeo, S. Rodino', T. D'Amico, L. Sebkova, N. Sacca', G.E. Di, F. Luzza, M. Imeneo, T. Larussa, R.S. Di, V. Annese, S. Danese and A. Gasbarrini. 2010. Treatment of relapsing mild-to-moderate ulcerative colitis with the probiotic VSL#3 as adjunctive to a standard pharmaceutical treatment: a double-blind, randomized, placebo-controlled study. Am. J. Gastroenterol. 105: 2218–2227.

Ukena, S.N., A. Singh, U. Dringenberg, R. Engelhardt, U. Seidler, W. Hansen, A. Bleich, D. Bruder, A. Franzke, G. Rogler, S. Suerbaum, J. Buer, F. Gunzer and A.M. Westendorf. 2007. Probiotic *Escherichia coli* Nissle 1917 inhibits leaky gut by enhancing mucosal integrity. PLoS. One. 2: e1308.

Ulisse, S., P. Gionchetti, S. D'Alo, F.P. Russo, I. Pesce, G. Ricci, F. Rizzello, U. Helwig, M.G. Cifone, M. Campieri and C. De Simone. 2001. Expression of cytokines, inducible nitric oxide synthase, and matrix metalloproteinases in pouchitis: effects of probiotic treatment. Am. J. Gastroenterol. 96: 2691–2699.

Veckman, V., M. Miettinen, J. Pirhonen, J. Siren, S. Matikainen and I. Julkunen. 2004. *Streptococcus pyogenes* and *Lactobacillus rhamnosus* differentially induce maturation and production of Th1-type cytokines and chemokines in human monocyte-derived dendritic cells. J. Leukoc. Biol. 75: 764–771.

Venturi, A., P. Gionchetti, F. Rizzello, R. Johansson, E. Zucconi, P. Brigidi, D. Matteuzzi and M. Campieri. 1999. Impact on the composition of the faecal flora by a new probiotic preparation: preliminary data on maintenance treatment of patients with ulcerative colitis. Aliment. Pharmacol. Ther. 13: 1103–1108.

Waidmann, M., O. Bechtold, J.S. Frick, H.A. Lehr, S. Schubert, U. Dobrindt, J. Loeffler, E. Bohn and I.B. Autenrieth. 2003. Bacteroides vulgatus protects against *Escherichia coli*-induced colitis in gnotobiotic interleukin-2-deficient mice. Gastroenterology. 125: 162–177.

Wehkamp, J., J. Harder, K. Wehkamp, B. Wehkamp-von Meissner, M. Schlee, C. Enders, U. Sonnenborn, S. Nuding, S. Bengmark, K. Fellermann, J.M. Schroder and E.F. Stange. 2004. NF-kappaB- and AP-1-mediated induction of human beta defensin-2 in intestinal epithelial cells by *Escherichia coli* Nissle 1917: a novel effect of a probiotic bacterium. Infect. Immun. 72: 5750–5758.

Williamson, E., J.M. Bilsborough and J.L. Viney. 2002. Regulation of mucosal dendritic cell function by receptor activator of NF-kappa B (RANK)/RANK ligand interactions: impact on tolerance induction. J. Immunol. 169: 3606–3612.

Wyatt, J., H. Vogelsang, W. Hubl, T. Waldhoer and H. Lochs. 1993. Intestinal permeability and the prediction of relapse in Crohn's disease. Lancet. 341: 1437–1439.

Yan, F. and D.B. Polk. 2002. Probiotic bacterium prevents cytokine-induced apoptosis in intestinal epithelial cells. J. Biol. Chem. 277: 50959–50965.

Young, S.L., M.A. Simon, M.A. Baird, G.W. Tannock, R. Bibiloni, K. Spencely, J.M. Lane, P. Fitzharris, J. Crane, I. Town, E. Addo-Yobo, C.S. Murray and A. Woodcock. 2004. Bifidobacterial species differentially affect expression of cell surface markers and cytokines of dendritic cells harvested from cord blood. Clin. Diagn. Lab Immunol. 11: 686–690.

Zhang, Z., D.J. Hinrichs, H. Lu, H. Chen, W. Zhong and J.K. Kolls. 2007. After interleukin-12p40, are interleukin-23 and interleukin-17 the next therapeutic targets for inflammatory bowel disease? Int. Immunopharmacol. 7: 409–416.

Zyrek, A.A., C. Cichon, S. Helms, C. Enders, U. Sonnenborn and M.A. Schmidt. 2007. Molecular mechanisms underlying the probiotic effects of *Escherichia coli* Nissle 1917 involve ZO-2 and PKCzeta redistribution resulting in tight junction and epithelial barrier repair. Cell Microbiol. 9: 804–816.

12

Mechanisms, Prevention and Management of Diarrhoea in Enteral Nutrition

Kevin Whelan[1,*] and *Stéphane M. Schneider*[2]

Introduction

Enteral nutrition (EN) is a method of artificial nutritional support commonly used in patients in the hospital or community setting. Diarrhoea can occur in 2%–95% of patients, the wide range resulting from differences in the patient populations and the definition of diarrhoea adopted (Whelan 2007). It may result in a number of negative clinical sequelae, including fluid and electrolyte abnormalities, faecal incontinence andpressure sores. This is a frequent reason for discrepancies between prescription and intake of EN in hospitalized patients (Martins et al. 2012) and may even result in the cessation of EN, which may exacerbate under nutrition and its associated consequences of increased morbidity and mortality. Importantly, diarrhoea is also distressing and burdensome for both patients and their carers (Majid

[1] Diabetes and Nutritional Sciences Division, King's College London, 150 Stamford Street, London, SE1 9NH, United Kingdom.
E-mail: kevin.whelan@kcl.ac.uk

[2] Nutritional Support Unit, Department of Gastroenterology and Clinical Nutrition, Archet University Hospital, 151 route de Saint-Antoine, CS 23079, 06202 Nice Cedex 3, France.
E-mail: stephane.schneider@unice.fr

* Corresponding author

et al. 2012). The chapter briefly reviews the mechanisms of diarrhoea during EN and then critically appraises recent and emerging evidence on the prevention and management of this distressing complication.

Pathogenesis of Diarrhoea

A number of factors contribute to the pathogenesis of diarrhoea in EN, including altered physiological response, antibiotics and enteropathogenic infections.

Altered physiological responses may occur during EN. Studies in healthy subjects found that intra-gastric EN results in abnormal water secretion into the ascending colon (Bowling et al. 1993). This may be exacerbated by suppression of distal colonic motor activity that accelerates colonic transit and reduces the opportunity for water absorption (Bowling and Silk 1998). If these occur in patients receiving EN, then in the absence of compensatory absorptive mechanisms, diarrhoea may result.

Many patients receiving EN also receive concomitant antibiotics, and a number of studies have shown that diarrhoea is associated with the prescription, number or duration of antibiotics (Halmos et al. 2010; Trabal et al. 2008). However, it is unclear whether this is merely antibiotic-associated diarrhoea, or whether there is a particular interaction between antibiotics, EN and diarrhoea. A number of studies support the latter hypothesis, including that diarrhoea during EN is in part associated with deleterious alterations in luminal microbiota. For example, in a small cohort study that monitored the faecal microbiota in patients during the first two weeks of EN, those who developed diarrhoea had higher clostridia and lower bifidobacteria, and interestingly, these alterations were present even at the start of EN (Whelan et al. 2009). Whilst there is convincing evidence for the role of antibiotics in predisposing to higher risk of diarrhoea during EN, it is unclear whether other factors associated with gastrointestinal dysbiosis (e.g., ageing, disease) may also predispose to such risks.

There is evidence of elevated risk of enteropathogenic infection in patients receiving EN. For example, one case-control study found that *Clostridium difficile* colonization was three-fold higher, and *C. difficile*-associated diarrhoea(CDAD) was nine-fold higher, in patients receiving EN, and this is despite similar antibiotic use (Bliss et al. 1998). More recently, in an analysis of 233 patients undergoing percutaneous endoscopic gastrostomy (PEG), 15 (6.4%) patients developed CDAD within one month, six (2.6%) of whom entered a cycle of recurrent CDAD, resulting in a major interruption to the delivery of EN (Yokohama et al. 2009). Prophylactic antibiotics are frequently given during PEG insertion, and following multivariate analysis, the duration of antibiotic prescription (but not actual antibiotic prescription) was an independent predictor of subsequent development of CDAD (Yokohama et al. 2009).

Prevention of Diarrhoea

A number of factors contribute to the pathogenesis of diarrhoea in EN, meaning that approaches to its prevention and management are necessarily multifaceted, many of which are nutritional interventions. For example, some fibres undergo fermentation and produce short-chain fatty acids (SCFA) that reverse the abnormal colonic water secretion in EN (Bowling et al. 1993) and some fibres can alter colonic motor activity (Silk et al. 2001). In addition, prebiotics and probiotics may suppress enteropathogenic colonization, alter colonic fermentation and modulate immune function (Table 1) and therefore minimize the risks from antibiotics and *C. difficile*. The evidence to support these nutritional interventions is reviewed.

Fibre

Fibres can exert a number of gastrointestinal effects, depending on their solubility in water and fermentability by the colonic microbiota, with some being categorized as prebiotics. They may prevent diarrhoea through reducing the rate of gastric emptying, improving gut barrier function, increasing epithelial cell turnover or regeneration, and increasing colonic fluid and electrolyte absorption (Elia et al. 2008). Previously, many fibre-enriched enteral formulas contained poorly fermentable soy polysaccharides, however these increased formula viscosity and tended to sediment. More recently the composition of fibre-enriched formulas has evolved towards the use of blends of soluble and insoluble fibres. Formulas containing such blends have been reported to increase faecal SCFA concentrations in adult patients receiving EN (Schneider et al. 2006), whereas this was not confirmed in one study in children, albeit perhaps due to the small numbers involved (Guimber et al. 2010).

Partially hydrolyzed guar gum is a soluble fibre added to enteral formulas and has the largest body of evidence supporting its use in diarrhoea prevention when compared against fibre-free formulas (Table 2).

A systematic review and meta-analysis of prospective randomised controlled trials (RCTs) of fibre-enriched formulas in the prevention of diarrhoea has been reported (Elia et al. 2008). Although it statistically aggregated the results of RCTs of different fibres, it reported a preventive effect of fibre-enriched EN on diarrhoea, with a significant reduction in the percentage of patients with diarrhoea (OR: 0.68, 95% CI: 0.48–0.96; 16 data sets) and a tendency for a reduction in the percentage of days with diarrhoea.

Table 1. Potential Mechanisms through which Prebiotics and Probiotics may Influence GI Physiology of Relevance to EN Diarrhoea (Whelan 2007).

Physiological effect	**Examples of prebiotic mechanisms**	**Examples of probiotic mechanisms**
Enteropathogenic suppression		
Luminal colonization resistance	Oligosaccharides reduce faecal pH in infants	Some bifidobacteria inhibit *S. typhimurium* infection in mice, whilst some lactobacilli decrease faecal pH
Mucosal competitive exclusion	Inulin may prevent recurrent CDAD	Some lactobacilli, and their products, inhibit adhesion of *E. coli* and *S. typhimurium* to cell lines
Colonic fermentation		
SCFA production	Inulin enhance SCFA production in pure culture and faecal culture	*S. boulardii* increases butyrate and total SCFA in patients receiving EN
Immune activity		
Immune regulation	Possible effects, limited human studies	Lactobacilli reduce IL-6 production in patients receiving nutrition support on the ICU
Immune stimulation	Possible effects, limited human studies	Some lactobacilli and bifidobacteria enhance phagocytosis and specific IgA production. One study has shown they may enhance IgA and IgG production in patients receiving EN on the ICU

CDAD: *Clostridium difficile*-associated diarrhoea; EN: enteral nutrition; GI: gastrointestinal; ICU: intensive care unit; SCFA: short-chain fatty acids

Table 2. Randomized Controlled Trials that Measure the Preventive Impact of Fibre on Stool Output or Diarrhoea.

Reference	Patient group	Number	Fibre studied	Results
(Vandewoude et al. 2005)	Geriatrics	155	Cellulose, hemicelluloses, pectin, inulin	Reduced number of stools, improved stool consistency
(Hart and Dobb 1988)	ICU	68	Ispaghula husk	No effect
(Homann et al. 1994)	Medicine/Surgery	100	Partially hydrolyzed guar gum	Reduced diarrhoea incidence
(Spapen et al. 2001)	ICU (sepsis)	25	Partially hydrolyzed guar gum	Reduced diarrhoea incidence
(Schultz et al. 2000)	ICU	44	Pectin/Fibre mixture*	No effect
(Belknap et al. 1997)	Medicine/Surgery/ICU	60	Psyllium	No effect
(Dobb and Towler 1990)	ICU	91	Soy polysaccharides	No effect
(Frankenfield and Beyer 1989)	ICU	9	Soy polysaccharides	No effect
(Guenter et al. 1991)	ICU	100	Soy polysaccharides	No effect
(Reese et al. 1996)	Postoperative	80	Soy polysaccharides	Reduced diarrhoea incidence in males only
(de Kruif and Vos 1993)	Postoperative	60	Soy polysaccharides	Reduced diarrhoea score
(Khalil et al. 1998)	Postoperative	16	Soy polysaccharides + oat fibre	No effect

*oat, soy, gum arabic, carboxymethylcellulose, fructo-oligosaccharides; ICU: intensive care unit

Few studies have directly compared different fibres for their impact on diarrhoea in EN. A 12-week RCT of EN in diabetic patients compared two formulas: a "conventional" fibre-enriched formula (55% of energy as carbohydrate; 30% lipids; 15% protein; and 2 g/L of undefined fibre) and a "diabetes-specific" fibre-enriched formula (45% of energy as carbohydrates; 38% lipids—mostly monounsaturated; 17% protein; and 1.5 g/L of a mixture of soluble and insoluble fibres) (Vaisman et al. 2009). Stool frequency during the first week was significantly reduced in the "diabetes-specific" fibre-enriched group (0.8 ± 0.3 stools/day) compared with the "conventional" fibre-enriched group (1.8 ± 0.3 stools/day, $p = 0.034$) and this did not significantly change over time.

Recent ASPEN Intensive Care Unit (ICU) guidelines highlight the interest in partially hydrolyzed guar gum and the lack of interest of insoluble fibre in preventing EN-associated diarrhoea. They go as far as to advise against the use of insoluble fibre, arguing that it may increase the risk of bowel obstruction in the critically ill. However, the evidence to support this recommendation is very limited (five cases of bowel obstruction in surgical and trauma patients) and it is largely based on expert opinion. Meanwhile, ESPEN guidelines on EN refer to fibres and their properties (or lack thereof), but fail to give any recommendation on the matter (Lochs et al. 2006).

Prebiotics and fermentable carbohydrates

Studies have shown that standard formulas (no fibre, no prebiotics) result in negative alterations to the colonic microbiota, including lower total bacteria (Whelan et al. 2005), higher aerobes (Schneider et al. 2000) and lower numbers of the butyrate-producing *Faecalibacterium prausnitzii* (Benus et al. 2010), with subsequent reductions in total SCFA and butyrate (Whelan et al. 2005; Benus et al. 2010) and higher faecal pH (Whelan et al. 2005). In view of this, some formulas now have prebiotics added to them. Prebiotics are defined as '*the selective stimulation of growth and/or activity of one or a limited number of microbial genera/species in the gut microbiota that confer health benefits to the host*' (Roberfroid et al. 2010). The most commonly used prebiotics are inulin-type fructans (inulin, oligofructose, fructo-oligosaccharides).

When added to enteral formulas, prebiotics have been shown to increase faecal bifidobacteria in healthy subjects (Whelan et al. 2005); however, whether such effects occur in patients receiving EN remains as yet unclear. For example, in one RCT of 15 patients receiving long-term home EN, a mixed fibre/prebiotic formula increased faecal SCFAs but had no effect on faecal bifidobacteria (Schneider et al. 2006). However, another RCT of 19 patients receiving long-term home EN investigated the impact of a standard formula (no fibre, no prebiotic) compared with a fibre/fructo-oligosaccharide formula on stool output (Wierdsma et al. 2009). Faecal microbiota were compared between a sub-group of patients (n=10) and there was shown to be a reduction in bifidobacteria in the standard formula, that did not occur in the fibre/prebiotic formula. Although diarrhoea was not measured *per se*, there were no differences in the number of soft stools between groups. Meanwhile, no studies have investigated the ability of prebiotic formulas to stimulate growth of bifidobacteria in in-patients with acute illness.

No studies have specifically investigated the effect of a prebiotic alone in the prevention of diarrhoea during EN, however, a small number have investigated the effect of a fibre/prebiotic formula (Elia et al. 2008). The most notable study is a RCT of 155 older patients who received a standard (no fibre, no prebiotic) or a fibre/prebiotic formula. In those receiving the prebiotic there was lower stool frequency (4.1 vs. 6.3 stools per week, P=0.008) and more formed stools (31% vs. 21% formed stools, P=0.001) than patients receiving standard formula (Vandewoude et al. 2005). In 2010, in a RCT of 34 surgical ICU patients with sepsis on EN, a standard formula and a formula with yellow pea fibre and fructo-oligosaccharides were compared; the authors did not find any significant difference in the frequency of diarrhoea between groups, albeit with a lower diarrhoea score in the fibre/prebiotic-supplemented group (Chittawatanarat et al. 2010).

A number of small cross-over RCTs comparing fibre/prebiotic formulas with standard formulas have been undertaken in children receiving long term EN. In one, there was no difference in stool consistency or frequency in 27 children irrespective of which formula was used, despite increases in number of bifidobacteria during the fibre/prebiotic formula (Guimber et al. 2010). In addition, a study of 25 children found no difference in the number of days with diarrhoea between patients when receiving a standard or a fibre/prebiotic formula (Evans et al. 2009). However, in a recent study that was limited by small numbers (n=14) and heterogeneous disease categories requiring sub-group analysis, children with neurological impairment experienced fewer watery stools when receiving a fibre/prebiotic formula (11%) compared with the standard formula (26%, P<0.05) (Khoshoo et al. 2010). Stool frequency did not differ between formulas. Studies investigating the effects of formulas enriched with prebiotics alone on the luminal microbiota and prevalence of diarrhoea are required. This is particularly important in the hospital setting, where the prevalence of diarrhoea is more of a burden.

Finally, a recent retrospective case-note review of 160 hospitalized patients who had received EN hypothesized that fermentable carbohydrates may actually be associated with the development of diarrhoea (Halmos et al. 2010). Following logistic regression analysis that adjusted for clinical factors associated with diarrhoea (length of stay, duration of EN, antibiotic use) it was shown that one formula was associated with a lower diarrhoea risk (OR=0.18, P=0.029) and that this was low in fermentable oligo-, di-, mono-saccharides and polyols (FODMAPs). The mechanism may relate to the osmotic potential of these carbohydrates in the small intestine, which have been shown to increase water delivery to the colon (Barrett et al. 2010), which in the absence of compensatory colonic absorption, may contribute to diarrhoea pathogenesis. The authors are careful to point out that the limitations of the study design impede a conclusion of a cause and effect relationship, but randomised controlled trials are to follow (Halmos et al. 2010).

Probiotics

Probiotics are *'live microorganisms which when administered in adequate amounts confer a health benefit on the host'* (FAO-WHO 2002). Commonly used strains include *lactobacilli, bifidobacteria, saccharomyces* (and commercial mixtures) and are available in capsules, powders, yoghurts and fermented milks.

Over the last 20 years a number of RCTs investigating probiotics in patients receiving EN have recorded diarrhoea as an outcome measure (Table 3).

Two small studies demonstrated no impact of probiotics on diarrhoea prevalence. In the first, 41 patients received *L. acidophilus/L. bulgaricus* or placebo and there were no significant differences in faecal weight or diarrhoea (Heimburger, Sockwell and Geels 1994). In the second, 28 patients on the ICU received VSL#3, sonicated VSL#3 or placebo, without any significant differences in diarrhoea duration between groups, although this study was probably under-powered to detect such outcomes (Alberda et al. 2007). For some time the largest study in the area has been a trial of 128 patients receiving EN on the ICU who received *S. boulardii* or placebo, and those in the probiotic group experienced 25% fewer diarrhoea days than placebo (Bleichner et al. 1997).

Over the last years, there has been rapid expansion in trials investigating probiotics in patients receiving EN where stool output and/or diarrhoea have been recorded as an outcome (Table 3). Firstly, a trial of 45 patients starting EN on the ICU demonstrated a reduction in liquid stools in patients receiving VSL#3 (mean 0.5 liquid stools/d) compared with placebo (1.1 liquid stools/d) (P=0.03) (Frohmader et al. 2010). Compared with the previous trial of VSL#3 (Alberda et al. 2007), this study was powered to detect differences in stool output. A small trial in 36 adults receiving EN on the ICU who currently had diarrhoea found that *L. rhamnosus* GG in addition to a small dose of inulin (560 mg/d) did not have benefit over the same dose of inulin alone in treating patients (Ferrie and Daley 2011). A synbiotic mixture (*L. paracasei* NCC 2461 and *B. longum* NCC 3601 along with fructo-oligosaccharides, inulin and acacia gum) was tested in a RCT in

Table 3. Randomised Controlled Trials that Measure the Impact of Probiotics on Stool Output or Diarrhoea in Patients Receiving EN.

Reference	Patient group	Probiotic intervention and placebo	Sample size	Results	*P* value
(Barraud et al. 2010)	167 adults starting EN on ICU	Ergyphilus ($2x10^{10}$cells/d) Placebo	87 80	55% developed diarrhoea 53% developed diarrhoea	NS
(Heimburger et al. 1994)	41 adults starting EN	*L. acidophilus* and *L. bulgaricus* (3 g/d) Placebo	18 23	31% developed diarrhoea 11% developed diarrhoea	0.21
(Ferrie and Daley 2011)	36 adults with diarrhoea during EN on ICU	*L. rhamnosus* GG ($2x10^{10}$cells/d) plus inulin (560 mg/d) Inulin (560 mg/d)	18 18	3.8 d duration of diarrhoea 2.6 d duration of diarrhoea	0.08
(Bleichner et al. 1997)	128 adults starting EN on ICU	*S. boulardii* (2 g/d) Placebo	64 64	14% of days with diarrhoea 19% of days with diarrhoea	0.0069
(Alberda et al. 2007)	28 adults starting EN on ICU	VSL#3—live cells ($9x10^{11}$ bacteria/d) VSL#3—DNA only ($9x10^{11}$ bacteria/d) Placebo	10 9 9	14% of days with diarrhoea 12% of days with diarrhoea 23% of days with diarrhoea	NS
(Frohmader et al. 2010)	45 adults starting EN on ICU	VSL#3 ($9x10^{11}$ bacteria/d) Placebo	20 25	0.5 liquid stools/d 1.1 liquid stools/d	0.03

VSL#3 consists of *L. casei, L. plantarum, L. acidophilus, L. delbrueckii, B. longum, B. breve,* Ergyphilus® consists of mainly *L. rhamnosus* GG, *L. casei, L. acidophilus* and *B. bifidum.*

1–3-yr-old children under mechanical ventilation in a RCT (Simakachorn et al. 2011); even though the probiotics were detected in the stools, there was no difference in stool frequency between groups. Finally, the largest trial to date is now an RCT of 167 patients receiving placebo or the commercial product Ergyphilus® (consisting mainly of *L. rhamnosus* GG) (Barraud et al. 2010). Although not the primary outcome, there were no differences in the proportion of patients developing diarrhoea (defined as 3 liquid stools/d) between the probiotic (55%) and the placebo (53%) group. Interestingly, the probiotic had differential affects on mortality between those with severe sepsis (probiotics reduced mortality, OR=0.38, P=0.035) and those with non-severe sepsis (probiotics increased mortality, OR=3.09, P=0.08).

Clearly, safety is an important consideration prior to using probiotics in the clinical setting. This is particularly relevant in patients receiving EN as increased gastric pH (due to enteral formula or gastric acid suppressing drugs) or post-pyloric administration (e.g., nasojejunal tube) will result in increased probiotic survival into the small intestine. Safety was highlighted following an RCT of probiotics in patients with severe acute pancreatitis that reported increased mortality in the probiotic group (Besselink et al. 2008). However, in that study the probiotic was a novel product with little previous data on human applications and was delivered mixed with the EN formula via nasojejunal tube.

Safety was investigated in a systematic review of case reports and clinical trials of probiotics in patients receiving nutritional support (Whelan and Myers 2010). Case reports of 32 patients receiving nutritional support (EN or parenteral nutrition) who developed an infection caused by probiotics were found. Risk factors included increased risk of bacterial translocation and the presence of a central venous catheter, which enables direct access to the systemic circulation when it is handled by contaminated hands. Despite this, the systematic review identified 50 trials of probiotics in patients receiving nutritional support (>4,000 patients) that did not report negative clinical sequelae. A risk-benefit analysis and routine monitoring for adverse events should therefore be undertaken when considering probiotic use in patients receiving EN.

There is contrasting evidence of efficacy of probiotics in preventing diarrhoea in patients receiving EN (Jack et al. 2010). This is likely due to variations in trial methodology, but in particular due to differences in functional characteristics of the various probiotics. Therefore, experts in nutritional support should be cautious when extrapolating the results from one probiotic to that of another.

Management of Diarrhoea

Most episodes of nosocomial diarrhoea are mild and will usually spontaneously abate after 72 hours, even though exacerbation of pre-existing diarrhoea may persist. Should diarrhoea continue for 72 hours or more, an abdominal exam should be performed, a stool sample should be tested for *C. difficile* enterotoxins, blood samples should be taken for a serum electrolyte panel (to evaluate excessive electrolyte losses or dehydration), and medications should be reviewed. As discussed, antibiotics (especially broad spectrum and those effective against anaerobes) are renowned for causing diarrhoea (Halmos et al. 2010; Trabal et al. 2008) while syrup or liquid preparations of medications often have a very high osmolar load or contain sorbitol leading to a cathartic effect (Klang et al. 2013). Lactose (as an excipient in numerous tablets) and FODMAPs in enteral formulas may have the same effect (Halmos et al. 2010).

Water and electrolyte replenishment is crucial and can be achieved orally, enterally or parenterally as appropriate. Antimotility agents, such as loperamide, loperamide oxide and codeine can be used, however this must wait until faecal impaction and, most importantly, *C. difficile* infection have been excluded (O'Keefe 2009). There is no evidence to support the use of probiotics (Ferrie and Daley 2011) or prebiotics in treating patients who already have diarrhoea. An example clinical algorithm for the management of diarrhoea in EN is shown in Fig. 1.

Research is required to investigate the effectiveness of a cadre of interventions to treat diarrhoea in patients receiving EN.

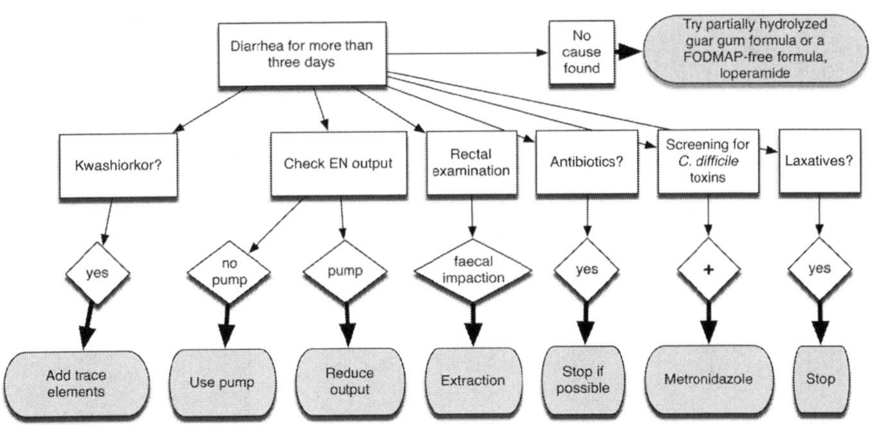

Figure 1. Algorithm for Clinical Management of Enteral Nutrition-associated Diarrhoea.

The composition of the enteral formula may also be adapted. Although only two studies are available, the use of soluble fibre may prove useful in treating diarrhoea in patients receiving EN. In an open-labelled study, galactomannans (7–28 g/d) were provided to 20 elderly in-patients with diarrhoea during EN and resulted in a significant decrease in the stool-water content and stool frequency (Nakao et al. 2002). Although this was not a controlled trial, the beneficial effects disappeared after fibre was discontinued. In a RCT in 20 ICU patients with diarrhoea during EN, the use of a formula containing partially hydrolyzed guar gum (compared to a standard formula) led to a significant decrease in the number of liquid stools, whilst allowing a higher volume of formula to be delivered (Rushdi et al. 2004). The addition of sodium chloride and trace elements to the formula may counteract the effects of active gastrointestinal water secretion and prevent specific deficiencies. Another approach to treating diarrhoea is to minimize the residue reaching the colon and some authors recommend pre-digested enteral formulas containing peptides and medium-chain triglycerides rather than whole proteins and long-chain triglycerides (McClave et al. 2009). Outside situations of major pancreatic exocrine insufficiency, this approach is not supported by any evidence. ASPEN combined these different approaches and stated that if there is evidence of diarrhoea, soluble fibre-enriched formulas or pre-digested formulas may be utilized. The lack of conclusive high quality evidence to support this recommendation resulted in it being given a grade E (McClave et al. 2009).

Despite these approaches to modifying the composition of the enteral formula, most importantly, EN should not be interrupted or stopped. In the clinical setting, diarrhoea is a frequently cited cause for EN to be interrupted or stopped (Meyer et al. 2009; Martins et al. 2012). However, continued bowel rest exacerbates bacterial overgrowth and gastrointestinal dysmotility, which perpetuates the diarrhoea. Furthermore, protein and energy goals will not be reached, which will predispose or exacerbate under nutrition and its associated consequences of increased morbidity, mortality, length of stay and of course cost. Reducing the delivery rate, whilst still achieving target volume is one possibility. Another option is to reduce EN provision and to replace protein and energy with supplementary parenteral nutrition, which can subsequently be tapered down once diarrhoea abates and the volume of EN is increased.

Conclusion

Diarrhoea is a major concern in EN and may represent a limitation to the broad use this therapy deserves. Recent research shows the main causal factor to be dysbiosis, related to the initial disease, antibiotics and also the formula composition (lack of certain fibres, presence of poorly-absorbed fermentable compounds). Future progress will come from new techniques in bacterial stool analysis as well as the progress in describing the microbiota, which may allow us to screen prospective EN patients in order to determine which ones are at risk for diarrhoea; this will allow us to start them on a specific formula, and such approaches need to be studied further (Schneider 2010). The priority in treating patients with EN-associated diarrhoea is to keep the provision of their water, electrolyte, protein and energy needs. For both prevention and treatment, a number of new fibre blends now available may be of interest; however, the need for clinical studies remains.

References Cited

Alberda, C., L. Gramlich, J. Meddings, C. Field, L. McCargar, D. Kutsogiannis, R. Fedorak and K. Madsen. 2007. Effects of probiotic therapy in critically ill patients: a randomized, double-blind, placebo-controlled trial. Am. J. Clin. Nutr. 85: 816–823.

Barraud, D., C. Blard, F. Hein, O. Marcon, A. Cravoisy, L. Nace, F. Alla, P.E. Bollaert and S. Gibot. 2010. Probiotics in the critically ill patient: a double blind, randomized, placebo-controlled trial. Intensive Care Med. 36: 1540–1547.

Barrett, J.S., R.B. Gearry, J.G. Muir, P.M. Irving, R. Rose, O. Rosella, M.L. Haines, S.J. Shepherd and P.R. Gibson. 2010. Dietary poorly absorbed, short-chain carbohydrates increase delivery of water and fermentable substrates to the proximal colon. Aliment Pharmacol. Ther. 31: 874–882.

Belknap, D., L.J. Davidson and C.R. Smith. 1997. The effects of psyllium hydrophilic mucilloid on diarrhea in enterally fed patients. Heart Lung. 26: 229–237.

Benus, R.F., T.S. van der Werf, G.W. Welling, P.A. Judd, M.A. Taylor, H.J. Harmsen and K. Whelan. 2010. Association between *Faecalibacterium prausnitzii* and dietary fibre in colonic fermentation in healthy human subjects. Br. J. Nutr. 104: 693–700.

Besselink, M.G., H.C. van Santvoort, E. Buskens, M.A. Boermeester, H. van Goor, H.M. Timmerman, V.B. Nieuwenhuijs, T.L. Bollen, B. van Ramshorst, B.J. Witteman, C. Rosman, R.J. Ploeg, M.A. Brink, A.F. Schaapherder, C.H. Dejong, P.J. Wahab, C.J. van Laarhoven, E. van der Harst, C.H. van Eijck, M.A. Cuesta, L.M. Akkermans and H.G. Gooszen. 2008. Probiotic prophylaxis in predicted severe acute pancreatitis: a randomised, double-blind, placebo-controlled trial. Lancet. 371: 651–659.

Bleichner, G., H. Blehaut, H. Mentec and D. Moyse. 1997. *Saccharomyces boulardii* prevents diarrhea in critically ill tube-fed patients. A multicenter, randomized, double-blind placebo-controlled trial. Intensive Care Med. 23: 517–523.

Bliss, D.Z., S. Johnson, K. Savik, C.R. Clabots, K. Willard and D.N. Gerding. 1998. Acquisition of *Clostridium difficile* and *Clostridium difficile*-associated diarrhea in hospitalized patients receiving tube feeding. Ann. Intern. Med. 129: 1012–1019.

Bowling, T.E., A.H. Raimundo, G.K. Grimble and D.B. Silk. 1993. Reversal by short-chain fatty acids of colonic fluid secretion induced by enteral feeding. Lancet. 342: 1266–1268.

Bowling, T.E. and D.B.A. Silk. 1998. Colonic responses to enteral tube feeding. Gut. 42: 147–151.

Chittawatanarat, K., P. Pokawinpudisnun and Y. Polbhakdee. 2010. Mixed fibers diet in surgical ICU septic patients. Asia Pac. J. Clin. Nutr. 19: 458–464.

de Kruif, J.T. and A. Vos. 1993. The influence of soyfibre supplemented tube feeding on the occurrence of diarrhoea in postoperative patients. Clin. Nutr. 12: 360–364.

Dobb, G.J. and S.C. Towler. 1990. Diarrhoea during enteral feeding in the critically ill: a comparison of feeds with and without fibre. Intensive Care Med. 16: 252–255.

Elia, M., M.B. Engfer, C.J. Green and D.B. Silk. 2008. Systematic review and meta-analysis: the clinical and physiological effects of fibre-containing enteral formulae. Aliment Pharmacol. Ther. 27: 120–145.

Evans, S., A. Daly, P. Davies and A. MacDonald. 2009. Fibre content of enteral feeds for the older child. J. Hum. Nutr. Diet. 22: 414–421.

Ferrie, S. and M. Daley. 2011. *Lactobacillus* GG as treatment for diarrhea during enteral feeding in critical illness: randomized controlled trial. J. Parenter. Enteral. Nutr. 35: 43–49.

Frankenfield, D.C. and P.L. Beyer. 1989. Soy-polysaccharide fiber: effect on diarrhea in tube-fed, head-injured patients. Am. J. Clin. Nutr. 50: 533–538.

Frohmader, T.J., W.P. Chaboyer, I.K. Robertson and J. Gowardman. 2010. Decrease in frequency of liquid stool in enterally fed critically ill patients given the multispecies probiotic VSL#3: a pilot trial. Am. J. Crit. Care. 19: e1–11.

Guenter, P.A.. R.G. Settle, S. Perlmutter, P.L. Marino, G.A. DeSimone and R.H. Rolandelli. 1991. Tube feeding-related diarrhea in acutely Ill patients. J. Parenter. Enteral. Nutr. 15: 277–280.

Guimber, D., B. Bourgois, L. Beghin, S. Neuville, P. Pernes, K. Ben Amor, A. Goedhart, J. Sijben, J. Knol and F. Gottrand. 2010. Effect of multifibre mixture with prebiotic components on bifidobacteria and stool pH in tube-fed children. Br. J. Nutr. 104: 1514–1522.

Halmos, E.P., J.G. Muir, J.S. Barrett, M. Deng, S.J. Shepherd and P.R. Gibson. 2010. Diarrhoea during enteral nutrition is predicted by the poorly absorbed short-chain carbohydrate (FODMAP) content of the formula. Aliment Pharmacol. Ther. 32: 925–933.

Hart, G.K. and G.J. Dobb. 1988. Effect of a fecal bulking agent on diarrhea during enteral feeding in the critically ill. J. Parenter. Enteral. Nutr. 12: 465–468.

Heimburger, D.C., D.G. Sockwell and W.J. Geels. 1994. Diarrhea with enteral feeding: prospective reappraisal of putative causes. Nutrition. 10: 392–396.

Homann, H.H., M. Kemen, C. Fuessenich, M. Senkal and V. Zumtobel. 1994. Reduction in diarrhea incidence by soluble fiber in patients receiving total or supplemental enteral nutrition. J. Parenter. Enteral. Nutr. 18: 486–490.

Jack, L., F. Coyer, M. Courtney and B. Venkatesh. 2010. Probiotics and diarrhoea management in enterally tube fed critically ill patients—what is the evidence? Intensive Crit. Care Nurs. 26: 314–326.

Joint Food and Agriculture Organization of the United Nations and World Health Organization Working Group. 2002. Guidelines for the evaluation of probiotics in food. London, Ontario, Canada.

Khalil, L., K.H. Ho, D. Png and C.L. Ong. 1998. The effect of enteral fibre-containing feeds on stool parameters in the post-surgical period. Singapore Med. J. 39: 156–159.

Khoshoo, V., S.S. Sun and H. Storm. 2010. Tolerance of an enteral formula with insoluble and prebiotic fiber in children with compromised gastrointestinal function. J. Am. Diet. Assoc. 110: 1728–1733.

Klang, M., V. McLymont and N. Ng. 2013. Osmolality, pH, and compatibility of selected oral liquid medications with an enteral nutrition product. J. Parenter. Enteral. Nutr. 37: 689–694.

Lochs, H., S.P. Allison, R. Meier, M. Pirlich, J. Kondrup, S. Schneider, G. van den Berghe and C. Pichard. 2006. Introductory to the ESPEN Guidelines on Enteral Nutrition: Terminology, definitions and general topics. Clin. Nutr. 25: 180–186.

Majid, H.A., P.W. Emery and K. Whelan. 2012. Definitions, attitudes, and management practices in relation to diarrhea during enteral nutrition: a survey of patients, nurses, and dietitians. Nutr. Clin. Pract. 27: 252–260.

Martins, J.R., G.M. Shiroma, L.M. Horie, L. Logullo, L. Silva Mde and D.L. Waitzberg. 2012. Factors leading to discrepancies between prescription and intake of enteral nutrition therapy in hospitalized patients. Nutrition. 28: 864–867.

McClave, S.A., R.G. Martindale, V.W. Vanek, M. McCarthy, P. Roberts, B. Taylor, J.B. Ochoa, L. Napolitano and G. Cresci. 2009. Guidelines for the Provision and Assessment of Nutrition Support Therapy in the Adult Critically Ill Patient: Society of Critical Care

Medicine (SCCM) and American Society for Parenteral and Enteral Nutrition (A.S.P.E.N.). J. Parenter. Enteral. Nutr. 33: 277–316.

Meyer, R., S. Harrison, S. Sargent, P. Ramnarayan, P. Habibi and D. Labadarios. 2009. The impact of enteral feeding protocols on nutritional support in critically ill children. J. Hum. Nutr. Diet. 22: 428–436.

Nakao, M., Y. Ogura, S. Satake, I. Ito, A. Iguchi, K. Takagi and T. Nabeshima. 2002. Usefulness of soluble dietary fiber for the treatment of diarrhea during enteral nutrition in elderly patients. Nutrition. 18: 35–39.

O'Keefe, S.J. 2009. A guide to enteral access procedures and enteral nutrition. Nat. Rev. Gastroenterol. Hepatol. 6: 207–215.

Reese, J.L., M.E. Means, K. Hanrahan, B. Clearman, M. Colwill and C. Dawson. 1996. Diarrhea associated with nasogastric feedings. Oncol. Nurs. Forum. 23: 59–66; discussion 66–58.

Roberfroid, M., G.R. Gibson, L. Hoyles, A.L. McCartney, R. Rastall, I. Rowland, D. Wolvers, B. Watzl, H. Szajewska, B. Stahl, F. Guarner, F. Respondek, K. Whelan, V. Coxam, M.J. Davicco, L. Leotoing, Y. Wittrant, N.M. Delzenne, P.D. Cani, A.M. Neyrinck and A. Meheust. 2010. Prebiotic effects: metabolic and health benefits. Br. J. Nutr. 104 Suppl 2: S1–63.

Rushdi, T.A., C. Pichard and Y.H. Khater. 2004. Control of diarrhea by fiber-enriched diet in ICU patients on enteral nutrition: a prospective randomized controlled trial. Clin. Nutr. 23: 1344–1352.

Schneider, S.M., P. Le Gall, F. Girard-Pipau, T. Piche, A. Pompei, J. L. Nano, X. Hébuterne and P. Rampal. 2000. Total artificial nutrition is associated with major changes in the fecal flora. Eur. J. Nutr. 39: 248–255.

Schneider, S.M., F. Girard-Pipau, R. Anty, E.G. van der Linde, B.J. Philipsen-Geerling, J. Knol, J. Filippi, K. Arab and X. Hébuterne. 2006. Effects of total enteral nutrition supplemented with a multi-fibre mix on faecal short-chain fatty acids and microbiota. Clin. Nutr. 25: 82–90.

Schneider, S.M. 2010. Microbiota and enteral nutrition. Gastroenterol. Clin. Biol. 34 Suppl. 1: S57–61.

Schultz, A.A., B. Ashby-Hughes, R. Taylor, D.E. Gillis and M. Wilkins. 2000. Effects of pectin on diarrhea in critically ill tube-fed patients receiving antibiotics. Am. J. Crit. Care. 9: 403–411.

Silk, D.B., E.R. Walters, H.D. Duncan and C.J. Green. 2001. The effect of a polymeric enteral formula supplemented with a mixture of six fibres on normal human bowel function and colonic motility. Clin. Nutr. 20: 49–58.

Simakachorn, N., R. Bibiloni, P. Yimyaem, Y. Tongpenyai, W. Varavithaya, D. Grathwohl, G. Reuteler, J.C. Maire, S. Blum, P. Steenhout, J. Benyacoub and E.J. Schiffrin. 2011. Tolerance, safety, and effect on the faecal microbiota of an enteral formula supplemented with pre- and probiotics in critically ill children. J. Pediatr. Gastroenterol. Nutr. 53: 174–181.

Spapen, H., M. Diltoer, C. Van Malderen, G. Opdenacker, E. Suys and L. Huyghens. 2001. Soluble fiber reduces the incidence of diarrhea in septic patients receiving total enteral nutrition: a prospective, double-blind, randomized, and controlled trial. Clin. Nutr. 20: 301–305.

Trabal, J., P. Leyes, S. Hervas, M. Herrera and M. de Tallo Forga. 2008. Factors associated with nosocomial diarrhea in patients with enteral tube feeding. Nutr. Hosp. 23: 500–504.

Vaisman, N., M. Lansink, C.H. Rouws, K.M. van Laere, R. Segal, E. Niv, T.E. Bowling, D.L. Waitzberg and J.E. Morley. 2009. Tube feeding with a diabetes-specific feed for 12 weeks improves glycaemic control in type 2 diabetes patients. Clin. Nutr. 28: 549–555.

Vandewoude, M.F., K.M. Paridaens, R.A. Suy, M.A. Boone and H. Strobbe. 2005. Fibre-supplemented tube feeding in the hospitalised elderly. Age Ageing. 34: 120–124.

Whelan, K., P.A. Judd, V.R. Preedy, R. Simmering, A. Jann and M.A. Taylor. 2005. Fructooligosaccharides and fiber partially prevent the alterations in fecal microbiota and short-chain fatty acid concentrations caused by standard enteral formula in healthy humans. J. Nutr. 135: 1896–1902.

Whelan, K. 2007. Enteral-tube-feeding diarrhoea: manipulating the colonic microbiota with probiotics and prebiotics. Proc. Nutr. Soc. 66: 299–306.

Whelan, K., P.A. Judd, K.M. Tuohy, G.R. Gibson, V.R. Preedy and M.A. Taylor. 2009. Fecal microbiota in patients receiving enteral feeding are highly variable and may be altered in those who develop diarrhea. Am. J. Clin. Nutr. 89: 240–247.

Whelan, K. and C.E. Myers. 2010. Safety of probiotics in patients receiving nutritional support: a systematic review of case reports, randomized controlled trials, and nonrandomized trials. Am. J. Clin. Nutr. 91: 687–703.

Wierdsma, N.J., A.A. van Bodegraven, B.M. Uitdehaag, W. Arjaans, P.H. Savelkoul, H.M. Kruizenga and M.A. van Bokhorst-de van der Schueren. 2009. Fructo-oligosaccharides and fibre in enteral nutrition has a beneficial influence on microbiota and gastrointestinal quality of life. Scand. J. Gastroenterol. 44: 804–812.

Yokohama, S., M. Aoshima, T. Asama, J. Shindo and J. Maruyama. 2009. Clostridium difficile-associated enteric disease after percutaneous endoscopic gastrostomy. J. Gastroenterol. 44: 121–125.

13

Influence of the Intestinal Microbiota on the Critically Ill Patient

Robert G. Martindale,[1,*] *Stephen A. McClave*,[2] *Malissa Warren*[3] and *Svetang Desai*[4]

Introduction

The gastrointestinal (GI) tract, from mouth to anus, represents the most diverse, complex and fragile microenvironment in the mammalian body. The human host and microbiome within are so closely related that many consider it a single "super-organism" (Gill et al. 2006). It has now been well validated that the intestinal microbiota influences the physiology, nutritional status, immune function, and overall health status of the host (Morowitz et al. 2011; Bengmark 2013). The intensive care unit (ICU) is an

[1] Oregon Health & Science University, 3181 SW Sam Jackson Park Rd, L223A, Portland, OR 97239 USA.
E-mail: martindr@ohsu.edu
[2] University of Louisville, 401 East Chestnut St., Ste 310, Louisville, KY 40202 USA.
E-mail: Stephen.mcclave@louisville.edu
[3] Portland VA Medical Center, 3710 SW US Veterans Hospital Rd, PO Box 1034, Portland, OR 97207 USA.
E-mail: Malissa.warren@va.gov
[4] Duke University, Section of Gastroenterology, Duke Clinic 2G/2H, 40 Duke Medicine Circle, Durham, NC 27710.
E-mail: Svetang@gmail.com
* Corresponding author

extremely "artificial" environment. Invasive techniques (which eliminate or dramatically alter natural barriers) and the use of multiple medications (which affect GI motility, luminal pH, blood flow, and mucosal oxygenation) radically influence gut microbiota. The widespread use of broad spectrum antibiotics and the need to provide artificial nutrition therapy make the changes even more complex to understand. Within hours of the insult or injury which leads to admission to an ICU, the microbiome is dramatically and rapidly altered (Morowitz et al. 2011; Alverdy et al. 2005). These changes in the gut microbiota accompany the often major changes in intra-vascular volume and blood flow to the GI tract which commonly occurs in the ICU population. Therapeutic measures involved with critical care can cause dramatic changes in the mucosal redox potential and pH within the epithelial cell. Numerous secretory products of the GI tract (such as saliva, bile, mucous, immunoglobulins, and pancreatic enzymes) are all significantly altered in the ICU setting, changes which in turn will secondarily alter the gut microbiome.

Physiologic stress on the host induced by illness, surgery or trauma has been shown to cause significant changes on the mucosal microenvironment through alterations in pH, redox potential and mucosal energy supply (Alverdy 2012). Alverdy and his group have reported in a variety of stress models that a rapid depletion in luminal phosphate is observed. This in turn induces normally non-pathogenic bacteria to undergo a phenotypic change and become pathogenic. These stress-induced mucosal changes result in upregulation of virulence factors in the endogenous bacteria which facilitate attachment and invasion of the host epithelial barrier (Zaborina 2011; Long et al. 2007).

Changes in Technology—Application to the ICU Setting

Recent advances in technology, like the addition for high-throughput sequencing and other molecular taxonomic technologies, have led to a more complete understanding of the level of complexity of the microbiome. The huge gaps in knowledge that were present with the previous techniques of fecal and luminal cultures for bacterial species identification have been replaced with DNA high-throughput sequencing and culture independent 16S ribosomal subunit technology. These changes have created more unanswered questions and revealed an even greater complexity to the unique symbiotic relationship between the human host and the bacteria within us. Although the new technologies yield a more comprehensive catalog of the human intestinal microbiota, the changes which occur in these organisms in the ICU remain mysterious. A better understanding of these relationships and the possible manipulation of the gut microbiome via metagenomics, metatranscriptomics, and proteonomics should lead to improved care for the ICU patient.

The Gut in Critical Illness is a Black Box!

Awareness of the influence of the gut microbiota on various disease states has historically been focused on specific disease states such as sepsis, inflammatory bowel, antibiotic-associated diarrhea, and cancer. Two decades ago, the concept of the GI tract as the "undrained abscess" of multiple organ failure (Marshall et al. 1993) and the "Gut as the Motor of Multiple Organ Failure" (Nieuwenhuijzen and Goris 1999 alluded to the concept of the gut microbiome as the etiology of disease, with no mention of the potential benefit derived from intestinal bacteria (Taneja and Marshall 2000). The National Institutes of Health and European Union Microbiome projects have allowed for the this relative black box to be opened somewhat, showing the potential positive (as well as negative) influence of the gut microbiome. A complete understanding of the changes and contributions of the intestinal bacteria in the ICU setting, however, remains elusive as it is now believed that less than 50% of the bacterial species found in the gut are routinely culturable (Morowitz et al. 2011). Well over 1000 species have been identified with high throughput sequencing and multiple 16sRNA methods (Wang 2013). The diagnoses and the pathophysiology of the disease processes which warrant ICU admission are widely heterogeneous, making the complexity of understanding the microbiome exponentially more difficult. The changes in the individual bacterial colonies are rapid, having been shown to take place within minutes to hours of stress or injury to the host, as a result of the alteration in stress hormones, hemodynamic changes, and immune suppression that occurs in setting (Babrowski et al. 2013).

Beneficial Contributions of the Microbiota in the ICU Setting

Microbiota presiding in the host GI tract is felt to interact not only between species but also with the intestinal mucosa of the host. Many of these effects yield benefits. The ability to manipulate and potentially exploit these microbial changes noted in the ICU setting has become the focus of research only in recent years. Numerous mechanisms of these benefits have been elucidated (Morrow et al. 2012). Early work done in this area by Deitch et al. using a murine stress model reported significant changes in the gut bacteria, with a rapid decrease in *Lactobacillus* species and a rise in gram negative species (Katayama et al. 1997). The beneficial effects of the bacterial symbiotic relationships include production of bacterial antimicrobial peptides inhibiting overgrowth of pathogens, enhancement of the physical epithelial barrier by increasing production of mucins (which adhere to intestinal mucosa), concomitant inhibition of pathogen adhesion, and competitive exclusion of the pathogenic microorganisms (Turroni et al. 2013). Certain microbiota has been shown to aid specifically in maintaining integrity of gap junctions (Feng et al. 2012), while others enhance host-cell antimicrobial peptide production which reduces epithelial adherence of pathogens (Sherman et al. 2005). Manipulation of the host microbiota has been shown to modulate metabolic and immune responses (Morrow et al. 2012).

Influences of Nutrient delivery on Host Microbiome in the ICU

Nutrient contributions to the host are best illustrated by comparing germ-free animals to conventionally raised control animals. Studies involving several animal models have reported that animals raised in a germ-free environment require 20 to 30% more calories and nutrients to maintain normal weight and growth curves than controls (Hooper et al. 2002). The metabolism and absorption of all three macronutrients including carbohydrate, protein, and lipids are influenced by the host microbiome. This interaction between ingested or infused nutrients and the microbiota is altered in the ICU setting, where patients commonly have relative periods of fasting which may last for days. Surprisingly little is known about the influence of periods of fasting on gut microbiota. It is very well described that complex carbohydrates not absorbed in the proximal gut will pass to the colon, where a wide array of bacterial enzymes not present in the mammalian host will hydrolyze these carbohydrates and utilize them for metabolic support as energy substrates, in many cases yielding end-products such as short chain fatty acids (SCFA's). The primary SCFA's produced by bacteria in the colon include acetate, lactate, butyrate, and propionate. These SCFA's have been reported to radically change in the ICU setting. In a recent paper, Osuka, et al reported that patient outcome and ICU course could be predicted by changes in fecal pH and content of short chain fatty acids (Osuka et al. 2012). The SCFA's produced by fermentation serve as the primary fuel for the colonocyte, with butyrate being responsible for up to 70% of colonocyte energy utilization. In the ICU setting, these SCFA's can yield a valuable source of additional energy, especially in those patients who have reduced absorption because of intestinal resection or in those patients undergoing various periods of luminal nutrient deprivation.

Although less studied than carbohydrate fermentation by gut bacteria, the influence of gut microbiota on protein metabolism has recently been shown to be of value in the recycling of intestinal urea and even to some degree in the de-novo synthesis of essential and non-essential amino acids (Davila et al. 2013). Estimates from complex labeling studies report that in some circumstances up to 20% of circulating leucine, lysine, and threonine have their origin from gut bacteria (Raj et al. 2008; Torrallardona et al. 2003). Although the actual net impact on protein nutrition to the host appears to be limited in the ICU setting, this area is only beginning to be studied (Bergan and Wu 2009). Once better understood, the amino acid contribution to host protein metabolism could become increasingly important to the catabolic ICU patient during periods where the provision of exogenous enteral or parenteral amino acids is limited.

The recycling of urea in the gut is significantly altered in the ICU setting, primarily when broad spectrum antibiotics are given. Because antibiotics are almost universally given to all ICU patients, the significance of their impact on the interaction between intestinal organisms and gut urea cycling is important to understand (Perez-Cobas et al. 2012). In work done primarily in neonates (Bergan and Wu 2009), treatment with antibiotics results in a rapid reduction in Bacteroidetes, with a concurrent increase in Firmicutes. The clinical significance and influence of these changes on clinical outcome is not clear at this point, but these radical changes would be expected to play a significant role in overall gut and host protein metabolism in the ICU setting (Bergan and Wu 2009).

The influence of the microbiome on systemic lipid metabolism has only been partially elucidated. A recent series of experiments demonstrated that colonic microbes produce variable amounts of SCFA's (primarily butyrate), which can then serve as ligands for Gpr41, a G protein-coupled receptor expressed by a subset of enteroendocrine cells in the gut epithelium. With activation of this receptor via butyrate, the microbes essentially become a regulator of lipid metabolism, nutrient utilization, and inflammation (Samuel et al. 2008; Sommer and Backhed 2013). The microbiome's ability to alter lipid metabolism and attenuate inflammation would obviously have significant influence on fuel availability and the subsequent immune response for the hyperdynamic ICU patient (Ohira et al. 2013).

Influence of ICU Interventions and Routine Therapies on Gut Flora

The complexity of ICU care, with the heterogeneity of patient populations and disease states, makes the evaluation of any single specific healthcare intervention difficult and somewhat speculative. Currently, many of the studies reporting changes in the intestinal microbiome in the ICU setting and the relationship of those changes to alterations in the epithelial border or barrier functions, describe only associations and not proven cause and effect.

First and foremost is the influence of the route of feeding in the ICU on the gut flora. In a series of elegant animal studies, Feng et al. reported that PN elicits a rapid and profound increase in mucosa inflammation, changes which result in the loss of epithelial barrier function. The mucosal inflammation associated with PN results in an increase in the pro-inflammatory cytokines TNF-alpha and IFN-gamma, an upregulation of toll-like receptors (TLR) 2, 4 and 9, and ultimately mucosal atrophy (Feng et al. 2012). The mucosal atrophy induced by PN was significantly decreased when the experiments were repeated using an IFN knockout model or when TNF receptor blocking agents were used. Other groups have now reported similar cytokine changes at the mucosal level in response to PN, but in addition to the increase in pro-inflammatory cytokines, they report a marked decrease in the anti-inflammatory cytokine IL-10 (Sun et al. 2008). These inflammatory changes are clearly associated with a rapid shift in the intestinal microbiome from Firmicutes to Gram-negative proteobacteria. This constellation of events has led to the hypothesis that the PN-induced changes could be averted or attenuated by appropriate selection and delivery of probiotics (Feng et al. 2012).

Currently, the use of commercially produced enteral formulations is the most common source of nutrient delivery in the ICU, yet little is known about the influence of such formulations on the gut microbiome. The variety and content of enteral formulations can be vastly different depending on macronutrient or micronutrient content, and manufacturing methodology. In addition, many of these formulations have varying amounts of soluble and insoluble fibers (prebiotics). No consistent inference of the influence of commercial formulations on the microbiome can be made. The prebiotic content varies widely, as does the type and amount of SCFA's produced. The effects of different commercial formulae have best been evaluated in those studies comparing formula feeding to breast feeding in infants. The microbiome is dramatically different between these two sources of nutrient (Carlisle et al. 2013; Carbrera-rubio et al. 2012).

Pain control and sedation in the ICU setting essentially incorporates a variety of opioids. These agents induce motility disturbances and alter transit time. With prolongation of the transit time, the ability of the gut to clear bacteria is significantly decreased, resulting in segments of bacterial overgrowth and dramatic changes in the luminal flora.

Attempts to Preserve Healthy Intestinal Microbiota

A variety of ICU interventions can be performed in an attempt preserve a "healthy" gut microbiome. Such interventions would include maintaining vascular perfusion and minimizing use of agents which dramatically alter either pH or motility. The use of broad spectrum antibiotics should be limited to conditions in which they have shown confirmed benefit, and they should be used for the shortest duration as possible. Efforts to use probiotics to prevent the numerous microbiological changes noted in the "normal" microbiome have not been uniformly successful (Petrot et al. 2012). A recent systematic review of the literature reported that use of probiotics in the ICU was associated with a decrease in infectious complications, a reduction in ventilator-associated pneumonia, and a non-significant trend toward decreased mortality. This report illustrated the difficulty with deriving conclusions or making more definitive recommendations from the studies analyzed (Petrot et al. 2012). It is inaccurate and naive to generalize findings by grouping the effects observed from use of a single probiotic species to that seen with different species of probiotics in other studies, especially in the ICU setting. Differences in the colonizing abilities of native intestinal flora, variability in probiotic strains, doses, timing, and duration of delivery, and heterogeneity in patient populations, disease processes, and their treatment modalities in an ICU setting make generalizations impossible. Nonetheless, studies do suggest a role for probiotics in preserving the healthy intestinal microbiota in ICU patients.

Clostridium difficile Infections in the ICU

Clostridium difficile, an anaerobic gram-positive, spore-forming bacterium, is the most lethal bacterial-induced diarrhea in the western world. The pathophysiology of *C. difficile* has been shown to establish itself in the colon and proliferate in the setting of antibiotic-induced changes in the native flora. *C. difficile* bacterial overgrowth of the native flora causes fluid secretion and extravasation, inflammation, and damage to the mucosal barrier. A variable amount of diarrhea results, with pseudomembranous colitis, toxic megacolon, *C. difficile* sepsis, and death occurring in the more severe cases if not aggressively managed (Bouza 2012). Over 90% of *C. difficile*-associated diarrhea (CDAD) occurs during or after antibiotic use, and approximately 30% of cases will recur (Safdar and Maki 2002). Standard treatment for first episodes of *C. difficile* infections is oral vancomycin or metronidazole (Drekonja et al. 2011). More recent data has suggested that the optimal therapy for refractory cases may require fecal transplant with the delivery of multiple flora (van Nood et al. 2013).

Early work done with Sarcomyces boulardii reported that a serine protease produced by the yeast could cleave and inactivate the toxin A and B produced by *C. difficile* (Castagliuolo et al. 1996). In a study of patients receiving a four week treatment for CDAD with oral vancomycin or metronidazole, McFarland showed that those who received *S. boulardii* (N = 57) in addition, were significantly less likely to experience recurrence than those given placebo (N=67) (RR = 0.59; 95% CI = 0.35–0.98) (McFarland et al. 1994). In a similar study, Surawicz et al. analyzed 168 patients with active CDAD taking vancomycin or metronidazole with or without added *S. boulardii* (Surawicz et al. 2000), showing again a significant reduction in recurrent CDAD with the addition of *S. boulardii* when compared to placebo. In a very small randomized trial, Lawrence et al. treated 15 patients with known recurrent CDAD with oral vancomycin with or without *Lactobacillus rhamnosus* GG (LGG). A non-significant reduction in CDAD was seen in those receiving LGG (Lawrence et al. 2005). In a 2006 meta-analysis of the use of probiotics for the treatment of CDAD, aggregating the data from six randomized controlled trials showed that *S. boulardii* was the only effective probiotic agent for reducing recurrence of CDAD (McFarland 2006).

Regarding the prevention of *C. difficile* in high-risk populations (not ICU patients), Hickson et al. reported delivering 100 g of *L. casei*, *L. bulgaricus*, and *S. thermophilus* twice daily for the duration of (and for one week after) the course of the antibiotic therapy. This strategy resulted in a significant decrease in both antibiotic-associated diarrhea and *C. difficile* infection (Hickson et al. 2007). In the largest systematic review of the literature to date evaluating the use of probiotics for CDAD reported in over 3800 patients, Johnston et al. found that "moderate-quality evidence suggests probiotic prophylaxis results in a large reduction in CDAD…." (Johnston et al. 2012). Although these studies report no data on the changes in the gut flora, broad sweeping statements about maintaining "normal" flora in the ICU population cannot be made.

Treatment with a single or even multiple probiotics may be inadequate for refractory cases of *C. difficile*. The concept of healthy donor fecal transplantation via either enemas, colonoscopic delivery, or through duodenal infusions via a nasoenteric tube has recently been shown to be more effective. In an elaborate study, van Nood et al showed that patients who received an infusion of donor feces had > 95% cure in *C. difficile* infection (van Nood et al. 2013). In a subset of those study subjects, investigators were able to show that the donor fecal infusion resulted in a dramatic increase in patient fecal bacterial diversity. Within 2 weeks of the infusion, the microbiota in the patient stool samples was indistinguishable from that of the healthy donors.

Based on the success of the fecal transplantation studies and the rather limited success of the probiotic trials in management of ICU-related *C. difficile* infections, management may be evolving toward delivery a more complex bacterial community simulating normal feces from healthy individuals instead of infusion of a single or multiple probiotic organisms (van Nood et al. 2013; Brandt 2013; Damman et al. 2012).

Ventilator-Associated Pneumonia

Hospital-acquired pneumonia is the single most lethal hospital-acquired infection, with ventilator-associated pneumonia (VAP) causing the majority of the observed mortality (To and Napolitano 2003). VAP typically occurs within 48–72 hours of intubation and requirement for mechanical ventilation. The populations at highest risk include trauma and burn patients in the ICU, having an incidence of VAP of approximately 15% (To and Napolitano 2003). Studies suggest that probiotics delivered locally and systemically may help preserve or at least partially re-establish the endogenous aerodigestive flora and reduce the incidence of VAP. The exact mechanisms of action are not completely elucidated, but most likely the probiotics work through selective colonization, alteration in the pathogenic bacteria biofilm production, and possibly immunomodulation (Klarin et al. 2008).

Morrow et al. conducted a very high quality prospective randomized double-blinded placebo-controlled trial in 138 critically ill patients, mechanically ventilated for greater than 72 hours, at high risk for developing VAP (Morrow et al. 2010). Study patients in the treatment arm received 10^9 colony-forming units of *Lactobacillus rhamnosus GG* twice daily, delivered by swabbing the oropharynx and by infusion through a nasoenteric tube. Confirming the diagnosis of VAP by bronchoscopy, they reported that the experimental group sustained a 50% reduction in VAP compared to controls given placebo. Not only did they report less VAP, they also reported reduced quantity of antibiotics required during the ICU stay and less acquisition of *C. difficile*. Limited bacteriologic information is reported, but it appears that the probiotic had significant benefit in preventing development of pathogenic bacterial overgrowth. In an earlier study, Forestier et al. reported that twice daily delivery of the single probiotic agent *Lactobacillus casei*, started on day 3 of mechanical ventilation, delayed colonization of the respiratory tree with *Pseudomonas aeruginosa* (Forestier et al. 2008).

A meta-analysis by Siempos et al. conducted prior to the Morrow et al. study above, found the incidence of VAP significantly lower in patients undergoing mechanical ventilation treated with probiotics compared to controls receiving placebo (Siempos et al. 2010). The remaining secondary outcomes included non-significant decreases in all-cause mortality, duration of mechanical ventilation, and number of patients experiencing diarrhea. An additional review of the literature by Schultz et al. on the prevention of VAP, reported that three out of eight studies showed a significant reduction in VAP in the probiotic group, with no effect on mortality, compared to placebo controls (Schultz et al. 2011).

The majority of these studies show a consistent pattern of reduced risk from VAP in patients treated with probiotics. The primary mechanism appears to be related to preventing local oropharyngeal colonization by pathogens. A consensus as to which probiotic to use, the optimal timing and duration of therapy and methods of delivery has not been established. These factors need to be sorted out prior to the promotion of wide-spread use in all ventilated patients or to derivation of any specific guidelines for the prevention of VAP in critically-ill patients.

Pancreatitis

Severe acute pancreatitis (SAP) is a disease process characterized by necrosis of the gland and risk for infectious complications and organ failure. Increased gut permeability with seeding of the necrotic portion with pathogenic bacteria of gut origin is a key mechanism in abscess formation within the gland. Not surprisingly, early randomized trials of probiotic use in SAP showed promise in reducing morbidity and improving clinical outcome.

In two small consecutive randomized trials in patients with SAP, using first one strain and then four strains of Lactobacillus, Olah showed significant reductions in pancreatic abscess, hospital length of stay, organ failure, and complications requiring need for surgery (compared to controls receiving placebo) (Olah et al. 2002; Olah et al. 2007). However, in a larger Dutch multi-center trial, use of six probiotic organisms (four Lactobacillus and two Bifidobacteria) caused net harm with increased incidence of organ failure, need for surgical intervention, ICU length of stay, and mortality (compared to controls not receiving probiotics) (Besselink et al. 2008). The discrepancy between the beneficial results of the Olah studies and the deleterious effects seen in the Dutch ProPatria study most likely relates to an overly aggressive strategy of nutrition therapy. The Dutch study used higher doses for longer duration of therapy (four weeks vs one week) than the Olah studies. High doses of prebiotic fiber, formula, and the probiotic mixture were infused directly into the jejunum, even in patients concurrently receiving multiple vasopressor agents. Nine of the study patients had evidence of ischemic bowel (vs none of the controls).

Unfortunately, the ProPatria study led to setbacks in this field of investigation. Fear generated by this report led to early cessation of a similar randomized trial in SAP by Sharma, which at the time of discontinuation showed no signs of a deleterious effect from the probiotics (and a potentially less inflammatory immune response) compared to controls (Sharma et al. 2011). No other randomized trial of probiotics in critically ill patients has shown such an adverse effect. Hopefully SAP studies in the future can focus on a more conservative strategy for delivery of the probiotics, which would help define the most appropriate dose, number of organisms, duration of therapy, and method of delivery.

Conclusions

The potential for manipulation of the intestinal flora in the ICU setting to improve recovery, decrease infections, and possibly improve mortality is on the horizon. The technology is now available for high quality studies that will elucidate the changes observed and the potential outcome benefits from replacing healthy flora. Such studies should allow the clinician to capitalize on the beneficial influences of the microbiome, while attenuating the detrimental effects. It appears that use of multiple strains of probiotics or even replacement of "healthy" fecal flora may be more beneficial than provision of any single probiotic. Novel investigative approaches may help shift the paradigm for the future; that the best clinical outcome results are not from killing bacteria in the ICU, but from bioecological control and manipulation.

References Cited

Alverdy, J., O. Zaborina and L. Wu. 2005. The impact of stress and nutrition on bacterial-host interactions at the intestinal epithelial surface. Curr. Opin. Clin. Nutr. Metab. Care. 8(2): 205–209.

Alverdy, J.C. 2012. During critical illness the gut does not pass the acid test. Crit. Care. 16(5): 150.

Babrowski, T., K. Romanowski, D. Fink et al. 2013. The intestinal environment of surgical injury transforms pseudomonas aeruginosa into a discrete hypervirulent morphotype capable of causing lethal peritonitis. Surgery. 153(1): 36–43.

Bengmark, S. 2013. Nutrition of the critically ill—a 21st-century perspective. Nutrients. 5(1): 162–207.

Bergen, W.G. and G. Wu. 2009. Intestinal nitrogen recycling and utilization in health and disease. J. Nutr. 139(5): 821–825.

Besselink, M.G., H.C. van Santvoort, E. Buskens et al. 2008. Probiotic prophylaxis in predicted severe acute pancreatitis: A randomised, double-blind, placebo-controlled trial. Lancet. 371(9613): 651–659.

Bouza, E. 2012. Consequences of clostridium difficile infection: Understanding the healthcare burden. Clin. Microbiol. Infect. 18 Suppl. 6: 5–12.

Brandt, L.J. 2013. American journal of gastroenterology lecture: Intestinal microbiota and the role of fecal microbiota transplant (FMT) in treatment of C. difficile infection. Am. J. Gastroenterol. 108(2): 177–185.

Cabrera-Rubio, R., M. Garcia-Nunez, L. Seto et al. 2012. Microbiome diversity in the bronchial tracts of patients with chronic obstructive pulmonary disease. J. Clin. Microbiol. 50(11): 3562–3568.

Carlisle, E.M., V. Poroyko, M.S. Caplan, J. Alverdy, M.J. Morowitz and D. Liu. 2013. Murine gut microbiota and transcriptome are diet dependent. Ann. Surg. 257(2): 287–294.

Castagliuolo, I., J.T. LaMont, S.T. Nikulasson and C. Pothoulakis. 1996. *Saccharomyces boulardii* protease inhibits clostridium difficile toxin A effects in the rat ileum. Infect. Immun. 64(12): 5225–5232.

Crooks, N.H., C. Snaith, D. Webster, F. Gao and P. Hawkey. 2012. Clinical review: Probiotics in critical care. Crit. Care. 16(6): 237.

Damman, C.J., S.I. Miller, C.M. Surawicz and T.L. Zisman. 2012. The microbiome and inflammatory bowel disease: Is there a therapeutic role for fecal microbiota transplantation? Am. J. Gastroenterol. 107(10): 1452–1459.

Davila, A.M., F. Blachier, M. Gotteland et al. 2013. Re-print of "intestinal luminal nitrogen metabolism: Role of the gut microbiota and consequences for the host". Pharmacol. Res. 69(1): 114–126.

Drekonja, D.M., M. Butler, R. MacDonald et al. 2011. Comparative effectiveness of clostridium difficile treatments: A systematic review. Ann. Intern. Med. 155(12): 839–847.

Feng, Y., M.W. Ralls, W. Xiao, E. Miyasaka, R.S. Herman and D.H. Teitelbaum. 2012. Loss of enteral nutrition in a mouse model results in intestinal epithelial barrier dysfunction. Ann. N Y Acad. Sci. 1258: 71–77.

Forestier, C., D. Guelon, V. Cluytens, T. Gillart, J. Sirot and C. De Champs. 2008. Oral probiotic and prevention of pseudomonas aeruginosa infections: A randomized, double-blind, placebo-controlled pilot study in intensive care unit patients. Crit. Care. 12(3): R69.

Gill, S.R., M. Pop, R.T. Deboy et al. 2006. Metagenomic analysis of the human distal gut microbiome. Science. 312(5778): 1355–1359.

Hickson, M., A.L. D'Souza, N. Muthu et al. 2007. Use of probiotic lactobacillus preparation to prevent diarrhoea associated with antibiotics: Randomised double blind placebo controlled trial. BMJ. 335(7610): 80.

Hooper, L.V., T. Midtvedt and J.I. Gordon. 2002. How host-microbial interactions shape the nutrient environment of the mammalian intestine. Annu. Rev. Nutr. 22: 283–307.

Johnston, B.C., S.S. Ma, J.Z. Goldenberg et al. 2012. Probiotics for the prevention of clostridium difficile-associated diarrhea: A systematic review and meta-analysis. Ann. Intern. Med. 157(12): 878–888.

Katayama, M., D. Xu, R.D. Specian and E.A. Deitch. 1997. Role of bacterial adherence and the mucus barrier on bacterial translocation: Effects of protein malnutrition and endotoxin in rats. Ann. Surg. 225(3): 317–326.

Klarin, B., G. Molin, B. Jeppsson and A. Larsson. 2008. Use of the probiotic lactobacillus plantarum 299 to reduce pathogenic bacteria in the oropharynx of intubated patients: A randomised controlled open pilot study. Crit. Care. 12(6): R136.

Lawrence, S.J., J.R. Korzenik and L.M. Mundy. 2005. Probiotics for recurrent clostridium difficile disease. J. Med. Microbiol. 54(Pt 9): 905–906.

Long, J., O. Zaborina, C. Holbrook, A. Zaborin and J. Alverdy. 2008. Depletion of intestinal phosphate after operative injury activates the virulence of P aeruginosa causing lethal gut-derived sepsis. Surgery. 144(2): 189–197.

Marshall, J.C., N.V. Christou and J.L. Meakins. 1993. The gastrointestinal tract. the "undrained abscess" of multiple organ failure. Ann. Surg. 218(2): 111–119.

McFarland, L.V. 2006. Meta-analysis of probiotics for the prevention of antibiotic associated diarrhea and the treatment of clostridium difficile disease. Am. J. Gastroenterol. 101(4): 812–822.

McFarland, L.V., C.M. Surawicz, R.N. Greenberg et al. 1994. A randomized placebo-controlled trial of *Saccharomyces boulardii* in combination with standard antibiotics for clostridium difficile disease. JAMA. 271(24): 1913–1918.

Morowitz, M.J., E.M. Carlisle and J.C. Alverdy. 2011. Contributions of intestinal bacteria to nutrition and metabolism in the critically ill. Surg. Clin. North Am. 91(4): 771–85, viii.

Morowitz, M.J., T. Babrowski, E.M. Carlisle et al. 2011. The human microbiome and surgical disease. Ann. Surg. 253(6): 1094–1101.

Morrow, L.E., M.H. Kollef and T.B. Casale. 2010. Probiotic prophylaxis of ventilator-associated pneumonia: A blinded, randomized, controlled trial. Am. J. Respir. Crit. Care Med. 182(8): 1058–1064.

Morrow, L.E., V. Gogineni and M.A. Malesker. 2012. Probiotics in the intensive care unit. Nutr. Clin. Pract. 27(2): 235–241.

Nieuwenhuijzen, G.A. and R.J. Goris. 1999. The gut: The 'motor' of multiple organ dysfunction syndrome? Curr. Opin. Clin. Nutr. Metab. Care. 2(5): 399–404.

Ohira, H., Y. Fujioka, C. Katagiri et al. 2013. Butyrate attenuates inflammation and lipolysis generated by the interaction of adipocytes and macrophages. J. Atheroscler. Thromb.

Olah, A., T. Belagyi, A. Issekutz, M.E. Gamal and S. Bengmark. 2002. Randomized clinical trial of specific *lactobacillus* and fibre supplement to early enteral nutrition in patients with acute pancreatitis. Br. J. Surg. 89(9): 1103–1107.

Olah, A., T. Belagyi, L. Poto, L. Romics, Jr. and S. Bengmark. 2007. Synbiotic control of inflammation and infection in severe acute pancreatitis: A prospective, randomized, double blind study. Hepatogastroenterology. 54(74): 590–594.

Osuka, A., K. Shimizu, H. Ogura et al. 2012. Prognostic impact of fecal pH in critically ill patients. Crit. Care. 16(4): R119.

Perez-Cobas, A.E., M.J. Gosalbes, A. Friedrichs et al. 2012. Gut microbiota disturbance during antibiotic therapy: A multi-omic approach. Gut.

Petrof, E.O., R. Dhaliwal, W. Manzanares, J. Johnstone, D. Cook and D.K. Heyland. 2012. Probiotics in the critically ill: A systematic review of the randomized trial evidence. Crit. Care Med. 40(12): 3290–3302.

Raj, T., U. Dileep, M. Vaz, M.F. Fuller and A.V. Kurpad. 2008. Intestinal microbial contribution to metabolic leucine input in adult men. J. Nutr. 138(11): 2217–2221.

Safdar, N. and D.G. Maki. 2002. The commonality of risk factors for nosocomial colonization and infection with antimicrobial-resistant staphylococcus aureus, enterococcus, gram-negative bacilli, clostridium difficile, and candida. Ann. Intern. Med. 136(11): 834–844.

Samuel, B.S., A. Shaito, T. Motoike et al. 2008. Effects of the gut microbiota on host adiposity are modulated by the short-chain fatty-acid binding G protein-coupled receptor, Gpr41. Proc. Natl. Acad. Sci. USA 105(43): 16767–16772.

Schultz, M.J. and L.E. Haas. 2011. Antibiotics or probiotics as preventive measures against ventilator-associated pneumonia: A literature review. Crit. Care. 15(1): R18.

Sharma, B., S. Srivastava, N. Singh, V. Sachdev, S. Kapur and A. Saraya. 2011. Role of probiotics on gut permeability and endotoxemia in patients with acute pancreatitis: A double-blind randomized controlled trial. J. Clin. Gastroenterol. 45(5): 442–448.

Sherman, P.M., K.C. Johnson-Henry, H.P. Yeung, P.S. Ngo, J. Goulet and T.A. Tompkins. 2005. Probiotics reduce enterohemorrhagic *Escherichia coli* O157:H7- and enteropathogenic E. coli O127:H6-induced changes in polarized T84 epithelial cell monolayers by reducing bacterial adhesion and cytoskeletal rearrangements. Infect. Immun. 73(8): 5183–5188.

Siempos, I.I., T.K. Ntaidou and M.E. Falagas. 2010. Impact of the administration of probiotics on the incidence of ventilator-associated pneumonia: A meta-analysis of randomized controlled trials. Crit. Care Med. 38(3): 954–962.

Sommer, F. and F. Backhed. 2013. The gut microbiota—masters of host development and physiology. Nat. Rev. Microbiol. 11(4): 227–238.

Sun, X., H. Yang, K. Nose et al. 2008. Decline in intestinal mucosal IL-10 expression and decreased intestinal barrier function in a mouse model of total parenteral nutrition. Am. J. Physiol. Gastrointest. Liver Physiol. 294(1): G139–47.

Surawicz, C.M., L.V. McFarland, R.N. Greenberg et al. 2000. The search for a better treatment for recurrent clostridium difficile disease: Use of high-dose vancomycin combined with *saccharomyces boulardii*. Clin. Infect. Dis. 31(4): 1012–1017.

Taneja, R. and J.C. Marshall. 2000. Vasoactive agents and the gut: Fueling the motor of multiple organ failure. Crit. Care Med. 28(8): 3107–3108.

To, K.B. and L.M. Napolitano. 2012. Common complications in the critically ill patient. Surg. Clin. North Am. 92(6): 1519–1557.

Torrallardona, D., C.I. Harris and M.F. Fuller. 2003. Pigs' gastrointestinal microflora provide them with essential amino acids. J. Nutr. 133(4): 1127–1131.

Turroni, F., M.Ventura, L.F. Butto et al. 2013. Molecular dialogue between the human gut microbiota and the host: A lactobacillus and bifidobacterium perspective. Cell Mol. Life Sci.

van Nood, E., A. Vrieze, M. Nieuwdorp et al. 2013. Duodenal infusion of donor feces for recurrent clostridium difficile. N Engl. J. Med. 368(5): 407–415.
Wang, A.K.Y.Y. 2013. Upper gastrointestinal microbiota and digestive disease. World Journal of Surgery. 19(10): 1541–1550.
Zaborina, O., A. Zaborin, K. Romanowski, T. Babrowski and J. Alverdy. 2011. Host stress and virulence expression in intestinal pathogens: Development of therapeutic strategies using mice and C. elegans. Curr. Pharm. Des. 17(13): 1254–1260.

14

Gut Microbiota in Obesity and Type-2 Diabetes: Links with Diet and Weight Loss Intervention

Judith Aron-Wisnewsky[1,2] and *Karine Clément*[2,*]

Introduction

Microbiota colonizes the gut soon after birth. Although its initial composition rapidly varies, influenced by numerous factors such as mother's microbiota, mode of delivery, breast feeding, environmental exposure, early antibiotic administration.... (Penders, Thijs et al. 2006; Palmer, Bik et al. 2007), it becomes relatively stable after the age of two and onward to the adult age (Marchesi 2011; Yatsunenko et al. 2012). The understanding of gut microbiota's composition and role was initially limited by the inability to cultivate most microorganisms colonizing the digestive tract. The field has taken another dimension with the availability of new culture-independent tools, such as the 16SrRNA gene sequencing

[1] Institut de Cardiometabolisme et Nutrition, ICAN, Assistance Publique-Hôpitaux de Paris, Hôpital Pitié-Salpêtrière, Département cœur et métabolisme, Paris, F-75013 France.
[2] INSERM, U872, équipe 7 Nutriomique, Paris, F-75006 France; Université Pierre et Marie Curie-Paris6, Centre de Recherche des Cordeliers, UMR S 872, Paris, F-75006 France.
* Corresponding author: Karine.clement@psl.aphp.fr

inventory (Shendure and Ji 2008). Moreover, metagenomic sequencing (i.e., characterisation of microbiota genome) represents high-throughput analyses of the microbiome that enable even broader insights into the gut microbiota composition (microbiome, i.e., collective microbes, their genomes and environmental interactions) (DiBaise et al. 2008; Turnbaugh et al. 2009). Obviously, these various methodologies used to characterize gut microbiota might induce some variability and may account for the diverging, or at least the different results that have been reported across the studies in the literature. Recently, functional analysis using metabolomic techniques sometimes combined with metagenomic approaches are now available to deepen even more our current understanding of the contribution of gut microbiota in many processes related to human diseases. These major technical breakthroughs have enabled the scientific community to better characterize gut microbiota in healthy humans (Turnbaugh et al. 2009). Additionally international studies, such as the Human Microbiome Project and the MetaHIT consortium are currently on going to improve our understandings in the links between microbiota and normal physiology on the one hand and in pathological conditions such as obesity and intestinal diseases on the other hand.

The epidemic of obesity has become a worldwide concern , particularly in countries that have adopted a western diet and sedentary mode of living (Flegal et al. 2008), resulting in increased prevalence of associated disorders such as metabolic syndrome, type-2 diabetes, nonalcoholic steato-hepatitis (NASH), cardiovascular diseases and some types of cancer (Kopelman 2000). The causes responsible for obesity development and progression are complex and involve factors as diverse as lifestyle habits, exposition to toxics, genetics/epigenetics, neural and endocrine hormones (Mutch and Clement 2006). Recent data in mice models have suggested an additional pathophysiological component with the contribution of gut microbiota not only in the development of obesity, but also in its related complications (Henao-Mejia et al. 2012; Serino et al. 2012) that contribute to make obesity a complex and chronic condition.

Thus these key discoveries in the past eight years created a major burst in the number of scientific publications enlightening the role of microbiota dysbiosis (i.e., imbalanced microbiota compared to the one of healthy individuals) in the complex facets of obesity. Although convincing in mouse models, it remains difficult to conclude in humans whether microbiota dysbiosis is the cause or consequence of obesity, most particularly, since major environmental confounding factors, including dietary changes, are known to play a role both in the modulation of microbiota (Hildebrandt et al. 2009) and obesity development and maintenance during lifespan (Astrup et al. 2008).

Microbiota in Obesity

Ninety percent of the mammalian gut microbiota belongs to two most prevailing phyla, Gram-negative Bacteroidetes and Gram-positive Firmicutes (Tilg and Kaser 2011). The first pioneering experiments displaying a role of microbiota in weight regulation originally came from mice studies. Germ-free mice, compared to conventional animals, had reduced adiposity despite increased food consumption. After colonization with cecal content from conventional mice, they rapidly gained weight, in particular fat mass, although their food intake decreased. This observation suggests an important role of microbiota in the regulation of energy storage (Backhed et al. 2004; Turnbaugh et al. 2006; Turnbaugh et al. 2008). Furthermore, the rational to suggest that microbiota could contribute to obesity came from gut microbiota manipulation in mice, illustrating its role in energy regulation and increased capacity to process otherwise indigestible dietary polysaccharides (Sonnenburg et al. 2005; Gill et al. 2006; Turnbaugh et al. 2006). This leads to subsequent intestinal absorption of short-chain fatty acids (SCFA) (Topping and Clifton 2001). Moreover, microbiota directly impact on gene regulation to favor increased storage into adipose tissue (Backhed et al. 2004). The hypothesis was that gut microbiota from obese animals could be more efficient in extracting energy from the diet and subsequently promote fat storage better than that of lean individuals. The discovery that obese subjects displayed higher SCFA and lower residual calories from food in feces than lean subjects is in agreement with this hypothesis (Topping and Clifton 2001).

Considering the various bacterial groups contributing to obesity, many studies in mice models (Ley et al. 2005; Turnbaugh et al. 2006) using 16SrRNA sequencing method, observed that obesity was associated with increased Firmicutes and decreased Bacteroidetes levels compared to lean counterparts independently of food consumption as reviewed recently in Clarke et al. 2012. This latter observation was partially confirmed in human studies (Ley et al. 2006; Armougom et al. 2009; Turnbaugh et al. 2009). The comparison of microbiota composition of 12 obese patients and 3 lean individuals revealed that obese had significantly fewer Bacteroidetes and more Firmicutes than did lean counterparts (Ley et al. 2006). Although this study included a limited number of subjects, 18,348 bacterial 16SrRNA sequences were analyzed. Those results were confirmed in two larger cohorts of humans (Armougom et al. 2009; Turnbaugh et al. 2009). Yet this initial set of data is currently controversial since other well performed studies found diverging results (Duncan et al. 2007; Duncan et al. 2008; Schwiertz et al. 2010). Duncan et al failed to detect any difference in the proportion of Bacteroidetes in obese and non-obese individuals. Furthermore, these apparent conflicting results cannot derive either from different technique

(FISH) since they were validated with 16SrRNA inventory, nor by the lack of power since this is the second study they performed to confirm previous concordant results (Duncan et al. 2007). Likewise, Jumpertz found the same level of the three dominant phyla (namely Bacteroidetes, Fimicutes and Actinobacteria) in groups with different BMI levels (Jumpertz et al. 2011). The discrepancies observed between studies across the literature could suggest that differences in the phylum level might not be universally true as previously reviewed (Harris et al. 2012). It is also well known that BMI (which is the ratio of weight on $(height)^2$ given in kg/m^2) is not an optimal phenotype to characterize the complex phenotypes associated with obesity. Subject phenotypic heterogeneity could explain these discrepancies which led to further studies using large scale metagenomic tools.

Firstly, as displayed very recently by Gordon et al., bacterial diversity (Yatsunenko 2012) may be of greater relevance than the ratio between different types of phyla or taxa since it greatly differed in humans originating from different countries. Indeed, in population with severe obesity and prevalent obesity related-diseases (namely the United-States), the overall diversity and richness of gut microbiota is decreased compared to Malawi and Amerindian populations. Likewise, children from rural Africa displayed increased microbiota richness and biodiversity than same age healthy children from Western Europe (De Filippo et al. 2010). Finally, these recent results confirm previous data in mice (Turnbaugh et al. 2008) but also in humans (Turnbaugh et al. 2009) where obesity was associated with reduced bacterial diversity. Similarly, our team has found in 2 cohorts in France and Denmark that decreased diversity during human obesity is strongly associated with metabolic complications (Cotillard et al. 2013; Le Chatelier et al. 2013). Most importantly, significant weight loss following bariatric surgery in morbidly obese subjects was associated with significant increased bacterial diversity (Kong et al. AJCN 2013 in press).

Secondly, the human microbiome has further been extensively studied in larger scale projects using Illumina-based metagenomic sequencing and uncovered interesting findings that might revise the conclusions of previous studies using 16SrRNA. Indeed 40% of the genes from each individual are shared with at least half of the individuals in the cohort (Qin, Li et al. 2010), suggesting both the existence of large inter-individual variability in gut microbiota composition but also a "core" functional microbiota shared by a large number of individuals. This was confirmed later in a large gut microbiome study including subjects of three different nationalities that enabled to identify three clusters called "Enterotypes" characterized by higher relative abundance of Bacteroides (Enterotype 1), Prevotella (Enterotype 2) or Ruminococcus (Enterotype 3) respectively (Arumugam et al. 2011). Finally, a very recent study on 145 Chinese subjects (71 with type-2 diabetes and 74 controls) applied the whole genome shotgun method to study gut microbiota. They confirmed, in this Chinese population, the presence of the same Enterotype, mentioned earlier, driven by the abundance of Bacteroides, Prevotella, Bifidobacterium and Ruminococcus (Qin et al. 2012).

Microbiota and Type-2 Diabetes

Mechanisms linking microbiota to insulin resistance phenotypes

Cani et al. cleverly demonstrated a convincing link between gut microbiota, glycemic status/insulin resistance and inflammation in rodents. A 4 weeks of high fat diet induced mice to develop a discrete increase in LPS, referred to as "metabolic endotoxemia", insulin resistance and a concomitant modification of their gut microbiota composition (i.e., a reduction of *E. rectal* and of some bacteria from the Bacteroidetes phylum) (Cani et al. 2007). The authors then used a model of CD14 KO mice which were submitted to a high fat diet. These genetically modified mice displayed a rather interesting phenotype: they gained less weight, were more insulin sensitive and had a decreased systemic and tissue inflammation compared to their wild type counterpart. CD14 KO mice when submitted to subcutaneous infusion of LPS lost their improved metabolic profile. Altogether, these results demonstrate that the LPS/CD14 system is mandatory to induce the metabolic effects of high fat diet, at least in mice (Cani et al. 2007).

Conversely, interventions which impact on both metabolic endotoxemia and gut microbiota composition (using antibiotics), associated with improved insulin resistance, decreased systemic inflammation and reduced weight gain both in ob/ob and high fat diet mice (Cani et al. 2008). Interestingly the authors suggested that high-fat diet induced increased intestinal permeability (as quantified by both the *in vivo* absorption of fluorescent dextran and the expression of occluding and ZO protein) that could partly be reversed by an antibiotic treatment. Furthermore, interventions using prebiotics led to reduce metabolic endotoxemia and improved intestinal permeability and integrity, respectively measured with tight junction protein expression and immunohistochemical staining in the jejunum (Cani et al. 2009). These authors linked high throughput microbiota analysis and metabolic profiling of gut microbiota and suggested that microbiota contributes to modulate host biology. They observed that ob/ob mice treated with prebiotics displayed improved metabolic phenotype with lower fasting glucose, reduced insulin resistance, decrease fat mass and increased fat free mass. Using two types of technologies to analyse microbiota (pyrosequencing and phylogenetic microarray), they demonstrated a shift in its composition (at the phylum level) in the prebiotic treated group: namely an increase in Bacteroidetes and decrease of Firmicutes. Most importantly they observed a large increase in the abundance of some bacteria (at the taxa level) in the treated mice (Everard et al. 2011).

Another very elegant study demonstrated that bacteria translocation in the blood was increased at the very early onset of insulin resistance. Indeed, after only one week of high fat diet, the systemic level of fluorescently labelled *Escherichia coli* given orally to mice was increased compared to that of mice fed a chow diet. This level continuously increased after 4 week of high fat diet even when the diabetic status is established. These results suggest that dietary factors impact on host biology (namely glucose homeostasis). The authors suspected an increased translocation of bacteria in agreement with observed increased intestinal permeability (Amar, Chabo et al. 2011). Interestingly, the level of bacterial DNA which was initially low in the mice tissues increased very significantly along the intervention both in the mesenteric adipose tissue (MAT) and in mesenteric lymph nodes. This was blunted in the CD14 KO mice confirming previous results (Cani et al. 2007).

The authors further studied mutant mice for Myd88 which is one of the major signalling molecules involved in TLR activation. These mice displayed a worsened metabolic phenotype (with increased fasting glycemia, weight, and reduced glucose tolerance) and concomitantly increased bacterial translocation (*E. coli*) as well as enhanced total content of bacterial DNA in MAT compared with wild type mice. Those results were mimicked in NOD2 mutant mice. Altogether these results demonstrate a link between dietary factor (such as fat content in the food), systemic inflammation, glucose homeostasis and microbiota which involves the Nod1/CD14/Myd88 system (Amar, Chabo et al. 2011). Finally, one month probiotic treatment with *Bifidobacterium animalis* subsp. *lactis 420 (B420)* enabled a decreased in bacterial translocation along with an improvement in insulin sensitivity suggesting that some probiotic intervention could reverse the adverse metabolic phenotype induced by high fat diet.

These results in mice concerning increased intestinal permeability in obesity needs to be deeper investigated in humans. However, recent observation suggests that abdominal adiposity indeed associates with increased intestinal permeability (Gummesson et al. 2011).

Type-2 diabetes bacterial signature

Qin et al. recently analyzed gut microbiota in a population of 345 Chinese subjects with or without type-2 diabetes (T2D) (Qin et al. 2012). Although the authors confirmed the presence of Enterotype in this population, they failed to observe any significant relationship between Enterotype and T2D disease status ($p=0.29$, Fisher's exact test), suggesting that these Enterotypes might not be causal drivers of metabolic pathologies. However they observed some peculiarities related to diabetic status. Type-2 diabetic patients' displayed a moderate degree of dysbiosis compared to controls. At the functional level, microbiota from diabetic patients is characterized by a decrease of butyrate producing bacteria (such as *Faecalibacterium prausnitzii*) and the increase in other bacteria involved in sulphate reduction and oxidative stress resistance. This observation confirms a previous one obtained in type-2 diabetic morbidly obese subjects in whom *Faecalibacterium prausnitzii* abundance was also significantly altered. Putative microbial functional roles were further investigated (Furet et al. 2010). Gut microbiota from type-2 diabetes patients displayed enrichment in membrane transport of sugars while it showed a decreased representation of functions related to vitamin metabolism and to cell motility (which seemed to be primarily driven by butyrate producing bacteria). Markers related to drug resistance were greatly enriched in diabetic patients. This last point, if confirmed in other studies, could open new areas of research and subsequently translate into practical clinical aspects. Indeed, microbiota could become a useful tool to predict which patient might respond to what treatment in a personalized medicine approach (Qin et al. 2012).

Likewise, obtaining a systemic bacterial signature to predict the future development of diabetes would be a useful tool to intensify care or introduce preventive treatment to those patients at risk. A first attempt has been proposed in the D.E.S.I.R cohort (Data from an Epidemiological Study of the Insulin Resistance Syndrome), where patients were followed for 9 years and the authors focused on who developed a diabetic status along the follow-up. Amar et al. demonstrated that the level of circulating bacterial genetic material (16SrDNA) in the blood at baseline predicted the onset of diabetes (OR=1.29 after adjustment for age, sex and fasting blood glucose) (Amar et al. 2011).

Microbiota and Food Intake

Long term food habits

There seems to be a strong interplay between food intake and microbiota composition. Enterotypes do not seem to be associated with body weight nor BMI or diabetes but appear to be mostly influenced by food habits. Each Enterotype is characterized by a network of dominant bacterial species that has specific abilities in deriving energy from specific types of nutrients. Furthermore, using a Food Frequency Questionnaire (FFQ) that enables to obtain detailed information on long term food habits, Wu et al. demonstrated that Enterotype 1 was highly associated with long term consumption of animal proteins and saturated fat while Enterotype

2 was associated with chronic carbohydrate-based diet (Wu et al. 2011). Importantly, this study highlighted that acute diet modifications induced rapid changes in the microbiome but not a proper switch from one Enterotype to the other. This latter result suggests that food habits and/or short term feeding constraints must be considered as key parameters linking Body Mass Index to microbiota composition.

Nutrient type and microbiota

When looked into close details, it seems that the nature of the nutrient load itself greatly impacts gut microbiota composition. A study in rodent model submitted to different diets differing in lipid density suggests that it is the fat content in the diet rather than weight variation that impacts on microbiota changes.

Mice on a relative high fat diet content displayed enriched Bacteroidetes microbiota whether they had lost weight or not (Ravussin et al. 2012). In humans, 14 overweight or obese patients with metabolic syndrome underwent a diet intervention and were followed for 10 weeks. The design was as follows: after a week of weight stabilization with a controlled diet, patients were divided in two groups and submitted, in a cross over model, to three weeks of diet containing either high content in resistant starch (RS) or non-starch polysaccharides (NSPs), and inversely for the next three following weeks. Then all patients were prescribed a three weeks caloric restriction diet to obtain weight reduction (Walker et al. 2011). Interestingly, changes in microbiota occurred all along the study intervention (whether on weight stabilization or weight loss). These results suggest that the change in diet content (herein the fiber content) was sufficient to induce microbiota composition modification, independently of weight variations. These results have then been confirmed using different technics to analyze gut microbiota (16S rRNA sequences analysis and RT-PCR method). Of note, these variations were not seen at the phyla level but rather at the taxa level for specific bacteria such as *Roseburia* and *Eubacterium(?) rectale* which are known to play a role in dietary carbohydrate digestion (Walker, Ince et al.). Another study included 20 obese subjects who received for 9 weeks in a cross over model, a diet with two different proportions of carbohydrate content (i.e., qualitative diet modification) (Duncan et al. 2007; Duncan et al. 2008). Likewise, no modification in microbiota composition was observed in relation with weight loss. However, depending on the carbohydrate richness in the diet, those same specific groups of bacteria, able to digest carbohydrate (*Roseburia* and *Eubacterium rectale*), were modified after the intervention. Again no changes were observed at the phylum level (Duncan et al. 2007). Whether this change is the cause or consequence of the diet intervention can of course not be deciphered.

Weight Loss and Microbiota Composition

Nutrient load and microbiota

Results diverge across the literature and some authors rather pointed out the effect of weight modifications (either weight gain or reduction) rather than the effect of nutrient itself on gut microbiota modulation (Clarke et al. 2012; Harris et al. 2012). The overall increase in dietary content rapidly associates with concomitant increase in Firmicutes and decrease in Bacteroidetes number in lean and obese humans (Jumpertz et al. 2011). In both mice and humanized mice models, the same pattern of changes was observed when submitted to high-fat (Cani et al. 2007; Hildebrandt et al. 2009) or western diet (Turnbaugh et al. 2008; Turnbaugh et al. 2009). By contrast, and in line with previous study in mice (Turnbaugh et al. 2008), Ley et al. observed that food restriction, which induced weight loss, was associated with increased Bacteroidetes and concomitant decreased Firmicutes level (Ley et al. 2006). Many other studies, evaluating microbiota modifications after different types of diet interventions also found an increase in Bacteroidetes number (Nadal et al. 2009; Santacruz et al. 2009). For example, after one year of this energy restricted diet, the levels of those phyla recovered that of normal weight subjects. Two studies in obese adolescents undergoing a 10 to 40% energy restricted diet for 10 weeks, also confirmed the increase in Bacteroidetes. The authors have also observed concomitant modifications at the taxa level, namely a reduction in *Eubacterium rectale* and an increase in *Lactobacillus-Enterococcus* (Nadal, Santacruz et al. 2009; Santacruz, Marcos et al. 2009).

Anyhow, one has to be cautious in jumping to conclusion since most studies explore short term interventions.

Bariatric surgery induced weight loss

The different weight-loss solutions to treat morbid obesity, be it either diet, physical activity or combined interventions, have proven disappointing in the long term (Wing and Phelan 2005) or insufficient regarding weight (Astrup et al. 2008) or risks reduction for massive obesity (Dyson 2010). Bariatric surgery, solely indicated in severe obesity, is now commonly used. It is actually the only mean to achieve major and sustainable weight reduction (Sjostrom et al. 2012; Buchwald, Avidor et al. 2004). Furthermore it yields a decrease in all-cause mortality and significantly improves most of the obesity-related diseases via various mechanisms extensively reviewed earlier (Dixon et al. 2011). Different surgical procedures are practiced that can fall into two main categories: purely restrictive (laparoscopic adjustable gastric banding (LAGB) and sleeve gastrectomy (SG)) or both restrictive and malabsorptive (Roux-en-Y gastric bypass (RYGB)) methods.

Bariatric surgery impact on microbiota

Limited data are available concerning the changes in microbiota composition after bariatric surgery, all of which have been studied in RYGB model, but it is anticipated that more and more information regarding change in gut microbiota after bariatric surgery will be provided in the near future. Although few data exist, these studies give a first interesting insight into microbiota changes after surgery. Up to now techniques used to characterize gut microbiota greatly differ from one study to the other. One has used large scale pyro-sequencing method but on a very small subset of patients (3 in each group), who were neither paired before and after RYGB nor investigated at the same time points inducing different weight responses (Zhang et al. 2009). Another one used metagenomic sequencing to study patients who were followed before and 3 months after RYGB. Yet again this study included only a small number of patients who presented mostly with type-2 diabetes (Graessler et al. 2012). We used a RTqPCR focused technique in a first study to investigate microbiota changes before, three and six months after RYGB, on a larger and precisely phenotyped cohort of 30 obese patients where clinical data as well as diet questionnaires were available and compared those analysis to that of 12 lean controls (Furet et al. 2010). We then analyzed gut microbiota using pyrosequencing methods in the same patients (Kong et al. AJCN 2013 in press) and related those data to host biological characteristics including adipose tissue gene expression.

Zhang et al. found that Firmicutes were dominant in the normal weight and obese individuals and decreased significantly after surgery (Zhang et al. 2009). Most importantly, microbiota functional differences were observed in obese versus normal weight or post-surgery individuals. Indeed, H2-producing bacterial groups (such as *Prevotellaceae*) as well as H2-using methanogenic Archea, both involved in energy extraction from indigestible polysaccharides, were present only in the obese individuals while no such bacteria were observed in lean subjects or after surgery. Gammaproteobacteria dramatically increased after surgery. Those results were further confirmed both by Graessler et al. (Graessler et al. 2012) who also pointed out a significant increase in proteobacteria and by Liou et al. (Liou et al. 2013). They observed a lower relative abundance of Firmicutes and Bacteroidetes after surgery (Graessler et al. 2012). By contrast, we pointed out that the *Bacteroides/ Prevotella* ratio was lower in obese than in control subjects and rose within 3 months after surgery to remain stable thereafter (Furet et al.). We also confirmed those results in our recent paper (Kong et al. 2013). Importantly this ratio negatively correlated with corpulence traits (body weight, body mass index and leptin), but the correlation was highly dependent on calorie intake. Actually 50% of associations found between gut microbiota components and obesity related traits were dependent on calorie intake. By contrast, bacteria of the species *Faecalibacterium prausnitzii* were directly associated with low-grade inflammation state reduction in obesity and with the metabolic status (in particular type-2 diabetes) independently of calorie intake, highlighting the fact that response to surgery greatly varies among individual and might also depend on pre surgery characteristics (Qin et al. 2010a). The level of *Faecalibacterium prausnitzii* increased after surgery (Furet et al. 2010). Our results contrast with those of Graessler, who found that *Faecalibacterium prausnitzii* decreased after surgery (Graessler et al. 2012). In our population, along with a significant improve of weight, inflammation and metabolic status, we observed a very significant increase in bacterial richness after surgery at 3 months that remained stable at latter follow-up. We observed some associations between adipose tissue gene expression and bacterial genes at baseline. Interestingly, we demonstrated a 10 fold increase in the number of these association 3 months after surgery suggesting a restored cross talk between gut microbiota and the host biology.

Although there seems to be some discrepancies between these studies, this may be related to the different number of patients studied, the different metabolic phenotype (severity of obesity and/or type-2 diabetes) and/or the differences in the technologies. Also, one cannot decipher whether those bacterial modifications are due to changes in food ingestion and digestion, to specific surgical procedure modifications or finally to metabolic improvements. There is a need to address the specific role of either energy restriction or surgery-induced modification on gut microbiota changes after bariatric intervention induced weight loss since as we pointed out some changes are highly dependent on food intake (Furet et al. 2010). Yet, gut microbiota seems to have some specific properties that can translate in metabolic phenotype. Indeed, Liou et al. transferred gut microbiota from RYGB operated mice. They observed that this transfer resulted in weight loss and decreased fat mass in the recipient animals (Liou et al. 2013).

Another study, in a rat model, linked metagenomic analysis and metabolic profiling to explore the impact of RYGB on gut microbiota composition. They observed a major disruption in gut microbiota ecology with a 52 fold higher concentration of Proteobacteria and a decreased concentration in both Firmicutes and Bacteroidetes as recently also observed by Graessler et al (Graessler et al. 2012). However, this last observation differs from previous results (Li et al. 2011b). To go further, functional studies in the same model have been performed, using both pyrosequencing methods and faecal water cytotoxicity assessment. Faecal waters were found to be highly cytotoxic after surgery, mainly due to an increase in faecal metabolites such as putrescine and other monoamines (Li et al. 2011b). This is consistent with the observed increase in Gammaproteobacteria (particularly *Enterobacter*) following RYGB in line with previous results (Graessler et al. 2012; Zhang et al. 2009). Undoubtedly, many future studies will use the RYGB model to confirm what is currently known. Most importantly, new integrative techniques will deepen the characterisation of gut microbiota composition and functional changes. Moreover this model could provide a strategic insight as to how microbiota changes can impact obesity-related diseases improvement.

Peculiar Model Effect

Literature seems to provide diverging results. How can we manage to make up with all of those data or at least explain these discrepancies?

Diet intervention models

Firstly, one needs to point out the great heterogeneity across studies. Models greatly differ since some authors used the rodent model (rats or mice) while other worked on humans who also differed as in regard to age, ethnicity, and clinical phenotype (obesity, Type-2 diabetes, metabolic syndrome, etc.). Clinical interventions vary across literature: some use diet interventions and some used surgical procedures (which are different in rodent model compared to humans). Also, studies use very diverse technologies (RT-PCR, FISH, pyrosequencing, metagenomic, and metabolomics). Finally, the comparative analysis is done at various time points in the follow-up of those interventions (either short or longer term). Indeed, Ley et al. observed humans individually submitted to a caloric restriction diet for a year (Ley et al. 2006). By contrast, Walker et al. and Duncan et al. studied the effect of a short diet with very small impact on weight (Walker et al. 2011; Duncan et al. 2008).

Secondly, there is still a controversy as to the reality of the changes in the Firmicutes/Bacteroidetes ratio after weight loss. However, the differences observed in the different studies could be due to the importance of the weight variations. Ley et al. demonstrated a strong association between the increase in Bacteroidetes and weight loss whatever the type of diet consumed by the subjects. This bacteroidetes increase was observed solely in patients whom had lost a significant amount of weight (Ley et al. 2006). Other studies also confirmed the positive correlation between microbiota composition and the importance of diet intervention induced weight loss (Nadal et al. 2009; Santacruz et al. 2009). However, most of the studies did not analyse their data as regard to the degree of weight loss and this might contribute to the diverging results across literature.

Thirdly, although there appear to be some discrepancies across the studies, one major element seems to be acknowledged by all: while there is a core metagenome shared by most individuals, there is also a great inter-individual variability in gut microbiota. Indeed, a study in healthy individuals using 16SrRNA to analyse gut microbiota, demonstrated that although large interindividual variability exists, 2.1% of the operational taxonomic units gathering 35% of the sequences were present in 50% of the sample suggesting the existence of a core microbiome (i.e., presence of identical bacterial sequences in unrelated individuals) (Tap et al. 2009).

Moreover, response to a diet intervention seems to be strongly linked to the initial microbiota composition. Altogether these results could suggest that microbiota composition and changes predicts the good or bad response to a diet intervention and could further be used in a personalized medicine approach. We subsequently need to take into account the precise individual phenotype (length of the disease, related comorbidities…).

Bariatric surgery models

Very importantly, along with weight loss, bariatric surgery induces subsequent changes in both environmental and systemic factors and most importantly in the digestive tract that were reviewed in (Aron-Wisnewsky, Dore et al. 2011). All of those changes might impact on microbiota composition after surgery (Sandoval 2011) and needs subsequently to be discussed.

During most abdominal surgery, it is highly recommended that patients undergo peri-operative antibiotic prophylaxis to prevent surgical site infection. Whatever the type of prophylaxis, an acute but probably transient effect on microbiota composition cannot be excluded that might potentially differ from one team to the other according to the antibiotic protocol (1999). After bariatric surgery, patients are advised to increase mastication (Favretti et al. 2002) for better food tolerance and comfort. This facilitates the initial steps of digestion by enhancing gastric motion, saliva production, and parasympathetic activity. Finally, RYGB procedure induces decreased acid production and altered and quickened transit time in the stomach (El Oufir et al. 1996). Although many variations of gastric bypass exist, those surgical procedures however share common modifications of the digestive tract and anatomical rearrangement. It consists in (i) Bile flow alteration, (ii) Restriction of the stomach size (iii) Altered flow of nutrients (iv) Vagal manipulation and (v) Enteric and adipose hormones modulations (Laferrere et al. 2007; Laferrere et al. 2008; Reed et al. 2011) resumed in the literature as the Brave effect (Ashrafian et al. 2011; Ashrafian et al. 2010;

Beckman et al. 2010). Furthermore, along with digestive tract modifications, the observed changes related to inflammation and the immune system could also impact microbiota composition. Importantly, weight loss and fat mass reduction (Ciangura et al. 2010) induced by bariatric surgery go along with decreased low-grade systemic (Dalmas, Rouault et al. 2011) and adipose tissue inflammation. Indeed macrophages infiltration within the adipose tissue decreases and their phenotype switch to a less inflammatory profile (Cancello et al. 2005; Aron-Wisnewsky et al. 2009). Also, RYGB induces profound decrease in calorie intake (Furet et al. 2010).

Finally although this field is currently being explored further, a number of food choice modifications have already been demonstrated. In particular, both in rat model and in humans, RYGB reduces high-fat or high-sugar containing food intake and preferences (le Roux, Bueter et al. 2011; Shin, Zheng et al. 2011). All of those modifications need to be taken into account when analyzing gut microbiota after bariatric surgery. Finally a recent paper on mice model compared gut microbiota composition in mice operated from RYGB, sham operated mice and sham surgery coupled with energy restriction. Interestingly, they pointed out that the change in gut microbiota was related to the surgery rather than weight loss or caloric restriction (Liou, Paziuk et al. 2013).

Conclusion

A large body of evidence now supports a role of gut microbiota in the development of obesity and its related-diseases in mice and maybe more indirectly in humans. Literature suggests that weight modifications are associated with variation in microbiota composition at the phyla level in some study and at the taxa level in others. However, these variations are impacted by both the food intake (and the nutrient content) and the degree of weight variations induced by the studied interventions. Also, some growing evidence already suggests that microbiota composition evolves after RYGB but this remains to be confirmed in larger cohorts and most particularly in the longer term, when patients regain weight. Also one has to take into account the numerous RYGB induced changes in the digestive tract, hormonal status, quantity and choice of ingested nutrients, all of which are prone to specifically modify microbiota composition. To date we are not able to decipher whether microbiota evolution is the cause or the consequence of weight loss, of the improvement of obesity-related disease or the surgical procedure (including all the above modifications).

Acknowledgement

The authors wish to thank support from Assistance Publique-Hôpitaux de Paris (PHRC), grants Appert, Fondation Coeur et Artère (Fondacoeur), Nestlé France as well as European Community (FP7 program, Metacardis) for studies related to gut microbiota exploration.

Reference Cited

ASHP. 1999. Therapeutic Guidelines on Antimicrobial Prophylaxis in Surgery. American Society of Health-System Pharmacists. Am. J. Health Syst. Pharm. 56(18): 1839–1888.

Amar, J., C. Chabo et al. 2011. Intestinal mucosal adherence and translocation of commensal bacteria at the early onset of type 2 diabetes: molecular mechanisms and probiotic treatment. EMBO Mol. Med. 3(9): 559–572.

Amar, J., M. Serino et al. 2011. Involvement of tissue bacteria in the onset of diabetes in humans: evidence for a concept. Diabetologia. 54(12): 3055–3061.

Armougom, F., M. Henry et al. 2009. Monitoring bacterial community of human gut microbiota reveals an increase in Lactobacillus in obese patients and Methanogens in anorexic patients. PLoS One. 4(9): e7125.

Aron-Wisnewsky, J., J. Dore et al. 2011. The importance of the gut microbiota after bariatric surgery. Nat. Rev. Gastroenterol. Hepatol. 9(10): 590–598.

Aron-Wisnewsky, J., J. Tordjman et al. 2009. Human adipose tissue macrophages: m1 and m2 cell surface markers in subcutaneous and omental depots and after weight loss. J. Clin. Endocrinol. Metab. 94(11): 4619–4623.

Arumugam, M., J. Raes et al. 2011. Enterotypes of the human gut microbiome. Nature. 473(7346): 174–180.

Ashrafian, H., T. Athanasiou et al. 2011. Diabetes resolution and hyperinsulinaemia after metabolic Roux-en-Y gastric bypass. Obes. Rev. 12(5): e257–272.

Ashrafian, H., M. Bueter et al. 2010. Metabolic surgery: an evolution through bariatric animal models. Obes. Rev. 11(12): 907–920.

Astrup, A., J. Dyerberg et al. 2008. Nutrition transition and its relationship to the development of obesity and related chronic diseases. Obes. Rev. 9 Suppl 1: 48–52.

Backhed, F., H. Ding et al. 2004. The gut microbiota as an environmental factor that regulates fat storage. Proc. Natl. Acad. Sci. USA 101(44): 15718–15723.

Beckman, L.M., T.R. Beckman et al. 2010. Changes in gastrointestinal hormones and leptin after Roux-en-Y gastric bypass procedure: a review. J. Am. Diet. Assoc. 110(4): 571–584.

Buchwald, H., Y. Avidor et al. 2004. Bariatric surgery: a systematic review and meta-analysis. JAMA. 292(14): 1724–1737.

Cancello, R., C. Henegar et al. 2005. Reduction of macrophage infiltration and chemoattractant gene expression changes in white adipose tissue of morbidly obese subjects after surgery-induced weight loss. Diabetes. 54(8): 2277–2286.

Cani, P.D., J. Amar et al. 2007. Metabolic endotoxemia initiates obesity and insulin resistance. Diabetes. 56(7): 1761–1772.

Cani, P.D., R. Bibiloni et al. 2008. Changes in gut microbiota control metabolic endotoxemia-induced inflammation in high-fat diet-induced obesity and diabetes in mice. Diabetes. 57(6): 1470–1481.

Cani, P.D., S. Possemiers et al. 2009. Changes in gut microbiota control inflammation in obese mice through a mechanism involving GLP-2-driven improvement of gut permeability. Gut. 58(8): 1091–1103.

Ciangura, C., J.L. Bouillot et al. 2010. Dynamics of change in total and regional body composition after gastric bypass in obese patients.. Obesity. (Silver Spring) 18(4): 760–765.

Clarke, S., E. Murphy et al. 2012. The gut microbiota and its relationship to diet and obesity: New insights. Gut Microbes. 3(3).

Cotillard, A., S.P. Kennedy et al. 2013. Dietary intervention impact on gut microbial gene richness. Nature 500(7464): 585–8.doi:10.1038/nature12480.

Dalmas, E., C. Rouault et al. 2011.Variations in circulating inflammatory factors are related to changes in calorie and carbohydrate intakes early in the course of surgery-induced weight reduction. Am. J. Clin. Nutr. 94(2): 450–458.

De Filippo, C., D. Cavalieri et al. 2010. Impact of diet in shaping gut microbiota revealed by a comparative study in children from Europe and rural Africa. Proc. Natl. Acad. Sci. USA 107(33): 14691–14696.

DiBaise, J.K., H. Zhang et al. 2008. Gut microbiota and its possible relationship with obesity.. Mayo. Clin. Proc. 83(4): 460–469.

Dixon, J.B., N.E. Straznicky et al. 2011. Surgical approaches to the treatment of obesity. Nat. Rev. Gastroenterol. Hepatol. 8(8): 429–437.

Duncan, S.H., A. Belenguer et al. 2007. Reduced dietary intake of carbohydrates by obese subjects results in decreased concentrations of butyrate and butyrate-producing bacteria in feces.. Appl. Environ. Microbiol. 73(4): 1073–1078.

Duncan, S.H., G.E. Lobley et al. 2008. Human colonic microbiota associated with diet, obesity and weight loss. Int. J. Obes. (Lond.) 32(11): 1720–1724.

Dyson, P.A. 2010. The therapeutics of lifestyle management on obesity. Diabetes Obes. Metab. 12(11): 941–946.

El Oufir, L., B. Flourie et al. 1996. Relations between transit time, fermentation products, and hydrogen consuming flora in healthy humans.. Gut. 38(6): 870–877.

Everard, A., V. Lazarevic et al. 2011. Responses of gut microbiota and glucose and lipid metabolism to prebiotics in genetic obese and diet-induced leptin-resistant mice. Diabetes. 60(11): 2775–2786.

Favretti, F., P.E. O'Brien et al. 2002. Patient management after LAP-BAND placement. Am. J. Surg. 184(6B): 38S–41S.

Flegal, K.M., M.D. Carroll et al. 2008. Prevalence and trends in obesity among US adults, 1999. 2010 JAMA. 303(3): 235–241.

Furet, J.P., L.C. Kong et al. 2010. Differential adaptation of human gut microbiota to bariatric surgery-induced weight loss: links with metabolic and low-grade inflammation markers. Dabetes. 59(12): 3049–3057.

Gill, S.R., M. Pop et al. 2006. Metagenomic analysis of the human distal gut microbiome. Science. 312(5778): 1355–1359.

Graessler, J., Y. Qin et al. 2012. Metagenomic sequencing of the human gut microbiome before and after bariatric surgery in obese patients with type 2 diabetes: correlation with inflammatory and metabolic parameters. Pharmacogenomics J.

Gummesson, A., L.M. Carlsson et al. 2011. Intestinal permeability is associated with visceral adiposity in healthy women. Obesity. (Silver Spring) 19(11): 2280–2282.

Harris, K., A. Kassis et al. 2012. "Is the gut microbiota a new factor contributing to obesity and its metabolic disorders?" J. Obes. 879151.

Henao-Mejia, J., E. Elinav et al. 2012. Inflammasome-mediated dysbiosis regulates progression of NAFLD and obesity. Nature. 482(7384): 179–185.

Hildebrandt, M.A., C. Hoffmann et al. 2009. High-fat diet determines the composition of the murine gut microbiome independently of obesity. Gastroenterology. 137(5): 1716–1724 e1711–1712.

Jumpertz, R., D.S. Le et al. 2011. Energy-balance studies reveal associations between gut microbes, caloric load, and nutrient absorption in humans. Am. J. Clin. Nutr. 94(1): 58–65.

Kong, L.C., J. Tap et al. 2013. Gut microbiota after gastric bypass in human obesity: increased richness and associations of bacterial genera with adipose tissue genes. Am. J. Clin. Nutr. 98(1): 16–24.

Kopelman, P.G. 2000. Obesity as a medical problem. Nature. 404(6778): 635–643.

Laferrere, B., S. Heshka et al. 2007. Incretin levels and effect are markedly enhanced 1 month after Roux-en-Y gastric bypass surgery in obese patients with type 2 diabetes. Diabetes Care. 30(7): 1709–1716.

Laferrere, B., J. Teixeira et al. 2008. Effect of weight loss by gastric bypass surgery versus hypocaloric diet on glucose and incretin levels in patients with type 2 diabetes. J. Clin. Endocrinol. Metab. 93(7): 2479–2485.

Le Chatelier, E., T. Nielsen et al. 2013. Richness of human gut microbiome correlates with metabolic markers. Nature 500(7464): 541–6. doi:10.1038/nature12506.

le Roux, C.W., M. Bueter et al. 2011. Gastric bypass reduces fat intake and preference. Am. J. Physiol. Regul. Integr. Comp. Physiol. 301(4): R1057–1066.

Ley, R.E., F. Backhed et al. 2005. Obesity alters gut microbial ecology. Proc. Natl. Acad. Sci. USA 102(31): 11070–11075.

Ley, R.E., P.J. Turnbaugh et al. 2006. Microbial ecology: human gut microbes associated with obesity.Nature. 444(7122): 1022–1023.

Li, J.V., H. Ashrafian et al. 2011a. Metabolic surgery profoundly influences gut microbial-host metabolic cross-talk. Gut. 60(9): 1214–1223.

Li, J.V., R. Reshat et al. 2011b. Experimental bariatric surgery in rats generates a cytotoxic chemical environment in the gut contents. Front Microbiol. 2: 183.

Liou, A.P., M. Paziuk et al. 2013. Conserved shifts in the gut microbiota due to gastric bypass reduce host weight and adiposity. Sci. Transl. Med. 5(178): 178ra141.

Marchesi, J.R. 2011. Human distal gut microbiome. Environ. Microbiol. 13(12): 3088–3102.

Mutch, D.M. and K. Clement. 2006 . Unraveling the genetics of human obesity. PLoS Genet. 2(12): e188.

Nadal, I., A. Santacruz et al. 2009. Shifts in clostridia, bacteroides and immunoglobulin-coating fecal bacteria associated with weight loss in obese adolescents. Int. J. Obes. (Lond.) 33(7): 758–767.

Palmer, C., E.M. Bik et al. 2007. Development of the human infant intestinal microbiota. PLoS Biol. 5(7): e177.

Penders, J., C. Thijs et al. 2006. Factors influencing the composition of the intestinal microbiota in early infancy. Pediatrics. 118(2): 511–521.

Qin, J., R. Li et al. 2010. A human gut microbial gene catalogue established by metagenomic sequencing. Nature. 464(7285): 59–65.

Qin, J., Y. Li et al. 2012. A metagenome-wide association study of gut microbiota in type 2 diabetes.Nature. 490(7418): 55–60.

Ravussin, Y., O. Koren et al. 2012. Responses of gut microbiota to diet composition and weight loss in lean and obese mice. Obesity. (Silver Spring). 20(4): 738–747.

Reed, M.A., W.J. Pories et al. 2011. Roux-en-Y gastric bypass corrects hyperinsulinemia implications for the remission of type 2 diabetes. J. Clin. Endocrinol. Metab. 96(8): 2525–2531.

Sandoval, D. 2011. Bariatric surgeries: beyond restriction and malabsorption. Int. J. Obes. (Lond.) 35 Suppl 3: S45–49.

Santacruz, A., A. Marcos et al. 2009. Interplay between weight loss and gut microbiota composition in overweight adolescents. Obesity. (Silver Spring) 17(10): 1906–1915.

Schwiertz, A., D. Taras et al. 2010. Microbiota and SCFA in lean and overweight healthy subjects.Obesity. (Silver Spring) 18(1): 190–195.

Serino, M., E. Luche et al. Metabolic adaptation to a high-fat diet is associated with a change in the gut microbiota. Gut. 61(4): 543–553.

Shendure, J. and H. Ji. 2012. Next-generation DNA sequencing. 2008. Nat. Biotechnol. 26(10): 1135–1145.

Shin, A.C., H. Zheng et al. 2011. Roux-en-Y gastric bypass surgery changes food reward in rats. Int. J. Obes. (Lond.) 35(5): 642–651.

Sjostrom, L., M. Peltonen et al. 2012. Bariatric surgery and long-term cardiovascular events. JAMA. 307(1): 56–65.

Sonnenburg, J.L., J. Xu et al. 2005. Glycan foraging *in vivo* by an intestine-adapted bacterial symbiont. Science. 307(5717): 1955–1959.

Tap, J., S. Mondot et al. 2009. Towards the human intestinal microbiota phylogenetic core. Environ. Microbiol. 11(10): 2574–2584.

Tilg, H. and A. Kaser . 2011. Gut microbiome, obesity, and metabolic dysfunction. J. Clin. Invest. 121(6): 2126–2132.

Topping, D.L. and P.M. Clifton. 2001. Short-chain fatty acids and human colonic function: roles of resistant starch and nonstarch polysaccharides. Physiol. Rev. 81(3): 1031–1064.

Turnbaugh, P.J., F. Backhed et al. 2008. Diet-induced obesity is linked to marked but reversible alterations in the mouse distal gut microbiome. Cell Host Microbe. 3(4): 213–223.

Turnbaugh, P.J., M. Hamady et al. 2009. A core gut microbiome in obese and lean twins. Nature. 457(7228): 480–484.

Turnbaugh, P., R.E. Ley et al. 2006. An obesity-associated gut microbiome with increased capacity for energy harvest. Nature. 444(7122): 1027–1031.

Turnbaugh, P.J., V.K. Ridaura et al. 2009. The effect of diet on the human gut microbiome: a metagenomic analysis in humanized gnotobiotic mice. Sci. Transl. Med. 1(6): 6ra14.

Walker, A.W., J. Ince et al. 2011. Dominant and diet-responsive groups of bacteria within the human colonic microbiota. ISME J. 5(2): 220–230.

Wing, R.R. and S. Phelan. 2005. Long-term weight loss maintenance.Am. J. Clin. Nutr. 82(1 Suppl): 222S–225S.

Wu, G.D., J. Chen et al. 2011. Linking long-term dietary patterns with gut microbial enterotypes. Science. 334(6052): 105–108.

yatsunenko, T.R.F., M. Manary, I. Trehan et al. 2012. Human gut microbiome viewed across age and demography. Nature.

Zhang, H., J.K. DiBaise et al. 2009. Human gut microbiota in obesity and after gastric bypass. Proc. Natl. Acad. Sci. USA. 106(7): 2365–2370.

Index

Color Plate Section

Chapter 1

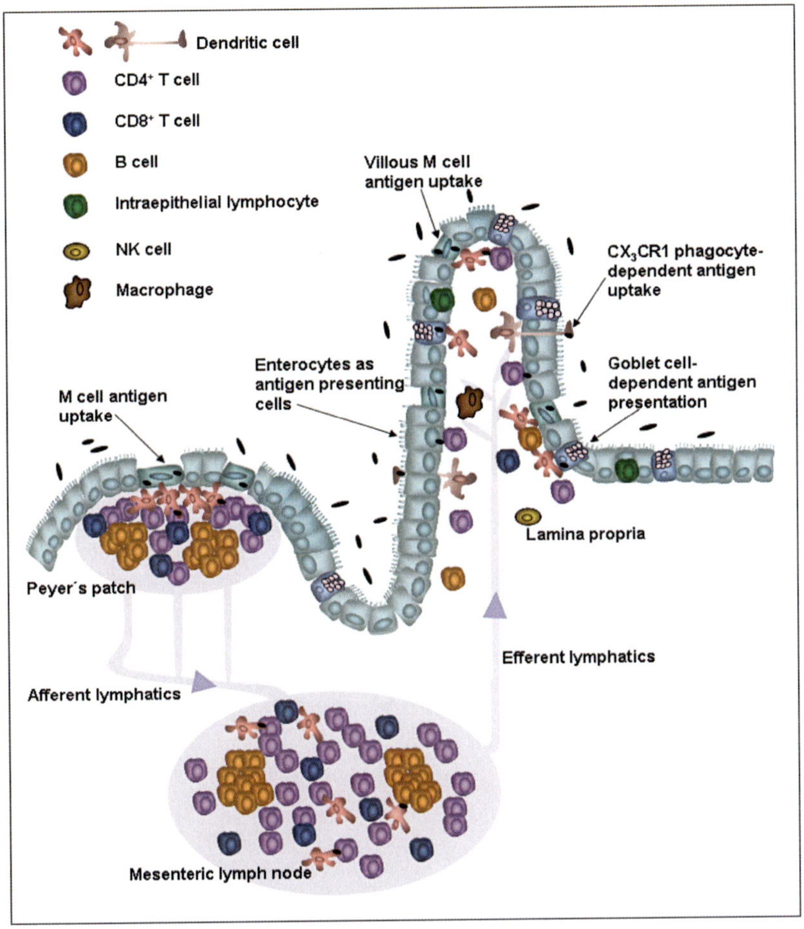

Figure 1. Antigen Recognition in the Intestine. Various cell types in the intestinal epithelium are able to uptake the luminal antigens and present them to the underlying immune system cells. Specialized antigen uptaking cells in the intestine are normal and villous micro-fold cells (M-cells) that transfer the antigen to the dendritic cells of the intestine. Enterocytes can function as the antigen presenting cells as they are able to process the luminal antigens and present them directly to the lymphocytes at the basal side. Goblet cells transfer the luminal antigens to the dendritic cells, while CX3CR1+ phagocyte can directly sample the luminal content using the transepithelial cellular processes.

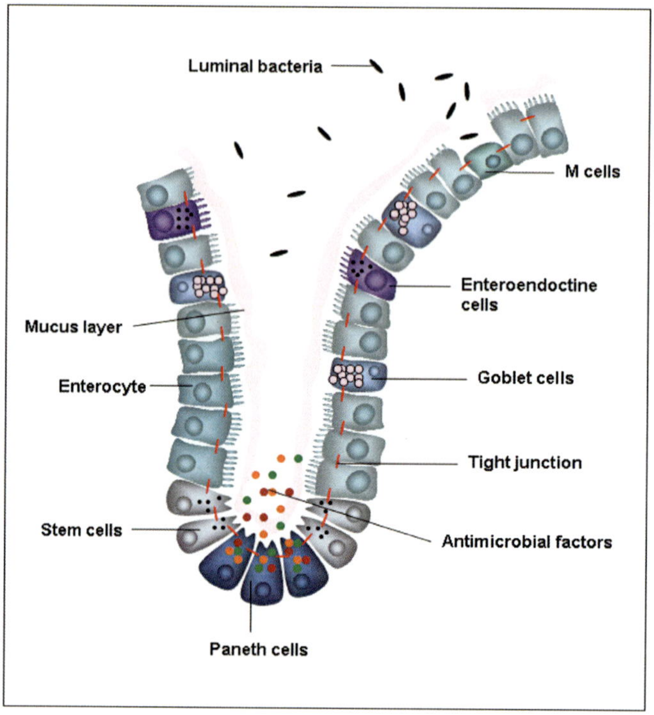

Figure 2. Mucosal Barrier in the Intestine. The intestinal epithelium consist of a single cell layer made of five cell types. All these cell types originate from stem cells located in the basis of intestinal crypts. The most numerous cells in the intestinal epithelium are enterocytes that are connected with the tight junctions. Intestinal epithelial cells are covered at the luminal side with mucus produced by Goblet cells. Mucus contains the antimicrobial factors secreted by Paneth cells. The mucus layer above micro-fold cells (M-cells) is disrupted allowing controlled luminal antigen uptake by these cells. Enteroendocrine cellssecrete the hormons that regulate the function of the intestinal epithelium.

Chapter 2

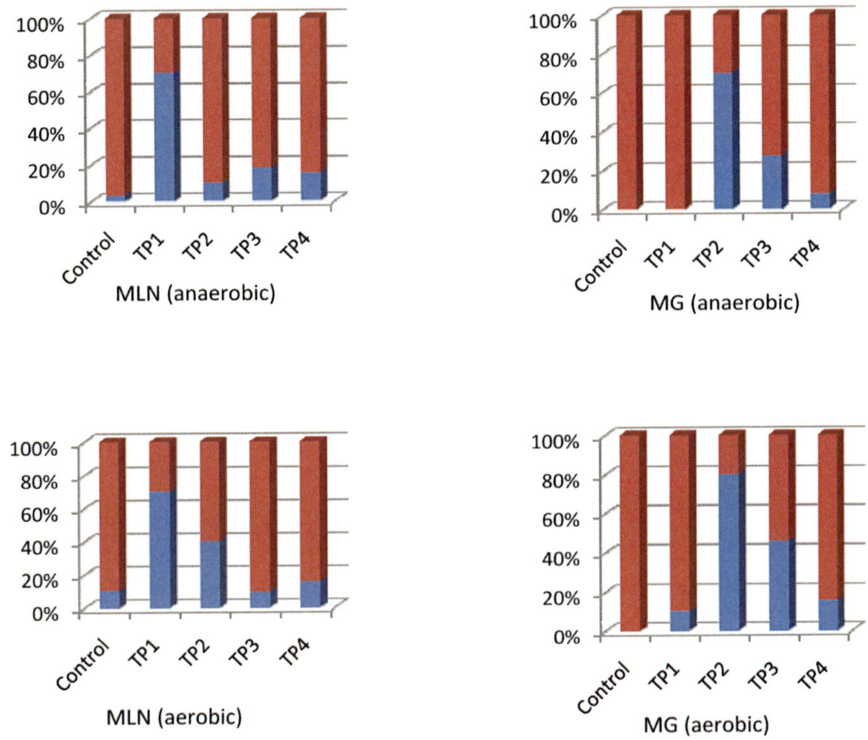

Figure 2. Microbiological analysis of mesenteric lymph nodes (MLN) and mammary glands (MG) of non-pregnantmice (control), pregnant mice (TP1) and lactating mice at 0–1 day (TP2); 3–4 days (TP3) and 14–15 days (TP4) after delivery. Red bars represent percentage of negative animals and blue bars represent percentage of positive animals.

Chapter 3

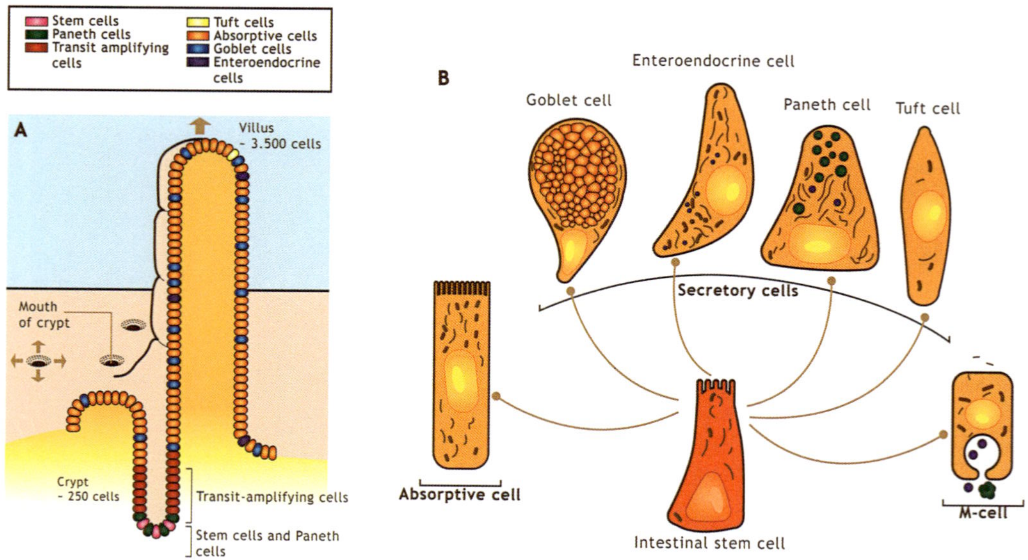

Figure 1. Structure of the Epithelia of the Small Intestine and Colon. The intestinal epithelium is spatially structured into invaginations called crypts, which extend into finger-like protrusions called villi in the small intestine and into the flat surface of the colon (A). Stem cells are found close to the base of the crypt. They generate highly proliferative cells that move upwards and subsequently differentiate into the various specialized lineages of the intestinal epithelium: absorptive cells (enterocytes or colonocytes), goblet cells, enteroendocrine cells, tuft cells, Paneth cells and the microfold (M) cells (B). The renewal process ends with the exfoliation of terminally differentiated cells at the tips of villi or from the surface of the colon epithelium. Paneth cells migrate towards the bottom of the crypt, where they remain for 6–8 weeks, supporting stem cell function. Cells expressing Paneth cell markers have been detected at the base of the crypts in the colon. These processes occur continuously and are highly coordinated, to allow tissue homeostasis.

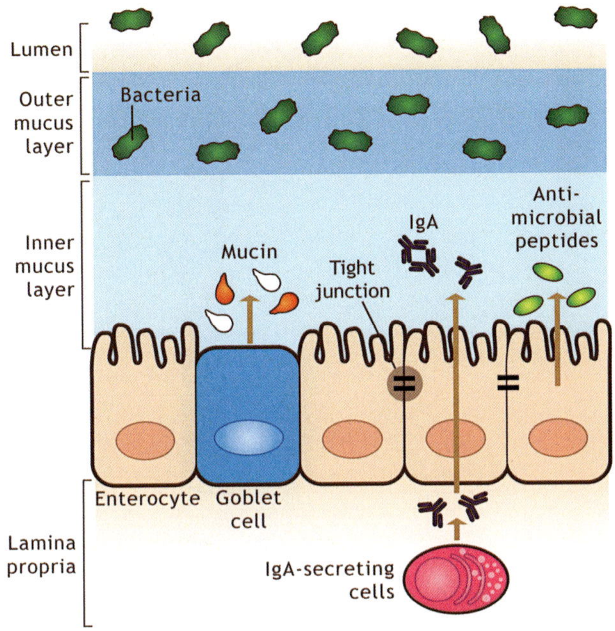

Figure 2. Mechanisms Minimizing Direct Contact Between Intestinal Bacteria and Epithelial Cells and Preventing Bacterial Penetration. Several mechanisms act in concert to limit direct interactions between the dense luminal microbial community and the intestinal epithelial cell surface. Goblet cells secrete gel-forming mucins, which assemble into a bilayer of intestinal mucus. The firm inner layer is dense and resistant to bacterial penetration, whereas the loose outer layer is a rich and reliable source of nutrients for the intestinal bacteria. The mucus layer concentrates antimicrobial peptides (AMPs), which are secreted principally by Paneth cells, but also by enterocytes. Some AMPs are produced constitutively, whereas others—such as RegIIIγ, a member of the C-type lectin family—are induced by commensal microorganisms. The epithelial cells are bound tightly together by intercellular junctional complexes, which regulate paracellular permeability. Another mechanism involve the transcytosis of IgA secreted by the underlying immune cells, which helps to limit the number of bacteria close to the epithelium and to prevent the penetration of host tissues by bacteria. This latter aspect is not considered in detail in this chapter.

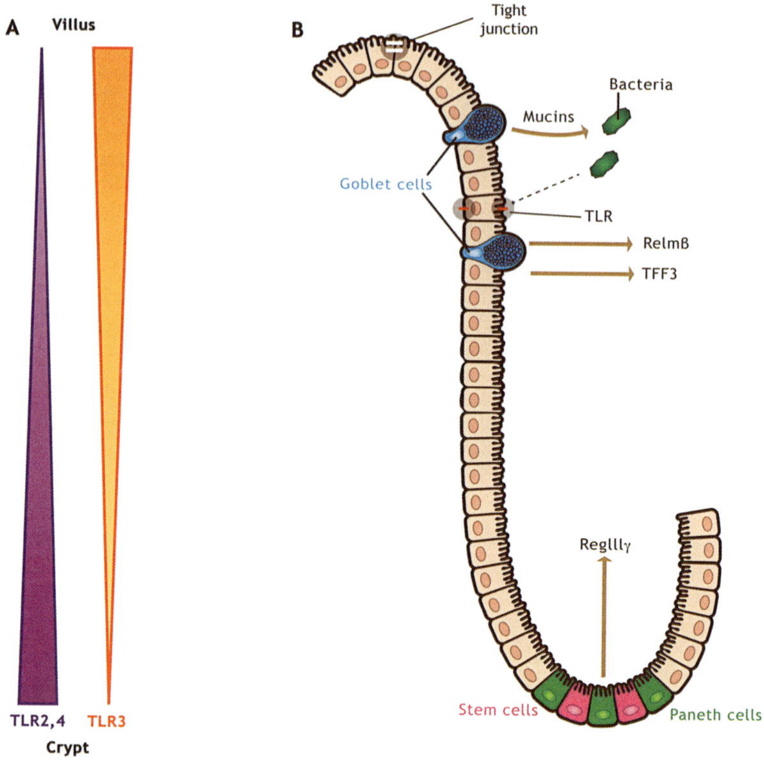

Figure 4. Epithelial Pattern Recognition Receptors for Perception of the Intestinal Microbiota. The sensing of microbe-associated molecular patterns is based on recognition by pattern recognition receptors (PRRs). Toll-like receptors (TLRs) and nucleotide oligomerization domain (NOD)-like receptors (NLRs) are the most important families of PRRs. TLR expression and distribution are controlled and signaling via these receptors is modulated so as to prevent deleterious inflammatory responses in the intestine. NOD2 is expressed predominantly in Paneth cells. PPR signaling plays a key role in epithelial homeostasis, as it stimulates the production of mucins, trefoil peptides (TFF3), resistin-like molecule β (RELMβ) and antimicrobial peptides, such as REG3γ, and strengthens tight junctions.

Chapter 4

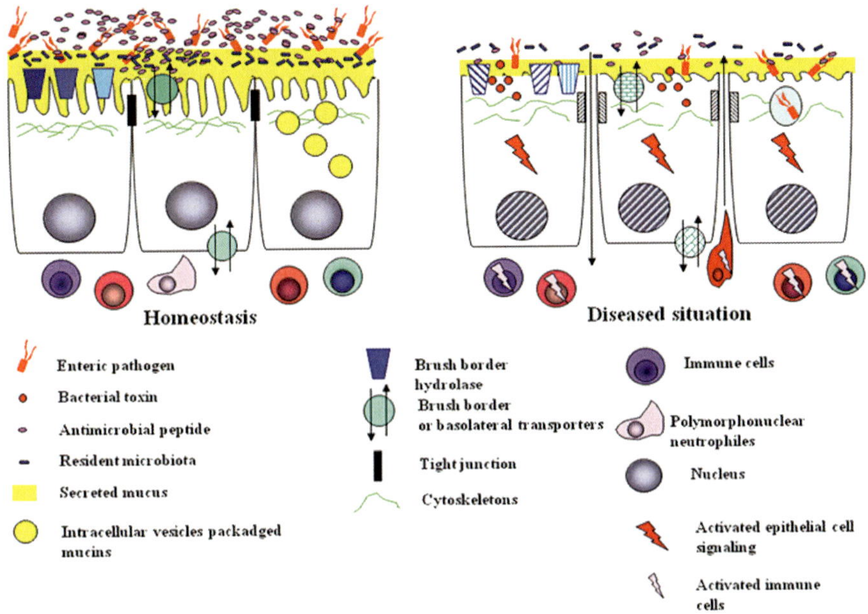

Figure 1. The Intestinal Epithelial Barrier. The intestinal epithelium consists of a single layer of highly polarized epithelial cells. The tight junction, a component of the apical junctional complex, seals the paracellular space between epithelial cells. Specific structural proteins composed the cytoskeleton and microtubule networks playing pivotal roles in the polarized organization of intestinal cells, the structural organization of the brush border, and the intracellular traffic of functional proteins to be addressed into the apical, lateral and basal domains of the polarized cells. The brush border at the apical domain and basolateral cell domain contain proteins and transporters exerting specific intestinal functions. The intestinal epithelial barrier plays an essential role in maintaining immune homeostasis. The lamina propria, located beneath the basement membrane, contains immune cells and neutrophils. Intestinal microbiota resides in the lumen, outside the mucus layer. Intestinal functions are modulated by quorum sensing-dependent cross-talk between the microbiota and host cells. Secreted mucins, combined with membrane-bound mucins, act as a physicochemical barrier and protect the epithelial cell surface against undesirable and harmful pathogens. Secreted antimicrobial peptides form the first chemical defense system against unwanted enteric pathogens.

Enteric pathogens use their adhesive factors to interact with the brush border membrane and hijack membrane-bound molecules as receptors for triggering signaling events. Invasive pathogens cross the epithelial cell membrane, penetrate into the host cells and pursue sophisticated intracellular lifestyles either in vacuoles or directly within the cell cytoplasm. Enteric toxins are secreted by both adhering and non-adhering pathogens, and either bind to membrane-bound receptors, thus activating signaling pathways, or are internalized. Enteric pathogens, acting either directly or via their toxins, disrupt the epithelial barrier, increasing paracellular permeability and inducing pro-inflammatory responses. Adhering and invading enteric pathogens trigger structural and functional lesions in epithelial cells, and even cell death.

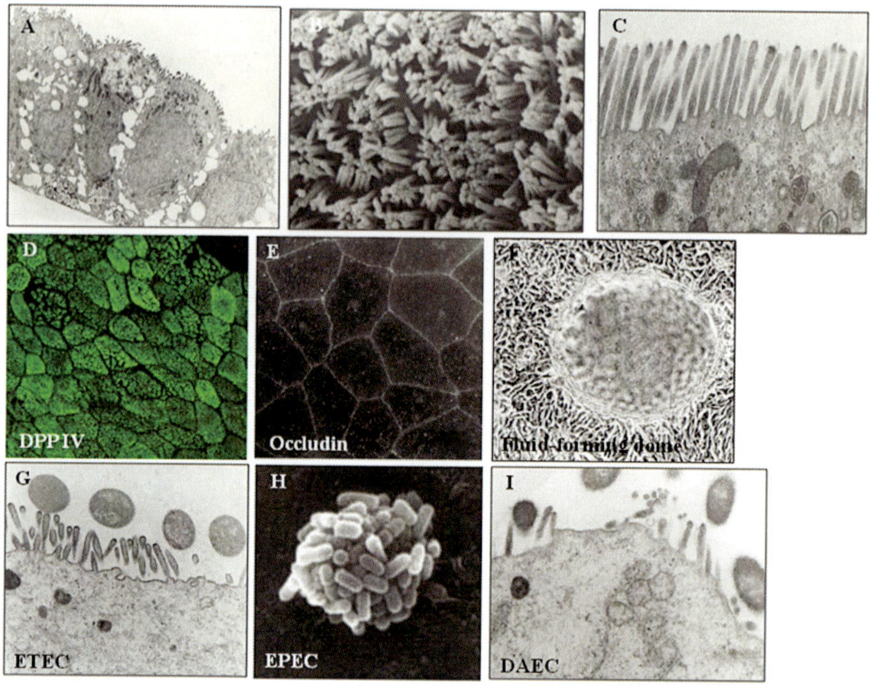

Figure 2. Human colon cancer Caco-2/TC7 Cells Spontaneously Differentiating in Culture into Enterocyte-like Cells of the Small Intestine. In A, transmission electron micrograph showing the polarized organisation of fully-differentiated cells forming a cell monolayer (15 days in culture). In B, scanning electron micrograph showing the well-organized and dense brush border. In C, transmission electron micrograph showing the well-ordered brush border microvilli. In D, immunofluorescent labeling showing the mosaic pattern of expression of functional brush border-associated DPP IV. In E, immunofluorescent labeling showing the tight junction-associated occludin localizing at the cell-to-cell contact of the cell monolayer. In F, microscopy phase-contrast micrograph showing a fluid-forming dome in the cell monolayer. In G to I, transmission (G and I) and scanning (H) electron micrographs showing interaction of ETEC with microvilli (G), typical, well-formed and compact EPEC microcolonies at the brush border (H), and brush border disappearance in Afa/Dr DAEC-infected cells (I).